Welsh Armorial Porcelain

Howell G. M. Edwards

Welsh Armorial Porcelain

Nantgarw and Swansea Crested China

Springer

Howell G. M. Edwards
Chemical & Forensic Sciences
University of Bradford
Bradford, UK

ISBN 978-3-030-97438-1 ISBN 978-3-030-97439-8 (eBook)
https://doi.org/10.1007/978-3-030-97439-8

Cover illustration: Dessert plate from the Nantgarw China Works, ca. 1817-1822, impressed NANT-GARW C.W., bearing the escutcheon, crest, torse and motto of Wyndham. The motto "Au Bon Droit" translates as "With Good Right". Henry Windham Wyndham-Quin (1782-1850) became Viscount Mount Earl of Adare in Co. Limerick, in February 1816 and the Earl of Dunraven in 1822. Family seats, Dunraven Castle, Glamorgan, and Uffords Manor, Norfolk. Reproduced with permission from Gwyn Jones, Director of the John Andrews Charitable Trust Collection, Oriel Plas Glyn-y-Weddw, Llanbedrog, Abersoch, Gwynedd.

This Springer imprint is published by the registered company Springer Nature Switzerland AG
The registered company address is: Gewerbestrasse 11, 6330 Cham, Switzerland

If I have seen further it is by standing on the shoulders of giants.
Sir Isaac Newton FRS (1642–1727)
Letter to Robert Hooke FRS (1635–1703), 5th February 1675.

Preface

Abstract Setting the scene for research into armorial porcelains: their purpose and the interdisciplinary approach required for their identification involving personal and manufactory records, armories and the social history of the period. The establishment of the methodology of the adopted approach for the determination of the identity of the entitled arms bearers from the full or partial coats-of-arms displayed on the porcelain. As this comprises the first dedicated text on the armorial porcelains of the two Welsh manufactories, Swansea and Nantgarw, a brief account of their anticipated number of exemplars is reviewed and their comparative rarity studied.

Keywords Swansea; Nantgarw; Porcelain; Coats-of-arms; Type of armorial

De omnibus dubitandum
Question everything.

Rene Descartes (1596–1650).

The depiction of coats-of-arms or heraldic crests on porcelain in the eighteenth and nineteenth centuries was undertaken by many china manufactories resulting from specific commissions received for porcelain services from royalty, the aristocracy and the landed gentry who had the right to bear arms. These commissions were generated often to celebrate a special event, perhaps a birthday, a marriage, the hereditary accession of a member of that family to a new title, the inheritance of an estate or an elevation in rank in the peerage. They could vary from the simple depiction of a gilt heraldic crest located centrally or in the reserve of an item of porcelain or earthenware to a fully decorated piece with a crest and a full or partial coat-of-arms placed strategically within the verge or centrally on the artefact. Armorial porcelains pose a rather different proposition for a ceramics historian from a research viewpoint compared with other porcelain artefacts produced at the same manufactories since they require the input of considerable additional and detailed genealogical and heraldic research to augment and ascribe their historical origin for the correct attribution of the identity of the bearer of the arms depicted on the porcelain. This genealogical research in particular can often illuminate the potential reason for the commissioning of the artefacts or services and historically this can give a new and often critically important dimension to the attribution of the porcelain. It is perhaps fair to say that armorial porcelains have usually not been appreciated so much by collectors generally for the artistic quality of the painting or gilding thereon as have other contemporary porcelain services, although there are some notable exceptions to this statement. A good example of this is the *Barry-Barry* or *Pendock-Barry* armorial service which was sumptuously decorated by William Billingsley, the master rose painter for the Derby China Works in the late eighteenth and early nineteenth centuries, which now provides one of the finest exemplars of his ceramic works of art (Edwards, *Porcelain Analysis and Its Role in the Forensic Attribution of Ceramic Specimens,* 2021; Denyer, Denyer and Edwards, *The Pendock-Barry Derby Porcelain Service: A Reappraisal*, to be published, 2022; Twitchett, *Derby Porcelain: An Illustrated Guide, 1748–1848*, 2002).

A factor which can detract somewhat from the full appreciation of an armorial piece of porcelain historically is the failure by researchers to identify the arms or heraldic device displayed thereon, which tends then to diminish its perceived importance and, therefore, downgrades perceptibly the armorial porcelain item alongside others of its genre that have been confirmed in their attribution. This certainly is the case for the armorial Welsh porcelains which form the subject of this book, especially the heraldically crested examples from the Swansea China Works, of which 50% remain unidentified currently after two centuries have elapsed since their creation.

Normally, the input of forensic scientific research principles into the regime of porcelain identification and attribution encompasses three distinct complementary components, namely, the weighted evaluation of any relevant historical documentation, especially the manufactory notebooks and the personal eye-witness recollections of people who actually had worked there for some time, the scientific data and the interpretation of the results from the chemical analyses (mineralogical, molecular and elemental oxide compositions) that have been undertaken on the body, glaze and pigments of the artefact concerned and the stylistic appreciation and measured opinion derived from an evaluation of the artefact by a ceramics connoisseur. The latter is normally afforded by a recognised expert and is based upon the translucency of the porcelain body, the artefact shape, the texture of the body and the glaze, the decoration if appropriate and the characteristic moulding of the porcelain piece in question in comparison with a range of established and accredited recognised factory exemplars. This synergistic approach is now considered essential for the modern holistic scientific assessment of the factory origin and attribution of a piece of porcelain, the chronological placement of the piece or artefact within its specific factory production period and the subsequent verification of its authenticity for its acceptance into the existing ceramic art genre. The applied decoration and enamelling of porcelain artefacts has recently been receiving considerable specific attention from the aspect of the elemental and molecular chemical analyses undertaken regarding the pigment composition of the enamels used and their characterisation for comparison within the chronological timeframe of pigment usage according to the factory workbooks and records, where these still exist. The use of out-of-context pigments chronologically that have been identified and exposed by chemical analysis on porcelain artefacts demands a reasonable and credible factual explanation and naturally gives rise to alarm relating to the presence of potential fakes or pieces that have been embellished in more modern times to make them more desirable and attractive for purchase by collectors because of their assumed historical or familial connections. Recent analytical work on porcelain glaze compositions has indicated the adoption of hitherto unknown procedures or recipes being used by specific manufactories, especially regarding the use of lead oxide or tin oxide in the glaze and the incorporation of other metal oxides, such as cobalt oxide, which may have been included as an additive to minimise the effect of a yellowing in the colour of the fired porcelain body due to the presence of an excessive iron oxide contamination of the clay or other raw materials used in porcelain synthesis. The application of analytical science to the pigment analysis of the coloured decoration can also reveal the presence of several novel and intriguing scenarios relating to the pigment raw material sourcing and its development, especially through the determination of trace elements in the enamels which can define some particular geographical locations and potential historical trade routes for their raw material supplies and their sourcing internationally by the manufactory. For example, the analytical discovery of the adoption of European arsenic-rich cobalt blue pigments from Saxony on Chinese porcelains decorated in underglaze blue enamel during the early eighteenth century is indicative of a reverse knowledge transfer exchange process that was occurring between European travellers and the manufacturers and decorators of the Chinese Imperial Qing Dynasty at Jingdezhen and Peking. This had hitherto only been suspected from historical documents and reports from the writings of the French Jesuit missionaries to Peking in the early 1700s and their letters to their superiors in Paris (Edwards, *Porcelain Analysis and Its Role in the Forensic Attribution of Ceramic Specimens,* 2021). It seems that both locally sourced cobalt oxide ores in China and their imported analogues from Europe (especially from Saxony) were utilised by Chinese decorators on their porcelains in the Kangxi (Xuanye) Dynasty period (1661–1722), particularly in the early years of the eighteenth century. It was found that the brighter blue colour thereby imparted to Chinese porcelains and its significantly improved adhesion to the substrate with little tendency to "run" upon the application and firing of the superficial glaze was much appreciated and admired by the manufacturers, the enamellers and by their prospective clients. In a recent research publication Colomban et al. (Non-invasive Raman analysis of 18th century chinese export/armorial overglazed porcelain: Identification of the different enamelling techniques, *Heritage*, 2022) have described their Raman and SEM/EDAXS analytical studies of several pieces of Chinese porcelain from the reigns of Kangxi and Qianlong which included porcelain commissioned by Philibert Orry, the Duc de Penthievre, and an armorial displaying the coat-of-arms of King Louis XV that was used at the Palace of Versailles. The analytical data identified the use of borax in the overglaze blue decoration as recommended in a French recipe of 1753, the use of manganese free-cobalt ores in the blue enamels and the use of arsenic in white enamels, which demonstrated the adoption of European ingredients and recipes in Chinese-decorated export porcelains.

Technically, the research approach that has to be adopted for armorial porcelains is in a rather different category to that undertaken for the majority of their other porcelain analogues, which can be termed "anonymous", since, having determined the source factory for the armorial porcelain, instead of defining the artist involved the key feature then becomes the identification of the coats-of-arms, crests or heraldic symbols which are displayed on the porcelain artefacts and arriving at a logical deduction therefrom as to the identity of the particular family bearer of the heraldic arms, which can be very important for the correct dating of the porcelain artefact in its attribution process. This confers an additional level of historical interest upon the ceramic artwork, especially if it can be associated chronologically with a particularly important event that occurred in history and for which the family celebration may have involved the commissioning of a special service to commemorate that event, whether this be an advancement in the peerage, a coming of age (for example, a 21st birthday), a marriage or the appointment to a local or national political position of the arms bearer. The right to bear arms and to display heraldic emblems is a personal one and is not conferred ubiquitously upon all family members; a noted exception to this rule is the membership of a Scottish Clan, but the entitled head of a Clan may also have the right to display their own independent arms aside from the others which may be also adopted by the Clan members.

The definition and identification of heraldic crests and coats-of-arms seems at the outset to be a straightforward task given the extensive armory literature that is available for consultation by the researcher, but it will transpire that some strict procedures need to be adopted to avoid several pitfalls that can be experienced in the assignment of the heraldic arms on armorial porcelain to a specific family bearer of those arms. It will be seen that even when the coat-of-arms or heraldic device has apparently been identified hitherto, through the careful perusal of the source armory literature their assignment to a particular family member is perhaps not unequivocal and can sometimes even be considered as being rather ambiguous or tentative, as the heraldic device, such as a crest or a motto, may have been in use by several different and related family arms-bearers and indeed by other unrelated families. In this case, special considerations and additional genealogical research then need to be applied to move to a potentially definitive and reasonable result that can be deduced for the proper attribution of the armorial devices, such as a crest or an emblazoned escutcheon. The adoption of the holistic analytical approach that has been demonstrably successful elsewhere for the attribution of porcelain artefacts and their source manufactories is also now necessary here as the examples in the following chapters will illustrate. Eventually, as found with the approved holistic analytical approach utilised for the attribution of other ceramic artefacts, its application extended to specifically armorial Welsh porcelains will also be seen to result in a measurable and quantifiable success in the assignment of several hitherto unidentified crests and partial coats-of-arms which will be a manifestly significant improvement upon that which can be realised now from a survey of the current situation in the ceramics literature (Edwards, *18th and 19th Century Porcelain Analysis: A Forensic*

Provenancing Assessment, 2020; Edwards, *Porcelain Analysis and Its Role in the Forensic Attribution of Ceramic Specimens,* 2021).

The number of armorial porcelain artefacts encountered for many early English porcelain factories in the eighteenth and nineteenth centuries is often found to be quite large, especially for the major factories such as Worcester, Coalport and Derby, which was a function both of their aristocratic and armigerous clientele and of their reasonably long porcelain manufacturing periods, although indisputably none of these can match the sheer output of their contemporary Chinese porcelain analogues in this respect. It has been estimated that approximately 70 million pieces of Chinese porcelain were imported into Europe by the British Honourable East India Company, the Dutch East India Company, the Swedish East India Company and their Portuguese affiliates from the Casa da India collectively during the seventeenth, eighteenth and nineteenth centuries (Vollker, *Porcelain and the Dutch East India Company*, 1954; Pinto de Matos, The Portuguese trade. *Oriental Arts XLV*/1 (Spring), 21–22, 1999; Pierson, The movement of Chinese ceramics: Appropriation in global history. *J. World Hist.* 23, 9–39, 2012; Graca, The Portuguese Porcelain trade with China. *Arts Asia* 7, 45–47, 1977; Finlay, The pilgrim art: The culture of Porcelain in world history. *J. World Hist.* 9, 141–187, 1998). The number of Chinese armorial porcelain artefacts encompassed in this estimated total is not quantifiable, but it must have been relatively significant. David Howard has recorded evidence for more than 5200 different individual English armorial services, some of which comprised several hundred pieces of porcelain, which were carried from China by the Honourable East India Company East Indiamen vessels alone during the eighteenth and early nineteenth centuries (see Chapt. 1) (Howard, *Chinese Armorial Porcelain,* Volume 1, 1974; Howard, *Chinese Armorial Porcelain,* Volume II, 2003; Howard, *A Tale of Three Cities – Canton, Shanghai and Hong Kong: Three Centuries of Sino-British Trade in the Decorative Arts,* 1997). Although we do not have a comparative figure for the total number of pieces of porcelain carried by the HEIC, an estimation of the quantity of armorial porcelain imported from China into Britain, with an assumption of perhaps 100 pieces per service, would then reasonably be given as about 10–20% of the total based on the figure given above – and this percentage only applies to the HEIC carriage of porcelain services. Other specific armorial commissions, outside of porcelain services, might involve an indefinable number of items such as punchbowls, vases, jars, containers and chamber, desk or boudoir sets. The research that has been undertaken into the origins and identification of the Chinese armorial porcelain services is significantly enhanced and assisted by the meticulous records that were maintained by the Honourable East India Company (HEIC) of their transportation of porcelain from China in their ships and of the individual commissions received by their captains and supercargoes from their British clientele. In many cases, this amounts to a researcher being able to assign the actual ship that was used to carry home to Great Britain the commissioned armorial china from Canton on a specific date and the name of the ship's captain or agent involved from a consultation of the works of Farrington (Farrington, *A Catalogue of the East India Ships' Journals and Logs, 1600–1834,* 1999; Farrington, *A Biographical Index of the East India Company Maritime Service Officers,*

1600–1834, 1999; Farrington, *Trading Places: The East India Company and Asia, 1600–1834*, 2002). Unfortunately, this wealth of historical detail that is accessible for Chinese porcelains is not normally available for the researcher of armorial porcelains that were commissioned directly from the contemporary English and Welsh porcelain manufactories, for many of which the manufactory records and workbooks no longer exist following the demise of most of these factories in the eighteenth and nineteenth centuries. In many cases, having identified the manufactory concerned, progress towards the attribution of the armorial bearings on armorial porcelains must be made, firstly, via a consultation of the armory literature, and then perhaps through the application of inductive reasoning and the assimilation of apparently unrelated clues in following potential leads to achieve a result. This aspect could well demand the adoption of novel approaches, which must be dissected carefully to reveal the factual basis from which an acceptable provenance of the true armorial bearings can then be established. Examples of this occurring historically will be revealed in the following text, where several Welsh porcelain armorials have been found to have been apparently incorrectly identified hitherto in the literature, through what can now perhaps be termed rather specious conclusions and the adoption of facile and erroneous assumptions which do not now stand up to a more rigorous forensic interpretation. Often, it is found that these original armorial assignments or attributions of armorial artefacts may not have any supporting argument for the justification of their conclusions other than perhaps someone believed it to be true many years ago, perhaps even from rather vague hearsay evidence provided by someone else historically. This is clearly an unsatisfactory procedure and, hence, nothing should now be assumed or taken for granted in this exercise, as will be seen in the examples cited later here, particularly referring to some tentative attributive assignments that were made historically and which have thereafter been assumed to be statements of fact which have become enshrined in the literature upon their repeated citation by later ceramic historians. Perhaps these attributions were originally made from a sole expert opinion, or from a misconceived interpretation of the relevant armory literature records at that time, and these are then no longer tenable when subjected to a realistic forensic scientific re-examination of the factual basis for those assignments having been made originally. It is a necessary part of the current study, therefore, to evaluate and check all armorial assignments for the Nantgarw and Swansea porcelains that have been made hitherto and not to assume that they have all been correctly attributed; in no way is this to be regarded as a diminution of the value of the earlier research into armorials, the majority of which stands firm under the present examination, but it is essential to establish the baseline determination for the armorial exemplars selected to which other unknowns presented for attribution can be added in the future.

In this book, the two Welsh porcelain factories of Swansea and Nantgarw have been selected for the current research project, which involves the detailing of the investigation into their armorial services, for three reasons. Firstly, both the Swansea China Works and the Nantgarw China Works had a strictly limited production period extending over a maximum of only five years, or rather less in reality, perhaps, during the second decade of the nineteenth century, extending between about

1815 and 1823 at the widest temporal range. This means that their porcelain production output was, therefore, quite limited in extent in comparison with some of their contemporary and much larger English manufactories, such as those at Coalport, Worcester and Derby. Secondly, they both had a restricted output of finished fine quality porcelains during their production periods because of particularly severe kiln firing losses, which at Nantgarw approached 90% of the pieces in the kiln firing charge that was matched perhaps by a still relatively large 70% estimated loss at Swansea; this meant that only one in every ten pieces of Nantgarw china actually survived the initial biscuit kiln firing process and subsequent glazing to be available for decoration and sale through an agent or directly from the factory to a demanding clientele. This was not just a problem for the Nantgarw and Swansea manufactories and it will be noted later in the text that the pre-French Revolutionary Sevres manufactory in 1778 recorded that they had experienced particularly severe kiln firing losses in the production of their esteemed soft paste porcelain service for the Empress Catherine the Great of Russia (Ekaterina II), which amounted to the total loss of 2200 pieces from a batch of 3000 that was specially fired in their biscuit porcelain kiln, a loss of 73%, which is comparative with the value mentioned for the Swansea China Works soft paste duck-egg porcelain noted by Lewis Weston Dillwyn some years later. Thirdly, unlike some of their contemporary English manufactories (for example, the Royal Worcester China Works), no previous book has devoted its theme to these Welsh armorial porcelains, which it will transpire have generally only been mentioned sporadically and *en passant* in most of the existing literature on Welsh porcelains extending over the past two hundred years and written since the manufactories both formally terminated their porcelain production in 1820. Some earlier historical accounts of the porcelains produced by the Nantgarw and Swansea China Works do not even mention the existence of armorial porcelains, for example Sir Arthur Church (Church, *English Porcelain: A Handbook to the China Made in England During the 18th Century as Illustrated by Specimens Chiefly in the National Collection*, 1885 and 1894) and William Turner (Turner, *The Ceramics of Swansea and Nantgarw. A History of The Factories with Biographical Notices of the Artists and Others, Notes on the Merits of the Porcelain, the Marks Thereon Etc.*, 1897). In contrast, the armorial porcelains from the Royal Worcester China Works are well represented in the historical literature and have been documented in several major collections nationally and internationally (*Armorial Worcester Porcelain of the First Period: Specimens from the Marshall Collection in the Ashmolean Museum and Other Sources*, 1964; Sandon, *The Ewers-Tyne Collection of Worcester Porcelain at Cheekwood*, 2007). Similarly, the armorials manufactured by the Derby China Works are also well illustrated, for example in John Twitchett's book (Twitchett, *Derby Porcelain: An Illustrated Guide, 1748–1848*, 2002). An exception in the literature for the Welsh porcelains is a brief, one-page article by David Phillips on heraldic emblems on Swansea porcelain (D. Phillips, "Heraldic Swansea Emblems", *Nantgarw & Swansea Newsletter*, p. 18, 2006) in which he cites nine examples of crested Swansea porcelain services, of which only three have been attributed to date, namely Clarke of Hereford, Lloyd of Bronwydd and Venn. Two of these armorials are now actually the subject of some conjecture and require a re-assignment, namely Clarke of Hereford and Venn, and no attribution was made for the other six "unidentified" examples cited in the article. In Fergus Gambon's authoritative and recent survey of the Andrews Collection of Welsh porcelain in Plas Glyn-y-Weddw (Gambon, *Porslen Abertawe a Nantgarw (Swansea and Nantgarw Porcelain)*, 2016), seven examples of crested porcelain are presented, comprising six from Swansea and one from Nantgarw. Of these, the Nantgarw example is shown to have been correctly attributed therein, but five of the six Swansea crested services are logged as being unidentified and only one has been attributed, namely Clarke of Hereford, about which there is now some re-attribution required, as we shall see later. In summary, therefore, this is an ongoing static but not very satisfactory situation and we should examine the potential reasons that contribute to it.

In the preparation of a companion volume to this book, which addressed the establishment of the principles of the holistic forensic scientific basis and the application of a proposed and proven scientific methodology for the examination and attribution of source factories for unknown porcelains (Edwards, *Porcelain Analysis and its Role in the Forensic Attribution of Ceramic Specimens*, 2021), it was realised that the current situation regarding the Welsh porcelain armorials was that they had simply largely been ignored or had been marginalised by ceramic historians hitherto. The author attributes this to the recognised failure historically to identify their family arms bearers, and an opportunity now existed using modern research methods for advancing our knowledge about these armorial porcelain exemplars, the people who commissioned them and whose arms they bore and the places that they occupied in the history of the two china manufactories at Nantgarw and Swansea. This situation has pertained into the present time and several ceramic historians of Swansea and Nantgarw china have occasionally noted this deficiency but have not addressed it in their writings, aside from commenting correctly that the armorials remain largely unidentified. In one of the most recent books to be written on the Swansea and Nantgarw porcelains by Fergus Gambon (Gambon, *Porslen Abertawe a Nantgarw (Swansea and Nantgarw Porcelain)*, 2016), Director of British Ceramics at Bonhams Fine Art Auctioneers and Valuers in London, in which he surveyed the premier Andrews Collection of Welsh porcelains at Oriel Plas Glyn-y-Weddw in Llanbedrog, Abersoch, Gwynedd, the following concise statements were made which summarise precisely the current perception among connoisseurs and historians regarding Welsh armorial porcelain artefacts and porcelain services:

> *A number of Swansea crested services are recorded but most of the crests have not been identified ...... full armorial bearings are a great rarity at Swansea. Those simply gilded with a family crest are more commonplace, although most have yet to be identified.*

And

> *Very little Nantgarw armorial porcelain is recorded.*

As stated above, Fergus Gambon noted the presence of six Swansea and one Nantgarw armorials in his survey of the collection at Plas Glyn-y-Weddw. The scene is now set, therefore, for some timely and extensive research to be undertaken in an attempt to better understand and to improve the current situation regarding these rather uncommon, but as yet little understood, examples of ceramic art pertaining to the Swansea and Nantgarw porcelain manufactories. In a parallel detailed research study, which has been undertaken by the author and two of his colleagues over several years recently (Denyer, Denyer and Edwards, *The Pendock-Barry Derby Porcelain Service: A Reappraisal*, to be published, 2022), the intricate and detailed genealogical, heraldic and historical research carried out on the important *Barry-Barry* armorial service from the Derby China Works, which was commissioned in the early nineteenth century, has now reached definitive conclusions which have paved the way for its re-assessment and correct placement in ceramics history. This study has demonstrated conclusively the pivotal role that can be played by genealogy in armorial research for the correct attribution and potential dating of a porcelain service commission. Until now, the *Barry-Barry* (alternatively known as the *Pendock-Barry*) armorial porcelain service has been the subject of much conjecture and discussion in the literature regarding the chronology of its commissioning from the Derby China Works in the early nineteenth century, and, hence, there has been much confusion and contradiction expressed about its dating and the identity of the artist involved in its superb decoration. As a result this particular research, the centre of gravity of which has required the detailed interpretation of the heraldic escutcheon displayed centrally on this armorial porcelain service related to the family coats-of-arms and using the funerary hatchments of the family displayed in the local parish church of St Peter's at Tollerton, has thereby provided a critical input to decisively ameliorate a highly unsatisfactory and uncertain situation which has pertained for many years in the literature and which was otherwise seemingly irresolvable (Denyer, Denyer and Edwards, *The Pendock-Barry Derby Porcelain Service: A Reappraisal*, to be published, 2022; Twitchett, *Derby Porcelain: An Illustrated Guide, 1748–1848*, 2002; John, *William Billingsley*, 1968). Basically, this had come down to a difference of opinion among several ceramics connoisseurs and historians historically regarding the precise chronology of commissioning of this armorial service, during which some seemingly incorrect assumptions about the operations of the Derby China Works and its enamellers had been made which fuelled the conjecture surrounding this service. The application of a forensic holistic approach wherein scientific analytical data and historical documentary information, especially concentrating upon the genealogical and heraldic studies, has resolved some conflicting issues, established the baseline and facilitated the correct attribution chronologically for this porcelain service and the consequential assignment of William Billingsley as its artist and enameller. The adoption of similar well-tried procedures and the assimilation of evidential and circumstantial material surrounding the armorial bearings on the Welsh porcelains should therefore perhaps set the basis for a better appreciation of the current Welsh armorial porcelain canon, the definition of which is an essential part of the application of the modern forensic scientific analytical research process. Having established this canonical baseline for the Swansea and Nantgarw armorial porcelains, future researchers shall then be in a much better position to extend the database for the evaluation of any newly discovered and pertinent material and exemplars that may surface in connection with the identification and attribution of Welsh armorial porcelain artefacts that remain yet to be unearthed. The author has surveyed the existing literature relevant to Welsh porcelain armorials and, hopefully, he has uncovered all currently known exemplars of the genre. If there are one or more additional Welsh porcelain armorials sitting unrecorded in private collections which have, therefore, not been featured here in this current research, the emergence of these as yet undiscovered examples will then provide a suitable cachet for adding to the database for the testing and evaluation procedures established herein for future research work that may be carried out in this fascinating area. Research is, by definition, an ongoing exercise, and each inductive iteration of the adopted forensic evaluation procedure for the identification of a particular armorial porcelain exemplar from the research database could possibly create a novel situation that may require some refinement of the originally established hypotheses or ideas. Thereby, this process could illuminate and inform the resolution of an existing conjectural situation that could have arisen elsewhere. This procedure was certainly adopted by that master practitioner of inductive reasoning in the Victorian era, Sherlock Holmes, and many of his statements made between one hundred and one hundred and forty years ago can be appropriately quoted with relevance to our task here today.

There is no doubt whatsoever that armorial porcelains from the Swansea and Nantgarw manufactories, despite their relative rarity compared with their more "anonymous" but nevertheless beautifully decorated analogues, deserve their rightful place in the Welsh ceramics canon and should therefore be assimilated properly into the existing genre of Welsh porcelains. This is the objective of the armorial research described here. Most of the crests and armorial bearings reproduced on Nantgarw and Swansea porcelain artefacts have now been identified as a result of the research outlined in this book, and this has, therefore, now removed these armorials as a class from their previous all-embracing category of "rare, but unidentified" in the Welsh ceramics genre which has been occupied hitherto by the Nantgarw and Swansea armorial porcelains. During this identification process, the attribution of unidentified armorials, and indeed several others that had been previously identified inappropriately, has occasionally revealed some interesting history either relating to the family arms bearers (armigerous families) or to some manufactory practices at Swansea and Nantgarw that were perhaps unsuspected.

Finally, during this research on Welsh armorial porcelains it would have been fallacious to have ignored the presence of armorial bearings found elsewhere, especially those comprising crests and escutcheons on silver plate and on earthenware ceramics, as these co-existed chronologically in the early nineteenth century with their presence on the porcelain artefacts that we are investigating here. A classic case would be the use of armorially engraved silver ladles for porcelain, creamware and earthenware sauce tureens, cream bowls and soup dishes in the dinner and dessert services used at Georgian dinner tables in the late eighteenth and early nineteenth centuries as well as silver cutlery, such as carving knives, serving spoons, knives and forks, which also might be expected to possess engraved armorial devices or crests. Hence, at the appropriate points in the text some lateral digressions will be made to illustrate armorial examples taken from engraved armorial silver plate and from ceramic armorial earthenwares. In the case of the latter, those from the Cambrian Pottery and, to a lesser extent, the Glamorgan Pottery in Swansea are particularly relevant as both were contemporaries of their neighbouring Swansea China Works where the porcelain was being manufactured. A novel approach adopted here was the comparison of the armorial bearings, crests and escutcheons which are found on selected examples of creamwares and earthenwares from the Cambrian Pottery with those unidentified crests appearing in the literature on Swansea porcelain. A second novel approach has been to investigate the role played by the local supporters of the Swansea and Nantgarw China Works and potential clients, many of whom were titled arms-bearers who undoubtedly would have possessed silverware and plate engraved with their crests and escutcheons and would have had bookplates displaying the same coats-of-arms in their library volumes, and who would also be prime candidates for the commissioning of armorial porcelain services from these manufactories to grace their tables.

In terms of setting a date for the appearance of the earliest armorial porcelains in Europe, historical evidence points to King Manuel I of Portugal being the first European monarch to commission a piece of Chinese porcelain which bears his coat-of-arms from a Portuguese expedition he despatched to China in 1517. However, the first record of a coat-of-arms being used to decorate European indigenous porcelain must reside with Duke Francesco de Medici of Tuscany, some years later, who offered several porcelain bottles of his own Florentine Medici porcelain as "diplomatic gifts" to King Felipe II of Spain in 1581, which bore the coat-of-arms and crest of the King. In the sixteenth century, porcelain was much admired for its novelty and beauty, and it possessed an associated alchemical legend regarding its protective ability to render harmless the toxins which tended to appear in contaminated food and drink. The first English porcelain armorial historically is recorded on Chinese porcelain and is believed to date from about 1698, but the first English armorial service (again on Chinese porcelain) for which examples of artefacts are still extant today is the service made for Sir William Pitt, a Governor of the HEIC in Madras, in 1705. There is apparently no record for the date of the first piece of English porcelain from a service to bear a crest or a coat-of-arms, but a diligent search of the literature assigns this probably to the Worcester China Works in 1767/8, which would have been preceded by the decoration in England of armorial porcelains imported in the white from China in London ateliers such as those of James Giles in the early 1760s. There is a rare Liverpool William Reid porcelain armorial punchbowl in existence, dating from about 1757/8, which represents probably the earliest single English porcelain armorial piece still extant, although, of course, this does not qualify technically as a porcelain armorial service for our study here. For the Welsh porcelains, the first armorial piece produced is difficult to date precisely, because all of the 18 different armorials identified thus far can be assigned to the fairly narrow date range between 1815 and 1823; it would be beneficial in this context if one could identify a potentially definitive reason for the commissioning of a Nantgarw or a Swansea armorial piece, using genealogical data in the absence of factory records which would have noted the precise date of such a commission. Potentially, the *Viscount Weymouth* armorial service for Lord Henry Thynne in Nantgarw porcelain is a contender for the earliest Welsh porcelain armorial, especially since it could have been commissioned to celebrate his coming of age in 1818. However, a combination of detailed heraldic and genealogical research has indicated that this armorial has been incorrectly ascribed hitherto and several important anomalies have arisen in the depiction of the coat-of-arms which are suggestive of a deeper meaning underlying the commission and potentially a slightly later date.

Chemical & Forensic Sciences
University of Bradford
Bradford, UK

Howell G. M. Edwards

Acknowledgements

In his research into the Welsh porcelains of Swansea and Nantgarw over a period of four decades the author acknowledges the efforts of previous writers on the ceramic history of these two porcelain manufactories of which several of the earliest had only a limited access to source documentation but were able to include information about their operations and personnel from eyewitness accounts. The porcelains from these manufactories were highly valued by a discerning clientele, but severe problems encountered in their production processes resulted in a limited lifetime of the manufactories of only a few years in the second decade of the nineteenth century between 1815 and 1820. It soon became apparent that a neglected area in previous accounts of Swansea and Nantgarw porcelains was the investigation of the special commissions of armorial porcelains placed by armigerous families which was not facilitated by the non-existence of factory work books and agency sales records. The author wishes to acknowledge the assistance and time given freely by the fol lowing colleagues, friends and museum curators for their informed discussions which have encompassed many years' experience in ceramics and documentary research. Several of these have kindly provided specimens for the studies described in this book whilst others have facilitated access to archives and papers and have given their permission for the reproduction of the appropriate photographs from the porcelain and ceramic specimens in their collections:

Dr Morgan and Mrs Rachel Denyer, Penrose Antiques, Thornton, West Yorkshire; Charles Fountain, Director, Nantgarw China Works Museum Trust, Tyla Gwyn, Nantgarw, Glamorganshire; Stuart Brown, Builth Wells, Powys; the late Dr Peter Bradshaw, Sheffield, South Yorkshire; Andrew Renton, Keeper of Art, Amgueddfa Cymru, National Museum of Wales, Cardiff, Glamorganshire; the late Dr William John, Newport, Gwent; Sarah George, Co-Faculty Librarian for Life Sciences and Subject Librarian for Archaeology, Chemistry and Forensic Sciences, University of Bradford; Gwyn Jones, Director, John Andrews Charitable Trust, Oriel Plas Glyn-y-Weddw, Llanbedrog, Gwynedd; Guest and Gray Antiques, Mayfair, London; Kingschina Antiques, Mayfair, London; Jonathan Gray, Swansea; Dr Luke Syson, Director, Fitzwilliam Museum, Cambridge; Indar and Aruna Pasricha, Kensington,

London; Fergus Gambon, Director of British Ceramics, Bonhams Fine Art Auctioneers and Valuers, New Bond Street, London.

In grateful appreciation to you all!

Finally, the author wishes to acknowledge the lifelong support and joint interest in so much of this research effort of his dear wife Gill who sadly passed away at the end of 2019 and the ongoing support of our daughter, Kate.

January 2022

Saltaire, West Yorkshire

Howell Edwards

Contents

List of Figures

List of Tables

About the Author

Howell G. M. Edwards was born in Skewen, South Wales, and is Professor Emeritus of Molecular Spectroscopy at the University of Bradford. He read Chemistry at Jesus College in the University of Oxford, and after completing his B.A. and B.Sc. degrees he studied for his doctorate in Raman spectroscopy at Oxford with Dr Leonard Woodward, *Aldrichian Praeelector in Chemistry*, and then became a Research Fellow at Jesus College, University of Cambridge, where he worked with Dr W. Jeremy Jones of Trinity College. He joined the University of Bradford as a Lecturer in Structural and Inorganic Chemistry, becoming Head of the Department of Chemical and Forensic Sciences, and he was awarded a Personal Chair in Molecular Spectroscopy in 1996. He has received several international awards (Sir Harold Thompson Award; Charles Mann Award; Emanuel Boricky Medal; Norman Sheppard Award) in a spectroscopic career which has resulted in the publication of over 1350 research papers in Raman spectroscopy and the characterisation of materials, along with six books on the application of this analytical technique to art, archaeology and forensic science. He has had a lifelong interest in the porcelains of William Billingsley, especially those from the Derby, Nantgarw and Swansea porcelain factories. He has authored five major books on Nantgarw and Swansea Porcelains: *Swansea and Nantgarw Porcelains: A Scientific Reappraisal*; *Nantgarw and Swansea Porcelains: An Analytical Perspective*; *Porcelain to Silica Bricks: The Extreme Ceramics of William Weston Young, 1776–1847*; *18th and 19th Century Porcelain Analysis: A Forensic Provenancing Assessment* and *Porcelain Analysis and Its Role in the Forensic Attribution of Ceramic Specimens*, all published by Springer-Nature Publishing, Dordrecht, The Netherlands; this particular book is his sixth on this subject. He has also produced several monographs on these manufactories: *William Billingsley – The Enigmatic Porcelain Artist, Decorator and Manufacturer*; *Nantgarw Porcelain – The Pursuit of Perfection*; *Swansea Porcelain – The Duck-Egg Translucent Vision of Lewis Dillwyn* and *Derby Porcelain: The Golden Years, 1780–1830*. He is currently preparing a research text for publication in 2022 on *Raman Spectroscopy in the*

Preservation of Cultural Heritage, for which porcelain artefacts will feature as artworks and a vital part of a nation's heritage, as does also the industrial archaeology, excavation and the preservation of the early porcelain manufactory sites. Howell Edwards is Honorary Scientific Adviser to the de Brecy Trust on the scientific evaluation of their artworks and paintings and is a Friend of the Nantgarw China Works Museum Trust. Howell Edwards was married to Gillian, who sadly passed away in 2019; they have one daughter Kate.

Chapter 1
The Birth of the Porcelain Trade with China

Science is the Highest Form of Creative Art.
Sir Chandrasekhar Venkata Raman, Bangalore, India.
Nobel Laureate in Physics, 1929.

Abstract A history of armorial porcelain imported into England from China through the Honourable East India Company (HEIC) and the oldest recorded English armorial service extant, namely the *Pitt* service from 1705. The change in dining fashion in early Georgian England that required large services and the emblazoning of armorial bearings upon these as evidence of status and social standing. Examples of finely decorated English armorial services for the aristocracy and the largest compound armorial service ever made for the HEIC. A consideration of what constitutes an armorial service, named services, pseudo-armorials, the *Frog* service for the Empress Catherine the Great of Russia. A survey of the literature on Swansea and Nantgarw china and the distillation of the distinct armorial exemplars, which number only 7 (or possibly 8) for Nantgarw and 11 for Swansea.

Keywords Chinese armorials · Pitt service · HEIC · Frog service · Nantgarw armorials · Swansea armorials

1.1 Chinese Armorial Porcelain – The Catalyst for the European Armorials

Undoubtedly, the largest generic collection of armorial porcelain which immediately comes to the attention of ceramics observers and historians is to be found in the Chinese export wares of the 18th and 19th Centuries, as can be evidenced from David Howard's comprehensive research on this topic which resulted in the publication of his two authoritative, scholarly and classic seminal texts (Howard, *Chinese*

H. G. M. Edwards, *Welsh Armorial Porcelain*,
https://doi.org/10.1007/978-3-030-97439-8_1

Armorial Porcelain, Volumes I and II, 1974 and 2003). In these books Howard cites details of over 5200 documented Chinese armorial services, all made for entitled armigerous British and American families in the 18th and early 19th Centuries, with over 3300 illustrations shown of their coats-of-arms and crests. Their unifying theme is that they were all transported from China into Great Britain as cargo in the Honourable East India Company (HEIC) East Indiamen ships from the port of Canton (Guangzhou). The two books of Anthony Farrington also prove to be an essential consultation source for cross-referencing and background information for the carrying out of detailed armorial research in this context (Farrington, *A Catalogue of the East India Ships' Journals and Logs, 1600–1834,* 1999a; Farrington, *A Biographical Index of the East India Company Maritime Service Officers, 1600–1834,* 1999b). The Farrington books provide specific details of the ships and their cargoes, their captains, agents and key personnel involved, the dates of their outward sailing and of their return to Britain with their porcelain artefacts and services. Considerable quantities of Chinese porcelain were imported into Europe by the HEIC and by the analogous Dutch East India Company (VOC), the Swedish East India Company (SOIC) and also by Portuguese vessels franchised through the British and Dutch East India Companies after the collapse of their own Portuguese Casa da India Company in the seventeenth Century (Edwards, *Porcelain Analysis and Its Role in the Forensic Attribution of Ceramic Specimens*, 2021). In contrast, the French East Indies Company (Compagnie Francaise pour le Commerce des Indes Orientales, 1664–1769) does not seem to have played a major role in the importation of Chinese porcelain into Europe; it was based at Pondicherry in India, which was governed by the French from 1674 until the British took over in 1760, where it primarily pursued an ongoing trade war commercially against its local British, Portuguese and Dutch competitors in India. In the books of Howard and Farrington are given details of the ships' captains, supercargoes, shipowners, names of the ships, dates of the voyages, the Governors of the three HEIC districts of Bombay, Madras and Bengal and the names of the influential merchant traders dealing in the shipments of the main cargoes of spices, tea, silks, and porcelain carried by the East Indiamen ships. These authoritative documentary histories are invaluable for tracing the commissioning of a Chinese armorial service and for establishing through an historical audit its provenance and the dates of its completion and eventual delivery to the client. As Anthony Farrington has noted (Farrington, *Trading Places: The East India Company and Asia, 1600–1834*, 2002), the HEIC soon became *"the greatest society of merchants in the universe"* and he also mentioned the following relevant observation in specific connection with the HEIC carriage of armorial porcelains on their East Indiamen ships:

> *Once the regular trade at Canton was established the company* (author: the HEIC) *came to concentrate upon three principal commodities – tea, textiles and inexpensive porcelain. All the fancy goods and special orders, for example, armorial porcelains or large decorative pieces, were left to the private trade of their servants and ships' officers.*

This is an important statement that is relevant to our theme in this book of armorial porcelain and reflects the unique place reserved for armorial porcelains and the special orders regarding their commissioning and shipment which were seemingly often undertaken effectively as a profitable "sideline" by the supercargoes and East Indiamen ships' captains. Whereas most of the porcelain shipped by the HEIC from China was stated to be of the "anonymous" and, therefore, not of the armorial and specific family-related category from the point of view of bearing coats-of-arms or heraldic devices, the reverse situation applied for the Dutch VOC, for which it has been recorded that their private commissions, including armorials, far outweighed in number their general shipments of porcelain (Vollker, *Porcelain and the Dutch East India Company*, 1954).

1.1.1 The Portuguese and China

In 1488, Bartolomeo Diaz rounded the Cape of Good Hope and entered the Indian Ocean, so starting the establishment of the Portuguese colonial empire, the Seculo de Oro, when followed by Vasco da Gama reaching India in 1498 and Pedro Alvarez Cabral reaching Brazil in 1500. The Portuguese initially discovered Chinese porcelain in India and the first pieces were brought back to Lisbon by Vasco da Gama in his flagship *Sao Gabriel* and presented to King Manuel I (reigned 1495–1521) on his return to Portugal in 1499 after navigating the Cape of Good Hope, crossing the Indian Ocean and making a landfall in Calicut. For his exploits Vasco da Gama was created "Admiral of the Seas of Arabia, Persia, India and All the Orient". King Manuel I was so pleased with his gift of porcelain that he ordered some more from a Portuguese expedition to China in 1517 to be decorated with a Portuguese armillary sphere device as exhibited on his personal standard and an ewer bearing his heraldic coat-of-arms, which probably represents the first specific European commission to be made for Chinese armorial porcelain historically (Rui 1990; Graca 1977). The fascination held for this novel ceramic material by the aristocratic rulers of Europe thereafter was almost immeasurable and was so labelled "*die Porzellankrankheit*" (the porcelain disease), or as Dr. Samuel Johnson described it somewhat later, "*the contagion of china-fancy*" (Finlay, 1998; Redford, *The Letters of Samuel Johnson,* 1992). King Manuel I commissioned Fernao Pires de Andrade to establish trade with China in command of a fleet of seven caravels, lateen-rigged and highly manoeuvrable small vessels of approximately 150 tons burthen, which became the first European vessels to officially reach the Chinese coast, landing first in Macao and then navigating the Pearl River estuary to reach Canton on the 15th August 1517, although there are several reports that forays were made earlier to Guangdong and Canton by Alvares and Perestrello in 1516. It is interesting that Andrade and his second-in-command, Tome Pires, were both apothecaries and this reinforces the idea that the main purpose of this Portuguese expedition was to evaluate the potential of Chinese herbal drugs and medicines for European adoption. The expedition had to wait 2 years in Canton before they could secure an audience with the Emperor Zhengde (Zhu Houzhao) to advance their trading protocols. In the meantime, diplomatic relations between the local Chinese and the Portuguese in Canton deteriorated and with the installation of a new Emperor Jiajing (Zhu Houcong) the visitors were incarcerated for several years. The situation did not

Fig. 1.1 Armillary sphere adopted as the personal standard of King Manuel I of Portugal in 1495 which was commissioned for his order of armorial Chinese porcelain in 1517 through the first Portuguese trade mission to China. *Public domain*

improve until Jiajing ceded Macao to the Portuguese in 1557 in recognition of their efforts to combat piracy in the China Sea and by that time most of the members of the Portuguese expedition had died. The next official European diplomatic mission to China was that of Matteo Ricci SJ, an Italian Jesuit, in 1598 who presented his credentials to the Emperor Wanli. It seems credible, therefore, that King Manuel never did receive his armorial Chinese porcelain commission, if indeed it had ever been ordered in China, as he had died in 1521 to be succeeded by King John III, and that the expedition did not return to Portugal (Brook, *Confusions of Pleasures: Commerce and Culture in Ming China*, 1998; Montalto de Jesus, *Historic Macao: International Traits in China, Old and New*, 1902). It is interesting that the armorial porcelain commission of King Manuel I would bear the armillary sphere, shown here on his personal standard in Fig. 1.1 which was adopted thereafter as a supporter of the Portuguese heraldic shield, which bore five escutcheons in cross azure, each charged with as many plates in saltire, all within a bordure gules charged with seven triple-towered castles (Fig. 1.2).

1.1.2 The First Porcelains

To begin at the beginning
Sherlock Holmes, A Study in Scarlet, 1887.

The discovery of porcelain in China is generally accredited by historians to belong to the Tang Dynasty, which ruled from 618 to 906 CE, although some historians have placed its manufacture even earlier than this, which may be attributed to a rather loose definition of what constitutes porcelain *versus* other glazed ceramic analogues; it is recognised that several attempts were made at its manufacture during this time, but with varying degrees of success, so one specific date cannot be identified for the start of porcelain manufacture in China. Some historians feel that porcelain started to appear during the earlier Sui dynasty and perhaps even during the divided Northern and Southern Kingdoms prior to that but the evidence is tenuous. Certainly, during the Song/Sung Dynasty, 960–1279 CE which followed a troubled Tang /Liang/Zhou Chinese interregnum, founded by the Emperor Taizu, quality porcelain ceramics were now appearing on the market and these wondrous

Fig. 1.2 Royal coat of arms of the Kingdom of Portugal from the sixteenth Century comprising five escutcheons contained within the shield and the armillary sphere of Fig. 1.1. adopted as the heraldic supporters. *Public domain*

new materials were being talked about enthusiastically by the first Western explorers and adventurers who had observed them in the Far East. Marco Polo (b. 1254, d. 1324), the Venetian adventurer and traveller (Marco Polo, *Il Milione*, 1928), who arrived in China with his merchant father in 1271 CE and was received at the Yuan Dynasty Court of the Mongol Emperors (Kublai Khan, Shizu, 1260–1294, and Temur Khan, Chengzong, 1294–1307), is believed to have been the first Westerner to describe Chinese porcelain he saw in Canton and to have brought a specimen back with him to Europe:

> *The most beautiful vessels and plates of porcelain, large and small, that one can describe are made here in great quantity .... more beautiful than can be found in any other place. And on all sides they are much valued for none of them are made in any other place but in this city and from there they are carried to many places throughout the world. And there is plenty there and the great sale, so great that for one goat you would actually have three bowls so beautiful that none would know how to name them better.*

However, the first actual literary record of the existence of Chinese porcelain was reported very much earlier by Suleiman, an Arabic merchant trading with China, who pre-dated Marco Polo by four centuries and who in 851 CE, during the preceding Chinese Tang Dynasty, said:

> *They have in China a very fine clay with which they make vases which are as transparent as glass; water is seen through them. These vases are made of clay.*

Although many ceramic historians believe that the ceramics of the Tang Dynasty were more like faience in character and really, therefore, comprised a predominantly non-translucent earthenware or stoneware substrate with an applied vitreous glaze, clearly some production of a translucent ceramic had been undertaken successfully during this period as described above by Suleiman, but perhaps this had been achieved only in rather limited quantities. A recently excavated wreck of an Arab dhow which sank in 830 CE in the Singapore Straits, off Belitung Island, with a cargo of ceramics *en route* for the Abbasid Dynasty in Arabia revealed that she carried a mixed ceramic cargo of 60,000 pieces. The cargo comprised mostly Tang Dynasty glazed stonewares, with some Yue celadons and about 300 pieces of a very fine Xing white porcelain from Heibei, which indicates that even at this early period the Chinese were manufacturing and exporting large quantities of novel glazed ceramics and some true porcelain pieces. Guo Yanyi (1987) has reviewed the raw materials that were available for porcelain production in China during the Yuan Dynasty (1271–1368 CE) and discussed the different characteristics that arose between the Southern kilns based around Jingdezhen and the Northern kilns based around Dehua. Basically, the porcelain stone that comprised a significant proportion of the porcelain paste in the Southern Chinese kilns was predominantly composed of quartz and sericite which conferred a highly siliceous nature upon the Chinese porcelains made there and the glazes were mixtures of porcelain stone and wood ash with some limestone added. In contrast, the Northern sites used more clay in their mixtures for their ceramic body and a feldspar and calcined wood ash for the glaze. The kiln design also differed between the Northern and the Southern sites: the *mantou* kiln (a horseshoe shaped kiln) was used in the North whereas the *dragon* kiln (of a linear sloping design built onto the surface contour of a hillside) was used in the South and both were believed capable of achieving a maximum operational temperature of near 1400 °C. From the Yuan to the Qing Dynasties a progression involving the addition of more kaolin to the clays or china stone raw materials was made to thereby improve the porcelain quality (Medley, *Yuan Porcelain and Stoneware*, 1974).The later Ming Dynasty, 1368–1644 CE, established the basis for the creation of the very fine quality Chinese export porcelains of exceptionally good translucency with applied blue and white decoration and whose imports by the Honourable East India Company (HEIC; The British "John" Company, founded in 1600), the Dutch East Indies Company (VOC; Veerenigde Oost-Indische Compagnie, founded in 1602) and independent or chartered Portuguese navigators in large quantities took Europe by storm in the 17th and 18th Centuries (Carswell, *Blue and White Chinese Porcelain*, 2000). A special landmark in the history of Chinese porcelains was the preparation and adoption of a compatible blue enamel which would not run into the glaze and thereby spoil the applied enamelling decoration when fired onto the porcelain substrates: this was created during the Ming Emperor Xuande's reign (Zhu Zhanji), the fifth Emperor of the Ming Dynasty, from 1424–1435 CE, and he was also responsible for the construction of the very large innovative dragon kilns at Jingdezhen and for the location of his Imperial porcelain manufactory there. Although the Ming period is justifiably famous for its blue and white porcelain designs and decoration, the later reigns of Kangxi (Xuanye) (1661–1722 CE), Yongzheng (1723–1735 CE) and Qianlong (1735–1796 CE) of the Qing Dynasty undoubtedly increased the flow of Chinese export porcelains into Europe with their use of polychrome enamel decoration, which was very much admired in the West. During the later Kangxi period, these polychrome porcelains acquired the names in the West of *famille jaune* (yellow), *famille noire* (modi susancai, black), *famille rose* (fencai, ruancai, yangcai, pink) and *famille verte* (wucai, green) (Kerr & Wood, *Science and Civilisation in China*, 2004). It is believed that the transition to polychrome enamelling from the standard blue and white decoration began in earnest in 1729 when the Emperor Yongzheng instructed the enamellers at Jingdezhen to utilise some new pigments and enamels that he had acquired for the decoration of their fine porcelains, the catalyst for this being a perceived increase in the export trade as the Emperor collectively termed these decorations *yangcai* – which literally translated means *foreign overseas painting*. The selection of Jingdezhen as the site for the Imperial porcelain manufactory was based upon the supplies of kaolin and *petuntse* that were readily available locally there: the surrounding hills lent themselves well to the construction of the famous dragon kilns, utilising the natural slopes of the land, and which it was said could hold up to 25,000 items at each firing using pine wood as a fuel. A serious problem was believed to be the decrease in temperature over a range of several hundred degrees that was experienced vertically in the kiln which would have certainly resulted in a variable quality of the porcelain production achieved from each firing process. The kilns at Jingdezhen were destroyed in the civil war that raged in China after the collapse of the Ming Dynasty in the mid-seventeenth Century and the influx of alternative Japanese porcelains into Europe then started to make an impact, firstly through their blue and white Arita porcelain but more so with their polychrome, red, blue and gold, Imari and Kakiemon wares (Medley, *The Chinese Potter: A Practical History of Chinese Ceramics*, 1989). The Dutch were able to arrange the importation into Europe of Japanese Arita porcelain at this time as they had a special arrangement with the Japanese Edo Emperors to fill the void left by the curtailment of availability of the Chinese porcelains for export. In the late 1680s, with the return to stability in China after the civil wars, the re-establishment of the Jingdezhen manufacturing kilns and the creation of the new Imperial enamelling workshops in the Forbidden City in Beijing were undertaken by the Emperor Kangxi (Xuanye), the first Emperor of the Qing Dynasty. This then restored the supremacy that had been enjoyed by the Chinese porcelain exported hitherto into Europe. This was further enhanced around 1700 by the adoption of a new glassy and superior deep blue cobalt pigment for decoration and also by the use of other coloured enamel pigments which were revealed in China by the French Jesuit mission to Beijing; these bright enamel pigments had first appeared on the enamelled metal and glass gifts presented by the Jesuits from the King of France to the Emperor Kangxi (reigned 1661–1722), as recorded by Father Xavier d'Entrecolles SJ. The Jesuits were prime movers in this information technology exchange with China, encouraged by a mandate from King Louis XIV to discover more details about the Chinese porcelain production. These new pigments were quickly adopted by the Chinese enamellers for their porcelains and modern chemical analyses now confirm their presence through trace element detection and the discovery of the presence of arsenic in several cases (Colomban et al. 2021) which was indicative of the Chinese use of European cobalt ores, such as those from Saxony. As a result, this directly facilitated the introduction of the novel Chinese polychrome enamelled porcelain to compete with the Japanese polychrome Arita and Kakiemon wares that were then becoming so admired in Europe (Curtis, *Glass Exchange Between China and Europe, 1550–1800*, 2000; Xu Xiandiong, *Europe-China-Europe: The Transmission of the Craft of Painted Enamels in the 17th and 18th Centuries*, 2015).

1.2 The Honourable East India Company

The detailed history of the Honourable East India Company (HEIC) can be accessed through its meticulously kept archival records from the early seventeenth Century until it was closed down by Act of Parliament in 1873; these provide an invaluable source of information about the porcelain trade between China and Europe in the 17th, 18th and 19th Centuries (Keay, *The Honourable Company: A History of the English East India Company,* 1993; Howard, *A Tale of Three Cities, Canton, Shanghai and Hong Kong: Three Centuries of Sino-British Trade in the Decorative Arts,* 1997; Keble Chatterton, *A World for the Taking: The Ships of the Honourable East India Company*, 2008; Lestock Reid, *Commerce and Conquest: The Story of the Honourable East India Company*, 1971; Morse, *The Chronicles of The East India Company Trading to China, 1635–1834*, 1926; Wild, *The East India Co.: Trade and Conquest from 1600*, 2000; Pierson 2012). Originally established through a Royal Charter by Queen Elizabeth the First on the 31st December, 1600 as *"The Governor and Company of Merchants of London Trading into the East-Indies"*, the powerful HEIC accounted for half the world's trade by the mid- to late-1700s, especially for commodities such as cotton, silk, indigo, sugar, salt, spices, tea, opium and porcelain.

Initially, a convoy of five Honourable East India Company ships, namely *The Scourge of Malice* (later renamed the *Red Dragon,* 38 guns, 600 tons, built in 1595 for the Earl of Cumberland), the *Hector* (300 tons), the *Ascension* (260 tons), the *Susan* (240 tons) and the *Guest* (a small pinnace), later to become known as East Indiamen vessels, under the command of Captain James Lancaster, departed Woolwich, London, in February 1601, bearing papers formalising trade with Sumatra, the Moluccas and the East Indies. In 1602, this fleet attacked and boarded the large Portuguese carrack *Sao Thome* (1200 tons) and sequestered her cargo of pepper and spices (including her extremely valuable cargo of nutmeg and cloves), thereby supplementing their trade mission to Batavia in the East Indies. The HEIC ships returned to London from Batavia in February 1603 with valuable cargoes of pepper, spices and tea which turned a healthy profit for their sponsors upon their re-sale in Britain, although it has been commented by contemporary observers that the profits actually achieved from these sales were not as great as those that had been anticipated since the quantities of the cargo offered for sale were so large! Captain James Lancaster was knighted for his exploits by King James in 1603. Later, the HEIC cargoes comprised mostly spices, tea, silk and porcelains as the trade expanded geographically to include China. Trading posts for the fledgling HEIC were initially established in India, from the earliest outpost at Surat in 1619, first at Fort St George in Madras in 1639, then at Fort William in Calcutta in 1640, at Bombay Castle in 1668 and then finally in Canton (Guangzhou) in China in 1684 (Beurdeley, *Porcelain of the East India Company,* London, 1962; Conner, *The China Trade, 1600–1860*, 1986). The Dutch VOC (Veerenigde Oost-Indische Compagnie) also were active in their privateering of Portuguese carracks and in 1602 the *Sao Caterina,* a large carrack of 1500 tons, was captured by an armed

Dutch East Indiaman and the 100,000 pieces of porcelain in her cargo (weighing some 60 tons!) were seized and later sold in Holland: it has been said that this incident catalysed and was responsible for the Dutch desire for the acquisition of Chinese porcelain (Borschberg, *The Seizure of the Santa Caterina Revisited*, 2002).

The HEIC was always a major force in global history and trade (Pierson 2019), as exemplified by the passing of the Tea Act of May 1773, which exempted the East India Company from paying import taxes to the British Government on goods that were destined for the American Colonies and instead the government recouped the financial deficit by taxing the colonists themselves. The colonists rebelled against this tea tax on the 16th December 1773 ("The Boston Tea Party") by offloading the tea overboard from the three East Indiamen vessels anchored in Boston harbour, namely the *Dartmouth*, the *Beaver* and the *Eleanor*. A fourth ship in this convoy of East Indiamen carrying tea from London to Boston, the *William*, was wrecked off Cape Cod on its outward journey and never reached Boston harbour. The colony of Massachusetts then lost its right to self-government in a retaliatory edict by the British Government, which catalysed the American Revolution and the Declaration of Independence from Great Britain in 1776. At one time in the late eighteenth Century, in addition to its large fleet of heavily armed East Indiamen, which were classed officially as fourth-rate ships-of-the-line, nominally they were listed as 30–40 gun frigates, for their procurement into the Royal Navy as and when required during the Napoleonic War, the Honourable East India Company (HEIC) maintained its own army of 70,000 troops, who also travelled as armed marines and gunners on board their East Indiamen ships. This is the reason that East Indiamen ships were formidable adversaries when unsuccessful attempts were made to pillage and board them by pirates in the China Sea. The East Indiamen ships were frequently used as outliers for the main battle fleets when incorporated into the Royal Navy, which relied upon the speed and manoeuvrability of these scouting vessels to outrun pursuers and thereby inform the parent fleet of the imminent approach of enemy ships and of the presence of blockade runners. The HEIC was formally dissolved in 1874, following the East India Stock Redemption Act passed by Parliament the year before, after which all of its assets were annexed by the British government, who then administered Imperial India until its Partition into India and Pakistan in 1947.

The coat-of-arms and heraldic crest of the Honourable East India Company is shown in Fig. 1.3; the crest is a *lion rampant gardant or holding between the fore-paws a regal crown proper*. The official East India Company mark was known as "*The Sign of Four*", shown here in Fig. 1.4, which depicts the figure 4 sitting on a heart divided by a horizontal line: in the upper section are the letters E and I, and the letter C sits in the lower section. This mark, known as a *bale*, was used as a form of kitemark or seal of quality on East India Company goods and especially on their local currency and early postage stamps, but it is strictly not an heraldic device. The motto of the Honourable East India Company is "*Auspicio Regis et Senatus Angliae*" which translates as "*By Command of the King and Parliament of England*". An interesting but related heraldic aside relates to the flag of the Honourable East India Company as adopted and flown by the East Indiamen ships (Fig. 1.5) between 1707 and 1801, which has more than a passing resemblance to the "Stars and Stripes" that

Fig. 1.3 Coat-of-arms of the Honourable East India Company (HEIC) granted by King William III in 1698: this is a full coat-of-arms with escutcheon, crest, helmet, torse, motto and supporters. Arms: Argent a cross gules, in the dexter chief quarter an escutcheon of the arms of France and England quartering the shield, ornamented and regally crowned or. Supporters: Two lions rampant guardant or each supporting a banner erect argent charged with a cross gules. Crest: A lion rampant guardant or holding between the forepaws a regal crown proper. Motto: "*Auspicio Regis et Senatus Angliae*", translating as "*By Command of the King and Parliament of England*". Torse: gules and argent. Mantling: erminois and or. *Private Collection—CC by SA.3.0*

was later adopted as their national flag by the emergent United States of America in 1776 after the Declaration of Independence (Fawcett 1937); its specific usage for this purpose was apparently personally endorsed by Benjamin Franklin to George Washington. The Honourable East India Company flag, a forerunner of the British Merchant Navy Red Ensign, bore 13 red and white alternating horizontal stripes (the actual number of these seemed to be variable and somewhat arbitrarily selected between 9, 11 and 13) and a Union flag canton in the upper left corner bearing the Crosses of St George and St Andrew. The appearance of a modified version of this flag with a circlet of white stars on a blue background replacing the British Union flag canton used on the East India Company flag was flown on board the American brig *Nancy* in June, 1776, in the harbour of St Thomas, British Virgin Islands, and this is believed to represent the first recognition of the United States as a nation just prior to the Declaration of Independence being signed (Johnson, *Saint Croix, 1770–1776: The First Salute to the Stars and Stripes,* 2006). In the early American flag, the thirteen red and white stripes represented the 13 secessionist colonies and this was flown on the *Alfred*, the flagship of the Congressional Navy, by Lt. Paul Jones. The reason for the adoption of the red and white stripes on its flag by the

Fig. 1.4 The "*Sign of Four*" logo of the HEIC, known as a bale, which was the equivalent of an early kitemark or sign of approved quality of manufacture. It appeared on coinage and articles marketed locally in India by the HEIC. Later versions had a quartered heart enclosing the letters VEIC for the United East India Company. Here it is on a postage stamp issued in the Sindh (Scinde) district in the Indus Valley by Sir Bartle Frere, Chief Commissioner of Sindh, using a system of postal runners (dawk) to deliver the mail from 1852–1854: each stamp had the fixed value of ½ anna. *Public domain*

Fig. 1.5 The East India Company Flag, 1707–1801, with the Cross of St George, the Cross of St Andrew and 13 red and white stripes; this provided a model for the Grand Union Flag of the emergent United States of America in 1776, in which white stars on a blue background replaced the British Union Flag, as advocated by Benjamin Franklin. *Public domain*

Honourable East India Company is lost in history but there is a theory that it represented the heraldic flag of the ancient Majapahit Empire of the East Indies with whom they initially traded, the precursor of modern Malaysia and Indonesia (Fawcett 1937).

In 1817 the HEIC placed an order for the largest quantity of armorial porcelain ever known in a single commission, comprising over 7000 items of combined breakfast, tea, dinner and dessert services from Richard Chamberlain's Worcester China Works, with a salmon pink border, sumptuously gilded and displaying the full coat-of-arms of the HEIC centrally for specific use by the Governor in their HEIC Fort St George headquarters in Madras. The cost of this commission at the time was

£4190 4 s (in pounds/£ shillings/s and pence/d, £sd, where 20s made £1, and where 12d made 1 s) which equates to approximately £420,000 today (~ $550,000).

The VOC, the Dutch East India Company, also commissioned their own porcelain services; in contrast with the HEIC these did not have armorial bearings but instead bore a monogrammed device, as shown in Fig. 1.6. Figure 1.6 is a porcelain plate ordered by the Dutch East Indies Company, dating from 1660, decorated with six geometric border panels in underglaze blue in the style known as "*kraak*" (derived from the Portuguese carracks which carried Chinese porcelain to Europe) bearing the VOC logo (Veerenigde Ost-Indisches Compagnie) as a cipher centrally. Strictly, as will be discussed later, the VOC logo on this plate is a cipher but it is certainly not an armorial device. This particular plate dates from the Japanese Edo period, and is known as Arita ware, when the Dutch were the only Europeans permitted to trade with Japan but they were confined to a small island, Deshima, off the coast of Japan near Nagasaki. This was at the time when the Chinese porcelain supplies to Europe were disrupted by strife and internal rebellions after the end of the Ming Dynasty in 1644 and before the Emperor Kangxi, the first Emperor of the Qing Dynasty, restored stability to the country in 1661. The Dutch persuaded the Japanese to simulate the earlier Chinese blue and white *kraak* porcelain wares for export as they already had a good following in the Netherlands. This plate (Fig. 1.6) now resides in the Metropolitan Museum of Art, New York (Brennan Ford and Impey, *Japanese Art from the Gerry Collection in the Metropolitan Museum of Art*, 1989).

Fig. 1.6 Dutch East India Company *kraak* style large porcelain plate, diameter 12.5″ (31.3 cm), Japanese Arita ware, bearing the logo VOC in underglaze blue with a six-panelled geometric border, Edo period, *ca.* 1660. Metropolitan Museum of Art, New York, the bequest of Dr. and Mrs. Roger G. Gerry, Roslyn, New York, 2000, Accession number 2002.447.40. *Public domain*

1.3 Early European Armorial Porcelains

The earliest European porcelain manufactory was established by the Medicis in Florence in the last quarter of the sixteenth Century. The artistic chronicler, Giorgio Vasari (*Le Vite de piu eccelenti pittori, scultori e architettori,* 1568) recorded that Bernardo Buontalenti was experimenting unsuccessfully with the art of manufacturing porcelain in Florence in the 1560s. However, in 1575, the Venetian ambassador to the Court of Tuscany, Andrea Gussoni, mentioned that Francesco Medici I, Grand Duke of Tuscany, had established a porcelain manufactory in the Casino of San Marco in Florence, having discovered the secret of the manufacture of the "*porcelain of India*" after 10 years of trials and experimentation. Gussoni said that the Medici porcelain was "*transparent, hard, light and delicate*" (Lane, *Italian Porcelain*, 1954). The generator of this interest in the Medici family that gave rise to their groundbreaking venture in this first synthesis of porcelain in Europe was undoubtedly Lorenzo de Medici, who received a Chinese porcelain vase in 1487 that he admired very much via the envoys of the Mameluk Sultan in Egypt and this triggered a desire amongst the successive Medici Grand Dukes to manufacture porcelain themselves after his death in 1492 (Lane, *Italian Porcelain*, 1954; Kerr & Wood, *Science and Civilisation in China*, 2004). Cosimo Medici I (reigned 1537–1574) was an avid collector of Chinese porcelain and a catalogue, dated 1553, listed over 400 pieces of Chinese porcelain in his personal collection. In 1574, Cosimo's son, Francesco Medici, became Grand Duke of Tuscany and he constructed new porcelain workshops in the Boboli Gardens and in the Casino di San Marco in Florence. Medici porcelain is ostensibly classified as a soft paste porcelain, which includes as a raw material a white clay component from Vicenza, fortuitously containing kaolin as a component. Francesco Medici used his porcelain artefacts as "diplomatic gifts" and his manufactory was never intended to be a commercial enterprise for the sale of porcelain: of relevance to the theme of this book, he presented several pieces of Medici porcelain to the powerful King Philip II of Spain (*Felipe Segundo*, 1527–1598; reigned 1556–1598) which bore the King's coat-of-arms in an underglaze blue decoration and the author maintains that these probably represent the first recorded examples, therefore, of European porcelain armorial artefacts, in contrast to the armorial pieces that had been commissioned in Chinese porcelain some 60 years hitherto by King Manuel I of Portugal (Garca 1977; Rui 1990). Francesco presented several "square" cross-section Medici porcelain bottles in 1581 to King Philip II of Spain, two of which are now in the ceramics collection of the Musee national de Sevres, France (acquisition numbers 5345 and 5778). These diplomatic gifts were much admired by Philip, who apparently used them frequently: in sixteenth Century Europe, porcelain was believed to possess a rather magical "material agency" which could act in response to the presence of toxins and could also heal minor skin afflictions – an idea which carried over from its mystical use as a "tempered earth" in alchemy (Bowen Backus, *Asia Materialized: Perceptions of China in Renaissance Florence,* 2014)! The Italian antiquarian and historian Guido Pancirolli was concerned at the loss of knowledge from the ancient world in the sixteenth Century and he wished to draw attention to these lost things (*deperdita*) during the advancement of learning in the Renaissance in a book that was published in Latin (*Rerum Memorabilium Iam, Olim Deperditarium, Libro Duo*, 1599 and 1602) by his student posthumously (and translated into English in 1715 by John Nicholson as *The History of Many Memorable Things Lost*, 1715) (Keller, *Accounting for Invention: Guido Pancirolli's Lost and Found Things and the Development of Desiderata,* 2012*;* Keller, *Knowledge and the Public Interest, 1575–1725,* 2015). One of these *deperdita* selected by Pancirolli was porcelain and his description reveals how Chinese porcelain was considered as a "magical" creation in the mid- to late- sixteenth Century:

> *Never has porcelain been seen before, it consists of a paste of plaster, eggs and shells of marine locusts and of similar species, which after being well mixed is secretly hidden in the ground by the father of a family who then acquaints his children with its hiding place. It remains there for 80 years without seeing the light of day, after which the heirs recover it and, finding it in a fit state for manipulation, make of it those precious transparent vases so beautiful in form and colour architects can find no fault in them; amongst their inestimable virtues is that of breaking should poison be put into them. He who knows the substance never recovers it himself but leaves it to his children, nephews and heirs as a rich legacy from which they derive much profit; it is far more precious than gold.*

Amusing though some of these allegations may now seem, this statement truly reflects the awe in which Chinese porcelain was held in Europe in the mid-sixteenth Century with its associated magical powers. The idea that the raw material of a kaolin paste was buried for a considerable time echoes the recent discovery that in parts of China the kaolin/petuntse mixture is buried in large dung heaps, where after some years it can be excavated and is found to have intercalated up to 70% w/w of absorbed urea into the layer silicates, with bonds formed between the hydroxyl groups of the kaolinite layers and the amide groups of the urea, which assists in the malleability of the kaolin/petuntse mixture and the achievement of its very fine potting capability (Seifi et al. 2016; Mako et al. 2013). The historical analogy being that Chinese "eggshell porcelain" thereby emulates the behaviour of snakes who use the warmth in dung heaps for the incubation of their very thin-shelled eggs; the earliest practical application of this procedure has been traced back (Weiss 1963; Lagaly 1984) to the Sung/Song Dynasty porcelain production (960–1279). So, perhaps not so far-fetched an allegation and observation after all and clearly it is based upon scientific fact!The substantial Medici bottles (height 28 cm, base 11 × 11 cm, volume 3.5 litres) presented to King Philip II and which bear his coat of arms, include heraldic devices to signify his several titles as King of Spain and Portugal, King of Naples and Sicily, King of England (through his marriage to Mary Tudor, Queen Mary I, *jure uxoris*, 1554–1558), Duke of Milan, Lord of The Netherlands, Duke of Burgundy and King of Jerusalem. The King's escutcheon is bordered with a fettered chain bearing a sheep in deference to his title of Knight of the Golden Fleece and replacing the English Order of the Garter that was bestowed upon him by Queen Mary. His heraldic supporters in the form of two beaked, sable eagles of St John are absent. The escutcheon is surmounted by the crown of a united Spain and Portugal. An interesting heraldic device can be seen beneath the escutcheon in the form of a crest that was occasionally used at this time with the Spanish coat-of-arms, depicting the Pillars of Hercules (*Fretum Herculeum*), which historically were believed to have been situated at Gibraltar and at Jebel el Musa on the North African mainland and unusually here are placed beneath the escutcheon. The normal crest accompanying these arms would have been the embattlements of Castile and Leon with a rampant lion, which would have been located in its normal placement above the escutcheon. The second bottle does not display this crest. The base of the first bottle depicts the cupola of the Duomo in Florence beneath which is an initial *F* for Francesco Medici – providing an example of a very early factory mark. Some estimate has been made of the total production of Duke Francesco's Medici porcelain at approximately 900 pieces and some 70 of these have survived in collections (Bowen Backus, *Asia Materialized: Perceptions of China in Renaissance Florence,* 2014), which is nevertheless a considerable number.

This use by Francesco Medici of fine quality armorial porcelain as a "diplomatic gift" followed on from the original idea of the Chinese Emperors, especially those of the Yuan Dynasty (1206–1368), such as the last Mongol Emperor, Togon Temur (also known as Huizong), who reigned from 1333–1368. He sent a Chinese delegation to the court of Pope Benedict XII in Avignon in 1338 bearing a very fine *Qingbai* porcelain vase with its characteristic *yingqing* bluish-green glaze – this being the earliest piece of Chinese porcelain that appears to have been specifically made for the European market. This artefact bore no armorial bearings and is shown in Fig. 1.7. Now known as the *Gaignieres-Fonthill Vase*, it resides in the National Museum of Ireland, Collins Barracks, Dublin, arriving there via a circuitous route through the King of Naples, then via the Kingdom of Aragon and thence to Jean, Duc de Berry and Dauphin of France, the son of King Louis XIV (Lane 1961). Although itself devoid of any armorial bearings or applied enamel decoration, the attractively moulded vase did seemingly acquire some additional silver mounts in the late fourteenth Century which bore the arms of the Angevin Kings and King of Poland, although these mounts are now missing (Edwards, *Porcelain Analysis and Its Role in the Forensic Attribution of Ceramic Specimens,* 2021). Other early Chinese diplomatic gifts of Chinese porcelain from the Chinese Emperors were made by delegations to the Ottoman sultans at the Topkapi Palace in Constantinople and to the Mongol Emperors in Samarkand in the mid-fifteenth Century.

The esteem in which porcelain artefacts were held at this time is aptly described by Frei Bartolomeo dos Martires, Archbishop of Braga, who waxed eloquently about this wondrous material at a dinner given by Pope Pius IV in 1563, just a few years prior to Francesco de Medici's successful experiments in Florence (Pinto de Matos 1999):

> *In Portugal we have a kind of tableware which being made of clay, may be compared advantageously with silver both in its elegance and its cleanliness. I would counsel all present to use it in preference to any other service and to banish silver from their tables. We call it porcelain. It comes from India* (Author: The Portuguese carracks brought Chinese porcelain that had been exported to India from China at this time and only rarely directly from China) *and is made in China. It is so fine and transparent that the whites resemble crystal and alabaster and the pieces that are decorated in blue dumbfound the eyes in a combination of alabaster and sapphire. They are esteemed by the greatest princes for their delightful curiosity.*

Fig. 1.7 The *Gaignieres – Fonthill Vase*, Yuan Dynasty Chinese porcelain, *ca.* 1338, as it is today, front view, minus its later addition of silver mounts, demonstrating the superb quality of its *Qingbai* porcelain; note that the vase itself which has a moulded floral decoration does not possess any armorial decoration, although it is recorded that silver mounts added later in the fourteenth century did have armorial devices engraved. *Copyright, The National Museum of Ireland, Collins Barracks, Dublin, and reproduced with permission*

This must indeed have been a revelation by Frei Bartolomei at the Pope's dinner party and one must infer that his knowledge of this new porcelain material was still rather restricted and not universal by any means. It was only a decade or two later that Duke Francesco de Medici was undertaking the manufacture of his porcelain for diplomatic gifts, which in this social climate must have been very well received and appreciated for their novelty.It seems then that the idea of using coats-of-arms on porcelain for creating diplomatic gifts of his own porcelain to other European rulers was generated by Duke Francesco de Medici in the 1580s; Francesco also used his own characteristic heraldic *alle*, comprising several balls mounted within an escutcheon from his own coat-of-arms, as his factory stamp as well as the cupola of the Duomo in Florence, as indicated earlier. Thereafter, some hundred years later, the first French porcelain manufactory at Rouen, which was founded under the patronage of King Louis XIV in 1674, started to produce luxury porcelain pieces from about 1690 (Grandjean 1999; Froissart 2008; Solon 1905) onwards: only some nine or ten Rouen soft paste porcelain artefacts have survived to this day and it is indeed fortunate that one of these is a *moutardier* of 1695, decorated in underglaze blue and bearing the coat-of-arms of a French legal advocate and counsellor to the parliament of the Departement de Normandie, Comte Jacques Asselin de Villequiers (1669–1728), whose family was ennobled in 1643. This *moutardier* is shown in Fig. 1.8, from which the coat-of -arms is clearly visible in its blue and white decoration. These arms comprise an escutcheon with a chevron containing three tall ewers (known as *burettes* in French heraldry) supported by two gryphons rampant regardant and were first attributed heraldically to Jacques Asselin de Villequiers by Andre Pottier at the Musee de Sevres in 1847. Above the escutcheon sits a coronet which displays nine balls on spikes (also known as "pearls") which denotes the rank of Comte in French heraldry– equivalent to that of a Viscount in the heraldic lists of the

British Isles. Undoubtedly, King Louis XIV embraced the emergence of porcelain as "the ambassadorial gift of preference in the eighteenth Century" (Walton, *The Era of Louis XIV: A Turning -Point in the History of Diplomatic Gifts*", 1992). Maureen Cassidy-Geiger and Guy Walton have reasoned that this idea was also behind the production of Meissen porcelain by Augustus, Elector of Saxony, destined as diplomatic gifts in the early half of the 18th Century (Cassidy-Geiger, *Fragile Diplomacy: Meissen Production for Foreign Courts, ca. 1710–1763,* 2007; Walton, *Diplomacy and Ambassadorial Gifts in the 16th and 17th Centuries,* 2001).

Philippe Colomban et al. (2018) have analysed early specimens of Rouen porcelain in the Musee national de Sevres collection and have determined that the *moutardier* mentioned here actually possesses a *hybrid* porcelain body, which is intermediate between *hard* and *soft paste* porcelains, but Rouen porcelain is still described as being a soft paste porcelain in most historical ceramic accounts. Clearly, the presence of this coat-of-arms on Rouen porcelain could indicate a possible divergence from the initial efforts of Francesco de Medici and his idea of using porcelain as diplomatic gifts, this being dependent upon whether the *moutardier* was presented to the Comte Jacques Asselin de Villequiers by the Rouen manufactory, or perhaps even by King Louis XIV, or whether it was purchased there personally by the noble client. At the contemporary French porcelain factory of St Cloud, the porcelain was manufactured and definitely sold commercially, as evidenced by the purchase in 1701/2 of some items of St Cloud porcelain by Ehrenfriede von Tschirnhaus, a visitor to the manufactory, who later joined Augustus the Strong, Elector of Saxony, to commence the manufacture of porcelain there at Meissen with Johann Bottger in the first decade of the eighteenth Century.

Despite this extensive documented history relating to early European porcelain manufactories, dating back to that of the Florentine Duke Francesco Medici in the 1570s, then through several start-ups believed to have spontaneously arisen, if only briefly, in Genoa, Parma and Milan, with a seemingly variable measure of success,

Fig. 1.8 Rouen porcelain mustard pot and lid, *moutardier*, *ca.* 1695, bearing the coat-of-arms of Comte Jacques Asselin de Villequiers, decorated in underglaze blue. *Musee national Ceramique de Sevres. CC by SA 3.0. Public domain*

then on to the porcelain manufactories of John Dwight at Fulham in England, and Rouen, St Cloud and Antoine Pavie (Paris) in France, some historians still tend to accord precedence for achieving the European challenge to Chinese imported porcelain to the factory of Augustus the Strong at Meissen in Saxony. The reason for this is uncertain but it seems rather incongruous to still perpetuate this "myth of primacy" to Meissen, as it has been termed, when there is now so much evidence to the contrary. One of the most recent scholarly books on early porcelain (Marchand, *Porcelain: A History from the Heart of Europe,* 2020) accords the recognition of the Meissen operation as being the first European firing of porcelain in 1708 under the supervision of Johann Friedrich Bottger. The Medici firing of porcelain in Tuscany some 130 years previously and the other later efforts elsewhere in the seventeenth Century described above are hence not acknowledged. The role of Bottger relative to von Tschirnhaus in the early formative years at Meissen up to 1708 is also rather diffuse and controversial and bears further objective analysis: it is clear that von Tschirnhaus was the driving force in the early experimentation and indeed it appeared that he had already synthesised several specimens of his own porcelain prior to his engagement by Augustus the Strong at Meissen, having travelled through England and France and meeting with porcelain manufacturers there (Edwards, *Porcelain Analysis and Its Role in the Forensic Attribution of Ceramic Specimens,* 2021). The precise date of synthesis of porcelain at Meissen is also open to debate as the earliest ceramics that were made there by Bottger after the death of von Tschirnhaus were certainly not porcelain as we know it but rather a type of stoneware: it seems that the first genuine hard paste porcelain to emerge from the Meissen factory actually appeared in 1713 at the Leipzig Trade Fair and the continuation of the operation was undoubtedly promoted by its commercial success in the second decade of the eighteenth Century and thereafter (Schulle and Goder 1982; Schulle and Ulrich 1982). For a picture of the earlier European porcelain manufactories, including the Fulham Pottery of John Dwight which has only been briefly mentioned here in passing as there is no evidence that armorial porcelains were ever made there, the reader is referred to the detailed summaries provided in the recent work of Edwards (Edwards, *Porcelain Analysis and Its Role in the Forensic Attribution of Ceramic Specimens,* 2021) and the Ramsays (Ramsay & Ramsay, *The Evolution and Compositional Development of English Porcelains from the sixteenth Century to Lund's Bristol c.1750 and Worcester c.1752—The Golden Chain,* 2017). A summary of the relevant data relating to several of these selected early European porcelain manufactories and their founders is provided in Table 1.1, along with information about their production of armorial porcelain artefacts, if these are known from surviving records or artefacts.

Table 1.1 Early European porcelain manufactories, their founders and incidence of armorial porcelain

Manufactory	Location	Founder	Date of manufacture	Type of porcelain	Armorial
Medici	Florence	Duke Francesco Medici	1578–1587	Hybrid paste	Yes
Fulham	London	John Dwight	1671–1703	Hard paste	No
Rouen	France	Louis Poterat	1693–1696	Soft paste	Yes
St Cloud	Paris	Duc d'Orleans	1702–1766	Soft paste	?
Pavie	Paris	Antoine Pavie	1703–1712	Soft paste	No
Meissen	Saxony	Augustus, Elector of Saxony	1708 – Present[a]	Hard paste	Yes
Ville d'Eveque	Paris	Marie Chicaneau	1711–1766	Soft paste	?

[a]Although the foundation date of the Meissen manufactory is given as 1708 officially, the year Ehrenfriede von Tschirnhaus died, the first porcelain did not appear from Johann Bottger at Meissen until 1713; the Meissen "porcelain" which was exhibited at the Leipzig Trade Fair in 1710 was a red glazed non-translucent stoneware, *Bottgersteinzeugt* (known as *Bottgerware*), and the first translucent porcelain from Meissen did not appear until 1713: W Schulle and W. Goder, "Die erfinding des europanischen porzellans durch Bottger – eine systematische schopferisch entwicklung", *Keramische Zeitschrift,* **34,** pp. 598–600, 1982. W. Schulle and B. Ulrich, "Ergebnisse gefugeanalytischer untersuchungen an Bottgerporzellan", *Silikatechnik*, **33**, pp. 44–47, 1982

1.3.1 Early English Armorial Porcelains

The earliest English porcelain manufactories, with the exception of the Fulham Pottery of John Dwight, which was founded in 1680, and other later smaller ventures at Pomona and Bovey Tracey in 1743/44 were those of Bow, Limehouse, Chelsea and Worcester, dating from 1744–1751, some of whose products do exhibit armorial work from albeit a rather later period. However, the earliest known recorded example of an English armorial porcelain service was commissioned in the late seventeenth Century by an English shipbuilder, Henry Johnson, of Blackwall, London, who built East Indiamen ships for the HEIC: it was of Chinese porcelain and was enamelled in China but, although documented, examples of this historic armorial porcelain service are not now believed to be extant (Howard, *A Tale of Three Cities: Canton, Shanghai and Hong Kong: Three Centuries of Sino-British Trade in the Decorative Arts,* 1997, p.95). Even though the earliest armorial commissioning of Chinese porcelain from Europe is historically documented for King Manuel I of Portugal in 1517, although there is some doubt as to whether this was ever delivered, it can be argued that the personal decoration of porcelain with one's own commissioned coat-of-arms or associated heraldic devices more generally occurred first in the early seventeenth Century, probably alongside the use of similar armorial bearings that were then also being applied to silverware goods and cutlery then being used for dining purposes. This, of course, is entirely independent of the much earlier examples known of armorial porcelains which were created for "diplomatic gifts" by the nobility who were sponsoring the manufactories at Florence and Rouen, namely Duke Francesco Medici I of Tuscany (reigned 1574–1587) and King Louis XIV (reigned 1643–1715), respectively, for which specimens of their artefacts (mostly non-armorial) are still extant.

The desire for armorial porcelain dinner, dessert, tea and breakfast services by wealthy families reflected a change in social attitudes through the seventeenth and eighteenth Centuries and the way that people met and dined together in Stuart and Georgian England. Hitherto, it was normal to have the meal at dinner served centrally and communally from large tureens or platters at table onto single plates, which were then removed and washed in the kitchen between courses and returned to the table for repeated use by the diners, so minimising the need for multiple sets of dishes or plates. The dishes and plates themselves were usually made of wood or pewter, or exceptionally of silver, which was of course very expensive (Davy, *Eat, Drink and Be Merry: the English at Table, 1600–2000*, 2000; Glanville & Young, *Elegant Eating: Four Hundred Years of Dining in Style*, 2002). As dining fashion moved onto the provision of multiple courses with individual servings for the seated guests and family the need arose for larger sets of dishes for the diners, along with comports, platters and tureens for the carriage of the food prior to its serving and for its display and serving on the table and sideboard, and the first large ceramic matched pattern dinner services hence appeared only in the early 1700s. Silver dishes in the seventeenth Century had engraved armorial bearings applied in addition to their English silver assay hall marks, so removing their anonymity and identifying the armigerous family ownership of the artefact. It was natural then to also treat porcelain services similarly and so the concept of armorial porcelains for everyday usage was born: dinner sets comprising multiple dishes decorated *en suite* were also found to be cheaper to commission in Chinese hard paste porcelain rather than silver, whether these were decorated in China or imported in the white for armorial enamelling in the ceramic ateliers and workshops here, such as those of James Giles in London, located in the fashionable Cockspur Street, Haymarket. James Giles was patronised by eminent personalities of the time, by royalty and the aristocracy. Additionally, the exhibition of a family's crest or coat-of-arms on a dinner service was a visual confirmation to guests of one's social status and ability to access desirable trade goods imported from overseas as well as advertising the established position in an elite society of the host and of their membership of a nobility or gentry with the officially registered entitlement to bear arms. David Howard is of the opinion (Howard, *Chinese Armorial Porcelain,* 1974) that large armorial services were not commissioned from China for use in England before the turn of the eighteenth century, which accords perfectly with the above scenario. In some manufactories, fragile porcelain ladles were not supplied except as optional extras for porcelain dinner or dessert services and these would normally then be otherwise made of silver, which would probably bear an engraved armorial device, so it would be natural to expect that similar armorial bearings on the porcelain dishes they accompanied would also be applied. Smith (Smith 2014) has also suggested that armorial porcelain dinner and dessert service ownership could be viewed as a gender specific masculine commodity at dinner tables in the eighteenth Century and provided an opportunity to advertise one's status in the community; the converse applied to armorial tea and coffee services, which were used at functions that were generally hosted by the ladies of the household.

The earliest example of a surviving armorial porcelain dinner service for an English family for which examples of the artefacts are still extant is that of Thomas Pitt, Governor of the HEIC Fort St George, Madras (1653–1726), who went to serve in India with the HEIC in the late seventeenth Century. This armorial service dates from 1705. It is of Kangxi, Qing Dynasty, Chinese porcelain decorated in the Imari style with predominantly red and blue pigments, in simulation of the imported Japanese Arita wares which were proving to be so popular in the late seventeenth Century and bears the Pitt arms centrally. The escutcheon is rather unusually placed heraldically above the crest of a stork and its torse; an example of a plate from this *Pitt* service is shown in Fig. 1.9. Kangxi (Xuanye) was the second of the Manchu Emperors and reigned from 1661–1722, who succeeded to the Imperial throne following the interregnal and rebellious period in China after the collapse of the Ming Dynasty in 1644. The heraldic supporters, helm, mantlings and family motto are also absent from this armorial service, so this then comes within our definition of being an armorial that possesses only a partial coat-of-arms (see later); the arms are contained in the form of an unusually shaped circular "escutcheon" surrounded by wreaths of blue foliage and comprise a chequy of azure and or with three gilt bezants. Thomas Pitt of Ewerne Stapleton, Dorset, made his fortune in India through the sale of a 127- carat diamond that he had acquired there to the French Regent of New Orleans, which was then incorporated as the Pitt Diamond into the Crown Jewels of the French monarchy. He later became Governor-General of Jamaica. The author has discovered an interesting situation in researching the *Pitt* service family credentials which has a bearing on most of the armorial research into Welsh porcelain artefacts that will be discussed here later, so it is worthwhile digressing in some detail to consider this particular aspect. In common with many noble arms-bearing individuals and families, reference to Burke's Armory (Burke, *Burke's General Armory—The General Armory of England, Scotland, Ireland and Wales: Comprising a Registry of Armorial Bearings from the Earliest to the Present Time*, 1884) and cross-referencing with the descriptions and pictorial crests displayed in Fairbairn (Fairbairn, *Crests of the Families of Great Britain and Ireland and Their Kindred in Other Lands,* Volumes I and II, 1905) reveals the presence of a number of possibilities for the assignment of the arms and crest displayed on this *Pitt* service armorial plate to a particular family arms bearer. In all, 15 different arms-bearing individuals of different familial descent are registered as being entitled to bear the arms of Pitt, but there are some significant and some subtle differences observed between their displayed individual armorial bearings and, for instance, the same crest and motto has also been adopted more widely by members of the Scottish Clan MacIntyre. Firstly, the award of arms to a particular individual or family may not match temporally with the established chronological date of the porcelain service and indeed it should be recognised that the title may have been an ancient one originally granted some centuries hitherto that is now deemed to be technically extinct and outside the time frame established for the commissioning of the porcelain service being investigated by the ceramic historian or analyst. The first entry in *Burke's Armory* that is

Fig. 1.9 Kangxi porcelain, Qing Dynasty, *ca.* 1705, Chinese armorial porcelain plate from the *Pitt* service, made for Sir Thomas Pitt of Ewerne – Stapleton, Dorset, the earliest surviving example of a Chinese armorial service ordered for the British aristocracy. Note the curious placement of the over-large crest below the escutcheon and the upwards curvature to the torse matching the curve of the cavetto of the plate, which should properly fit over the helm and require a downwards curvature. The oval escutcheon displays three bezants – a Byzantine gold coin discovered by the Crusaders *en route* to Jerusalem and in Latin called "*Bizantius aureus*", a gold bezant. The bezants depicted here are more akin to elliptical golden roses and should properly be shown as being plain and circular heraldically. *Private Collection. Reproduced with the permission of Kingschina Ltd, Mayfair, London*

relevant here is that of Thomas Pitt of Boconnock, co. Cornwall, who was created Baron Camelford in 1784. His arms are described as they should appear on the armorial plate: namely, sable a fess chequy argent and azure between three bezants or; it should be noted that the chequy in this case is tinctured argent and azure (white and blue), whereas in Pitt of Ewerne Stapleton it is or and azure (yellow and blue). The crest is a stork argent as shown in Fairbairn (Fairbairn, *Crests of the Families of Great Britain and Ireland and Their Kindred in Other Lands,* Volumes I and II, 1905) and illustrated here in Fig. 1.10. The supporters of the escutcheon are given as two Cornish choughs regardant elevant proper. His motto is "*Per ardua liberi",* which translates as "*Freedom through adversity*". The younger brother of this Thomas Pitt was William Pitt, Earl of Chatham, the British Prime Minister, who has the same coat of arms, but the crest is now a stork which has its dexter claw resting on an anchor erect cabled or (Fig. 1.11, Fairbairn: Plate 224/10). The supporters are a lion rampant gardant and a buck attired collared and chained with the motto "*Benigno numine",* which translates as "*By the Favour of the Heavens".* A third Pitt, Thomas Pitt, the Earl of Londonderry, was ennobled in 1726, but this became

Fig. 1.10 Stork crest for Sir Thomas Pitt of Ewerne-Stapleton, co. Dorset, the titled arms-bearer of the *Pitt* service shown in Fig. 1.9, as depicted in Fairbairn (*Crests of the Families of Great Britain and Ireland and Their Kindred in Other Lands,* Volumes I and II, T.C. and E.C. Jack, Edinburgh and London, 1905, Plate 33/9). Note the torse curving downwards properly as it should do to encompass the contour of the helm and the different portrayal of the open beak of the stork compared with the armorial plate in Fig. 1.9

Fig. 1.11 Stork crests for some other Pitt individuals showing small variations within the crest as represented in Figs. 1.9 and 1.10: left, the crest of William Pitt, Earl of Chatham (Fairbairn: Plate 224 /10) and right, crest of Thomas Pitt, Earl of Londonderry (Fairbairn: Plate 294/11). Depicted in Fairbairn (*Crests of the Families of Great Britain and Ireland and Their Kindred in Other Lands,* Volumes I and II, T.C. and E.C. Jack, Edinburgh and London, 1905)

extinct upon his death in 1764, and the arms are now quartered with those of Pitt having first and third and the second and fourth quarters being sable with two wings conjoined argent. The crest is now a stork argent, beaked and membered or, holding up its dexter claw (Fig. 1.11, Fairbairn: Plate 294/11). The supporters are two falcons sable with a chaplet of red roses barbed, with the motto: "*Amitie*" which translates as "*Friendship*". A fourth Pitt, from Ewerne Stapleton, co. Dorset, was granted arms in 1604 in the reign of King James I, who also has the same arms as Thomas and William Pitt and the same crest as Thomas Pitt, but not William Pitt, and has the motto "*Per ardua liberi*". There are eleven other Pitts with other notable differences in their escutcheons, supporters, crests and mottoes. Detailed examination of the Pitt armorial plate shown in Fig. 1.9 reveals several points which require amplification before an assignment can be made confidently to a particular Pitt arms-bearing individual:

- The escutcheon on the *Pitt* service plate is seen to match an assignment to several possible contenders, chiefly Pitt of Boconnock, co. Cornwall, Pitt (Earl of Chatham), Pitt (Earl of Londonderry) and Pitt of Ewerne Stapleton, co. Dorset. However, chronologically we can exclude the Earl of Chatham and the Earl of Londonderry from this discussion as they had not yet been granted arms in 1705, the date established historically for the commissioning of this armorial service.
- The crests of the Earl of Chatham and Earl of Londonderry differ from the other two Pitts described above in that the dexter claw of the stork is raised and in the Earl of Chatham's case it is also seen to be holding an anchor (Fig. 1.11)

The choice of assignment to a family arms bearer now comes down to that between Pitt of Boconnock and Pitt of Ewerne Stapleton: the former had arms granted in 1784 whereas the latter had arms granted in 1604. Hence, it can be deduced logically that Pitt of Ewerne Stapleton is the one whose arms are depicted on the plate when the armorial service was commissioned chronologically in 1705. An interesting point emerges when the crest of Pitt of Ewerne Stapleton as depicted in *Fairbairn* (Fairbairn, *Crests of the Families of Great Britain and Ireland and Their Kindred in Other Lands,* Volumes I and II, T.C. and E.C. Jack, Edinburgh and London, 1905) (Fig. 1.10) is compared in detail with that painted on the armorial plate as shown in Fig. 1.9: the beak of the stork in the Pitt crest as appearing in Fairbairn's armory is clearly open and the tongue is exposed whereas the beak of the stork depicted on the crest enamelled on the plate is firmly closed. Also, the torse is displayed incorrectly on the armorial porcelain plate and curves upwards, following the curvature of the edge of the plate base, known as the *cavetto*, whereas a properly executed torse heraldically, if it is shown curved at all as in the crests that appear in in Fig. 1.11 from the armory, should curve downwards to match the contour of the curvature of the top of the helm upon which it is designed to sit, as indeed it does as depicted in the armorial sketch from Fairbairn's armory in Fig. 1.11. Several ceramic and armorial historians have commented hitherto that the Chinese depiction of British armorials enamelled on their porcelains is sometimes lacking in exactitude and precision and that this would naturally depend critically upon the accuracy of the original copy drawing that was supplied by the arms-bearing individual or family to the HEIC agent for enamelling in the workshops in China and also upon the specific knowledge and attention to detail given by the HEIC supercargo who was responsible in Canton for inspecting and accepting the work carried out by the enamellers in China. This is probably the case here and a small difference in the enamelled crest and torse has escaped the attention of the supercargo who was monitoring the enamelling being done, possibly on site in Canton, or more likely at this time in Jingdezhen and then presented as the finished article for shipment in Canton. Before the Chinese Hongs or guilds were established on the waterfront in Canton as official agents of the Imperial porcelain manufactory at Jingdezhen for the local decoration of porcelain, when foreign nationals were excluded from the city of Canton and were confined to the port area where these negotiations and the deals were made, an arduous journey of over 1000 Km needed to be made by Chinese porters between Canton and Jingdezhen for this purpose (Edwards, *Porcelain Analysis and Its Role in the Forensic Attribution of Ceramic Specimens,* 2021). Much of this journey involved the hazardous manual porterage of the delicate enamelled porcelain items in traversing mountainous country passes and the porcelain was usually packed into tea chests with tea to minimise the potential damage incurred in transit. In this particular case and at this time, the armorial decoration of the Pitt service probably would have been accomplished in Jingdezhen from a template or drawing of the crest and associated arms. It is also clear that the presence of the motto in this case would have facilitated in the identification of the armorial service more easily by ceramic historians and would have acted as a confirmation of its attribution.

Although Pitt of Ewerne Stapleton has been identified unequivocally as the owner of this important historical service, it is found that he has also been matched elsewhere in some historical accounts with an incorrect motto – his arms there being accorded the motto of *"Amitie"* (*"Friendship"*), which is the actual motto of Pitt, the Earl of Londonderry, instead of his own true motto *"Per ardua liberi" ("Freedom through adversity")*! As will be seen later for a particular Swansea porcelain armorial service exemplar, in which the motto displayed also does not belong properly to the individual whose arms are represented, the armorial researcher should be aware of the occasional occurrence of this sort of discrepancy. In this case too, the alternative selection of an individual or family made on the basis solely of the motto does not correlate with their coat-of-arms as displayed on the escutcheon shown on the Swansea porcelain armorial plate, so the presence of a motto does not always assist in the correct and unequivocal attribution of the arms being made and can even be a misleading sidetrack to the armorial investigation. The heraldic background is that the motto, unlike the coat-of-arms or the crest, does not have to be formally registered with the College of Arms and can be assumed by usage.

Several porcelain armorial services from English manufactories are renowned for their sumptuous and opulent accompanying decoration, which often provide examples of the epitome and zenith of ceramic enamelling art: examples include the *Duke of Clarence* service from the Royal Worcester China Works (Fig. 1.12), the *Pendock-Barry* (*Barry-Barry*) service (Twitchett, *Derby Porcelain: An Illustrated Guide, 1748–1848*, 2002; Edwards, *18th and 19th Porcelain Analysis: A Forensic Provenancing Assessment*, 2020; John et al., *The Nantgarw Porcelain Album*, 1975, Illustration 65; Denyer, Denyer and Edwards, *The Pendock-Barry Derby Porcelain Service: A Reappraisal*, 2022) (Fig. 1.13) and the *Earl of Shrewsbury* service from the Derby China Works, the *Lord Nelson* service from Chamberlain's Worcester

Fig. 1.12 Armorial plate from the Duke of Clarence dinner-dessert service in Royal Worcester porcelain, Flight period, ordered by His Royal Highness Prince William in 1789 (who later became King William IV in succession to King George IV in 1830) to celebrate his accession to the royal Dukedom. Sumptuously decorated with his coat-of-arms centrally located and enclosed in the Order of the Garter with its motto, *Honi Soit Qui Mal y Pense*, and displaying the heraldic badges of the Order of the Thistle and Order of the Garter in the verge, which also shows St George impaling a dragon with a lance. *Reproduced from The Art Institute, Chicago (CCO; public domain): Joseph Maier and Arthur Lewis Liebman Memorial. Gift of Kenneth J. Maier MD, ref 1994.90*

China Works (Fig. 7.1) and the *King William IV* service from the Royal Rockingham China Works (Cox & Cox, *Rockingham Porcelain,1745–1842,* 2001). Examples of armorial services are known for several English manufactories dating from the early Worcester Dr. Wall or First Period Worcester (1751–1783) through to the Liverpool manufactories, which specialised in Masonic emblems and associated coats-of-arms between 1760 and 1770, and onwards with Derby from the 1780s into the nineteenth Century. It appears that several early crested and monogrammed, initialled or ciphered Derby services exist but some have not been fully researched: an example of one that has been provenanced and researched is the *Viscount Cremorne* service of 1788 described here later and illustrated in Fig. 1.19, which matches with the work records and correspondence in the Derby China Works archive. A second example of an early Derby armorial service is shown in Fig. 1.14, this being a tea cup and saucer bearing the arms of the See of Durham and ordered by Bishop John Egerton, who was Bishop of Durham between 1771 and 1787. An archival record exists from the Derby Covent Garden showroom for this tea service being purchased at a cost of £9 14 s in 1784. John Egerton married Mary Boughton in 1782 and this tea service could well have been made for the establishment of their new family home in Durham. The arms are azure a cross patonce or between four lions

Fig. 1.13 Dinner plate from the *Barry-Barry* (*Pendock-Barry*) armorial service supplied from the Derby China Works to Pendock Barry Neale of Tollerton Hall, Nottinghamshire, in 1806/7. It is a beautiful example of William Billingsley's rose painting and was probably executed externally for Derby when he was resident as a freelance enameller in Brampton-in-Torksey. Marked in gold with the crossed swords, crown and two batons of the Derby China Works manufactory. Note that this armorial exhibits the escutcheon only and much detailed genealogical research was undertaken from this to determine conclusively its correct attribution and chronology. *Private Collection*

rampant argent and the crest is a lion rampant in azure couped with a ducal coronet as shown in Fig. 1.14. It may seem rather incongruous that a Bishop should possess a ducal coronet exhibiting strawberry leaves and no pearls (see Chap. 2 for a study of heraldic devices and symbols), however, historically the Bishops of Durham have been granted uniquely the use of a ducal coronet in their arms as befitted their status and title of the Prince-Bishops of Durham.

For the Welsh porcelains, the only complete armorial representation is the *Viscount Weymouth* plate illustrated in John et al. (*The Nantgarw Porcelain Album*, 1975) and this is quite possibly the earliest armorial in the known range if the hypothesis is correct that it was produced to celebrate Lord Frederick Thynne's 21st birthday in 1818, although there may be serious complications with the acceptance of this in view of discrepancies arising from the genealogy and some anomalies that have been discovered in its armorial portrayal (see later in Chap. 4).

However, it is rather a daunting task to assign absolutely a date for the first armorial service produced by an English porcelain factory in the eighteenth Century. Stephen Hanscombe (2008) has uncovered and researched several items of "anonymous" early Worcester porcelain which were decorated in the atelier of James Giles in London in the mid-eighteenth Century as well as porcelain of Chinese origin bearing coats-of-arms for English individuals or families which were applied locally

Fig. 1.14 Tea cup and saucer, Derby porcelain, *ca.* 1782–4, bearing the arms of the See of Durham and commissioned by John Egerton, Bishop of Durham (1771–1787). An entry in the Derby sales records of their Covent Garden showroom in London gives an entry in 1784 for the purchase of this service at a cost of £9 14 s. *Reproduced with the Courtesy of Bonhams Auctioneers and Valuers, Bond Street, London*

in Giles' workshop around the same time. The early examples of indigenous English armorial porcelains are quite rare and several ceramics historians have expressed surprise that the Chelsea Porcelain Manufactory, in particular, did not seize upon the opportunity to use armorial devices on porcelain in the 1750s and 1760s especially since their manager Nicholas Sprimont, who was a silversmith by profession, designed elaborate Chelsea porcelain table pieces such as candelabra, comports and the famous goat-and-bee jugs which were clearly based upon their silver analogues and which were invariably accompanied by their appropriate armorial engravings. A typical Chinese hard paste porcelain item decorated in the James Giles workshop in London in the 1760s is shown in Fig. 1.15: this artefact is a teapot with an unidentified armorial escutcheon, dating from the period 1760–1770, which is now in the Helena Woolworth McCann Collection of the Metropolitan Museum of Art, New York (Winfield Foundation; Accession Number 51.86.394 a, b).

It has been claimed that the first English porcelain factory to bear armorial devices is Worcester and examples exist certainly from the 1760s, such as the Beaumont-Ayscough trio bearing a *faux* Meissen mark dating from 1765–68 (Rissik Marshall 1946), and the arms of Gavin impaling Heasey on a Worcester porcelain trio of a similar period, with an escutcheon comprising a vertical sword with a saltire and mullet (Gavin) and a fess argent with gules shield (Heasey) along with the crest of a full-rigged ship in sail (Rissik Marshall, *Coloured Worcester Porcelain of the First Period*, 1951). The production of this armorial artefact must therefore predate the death of Christina Heasey in 1767. Another early First Period Worcester example is a trio bearing the arms of Warwick Calmady of Cornwall, decorated in London, from about 1770. An interesting early Derby China Works porcelain tea

Fig. 1.15 A Chinese hard paste porcelain teapot decorated in the James Giles workshop in Haymarket, London, bearing an unidentified armorial escutcheon dating from the period 1760–1770, which is now in the Helena Woolworth McCann Collection of the Metropolitan Museum of Art, New York (Winfield Foundation; Accession Number 51.86.394 a, b). The escutcheon shape is of an unusual Italianate type and contains three wolf's heads couped armed sable on an erminois field, with the gilt crest of a wolf ambulant erminois on a sable and or torse. *Public domain*

bowl commissioned by a Mrs. Mary Bootle of Rode Hall in 1780 was made to be *en suite* with the earlier Chinese armorial service of Sir Thomas Bootle (1685–1753) and Robert Bootle (1694–1758) of Lathom House, Ormskirk, Lancashire (McKeown, *English Ceramics: 250 years of Collecting at Rode Hall*, 2006). Several items were made in armorial Derby porcelain to replace broken Chinese porcelain items from this earlier Chinese armorial porcelain service which had been commissioned in 1737. Rode Hall, Cheshire, was the seat of the Wilbraham family and Mary Bootle, daughter of Robert Bootle, married Richard Wilbraham, MP for Chester, in 1755, thereby taking the name of Wilbraham-Bootle and inheriting the estates of Lathom House and Rode Hall; her father was a sea captain and later a Director of the HEIC who specialised in the importation of Chinese porcelains. The Bootle coat-of-arms shown on the service comprises gules on a chevron engrailed between three combs argent with a cross patee-fitchee in the field, motto *"In Porto Quies"*, meaning *"In the Haven There is Rest"*.

A rare and early armorial product of the Bristol factory of Richard Champion is a hard paste porcelain teabowl with the arms of the influential politician Edmund Burke, commissioned upon his election as MP for Bristol in 1774 (Barker 1951).

An example of an early non-armorial but prestigious and sumptuously decorated porcelain service was ordered by King George III and Queen Charlotte in 1763 from the Chelsea Porcelain Manufactory as a "diplomatic gift" for Adolphus Frederick IV, the Duke of Mecklenburg -Strelitz (1738–1794), who was brother to the Queen; surprisingly perhaps, it does not contain armorial bearings despite its formal classification as a diplomatic gift. This service, which was completed in March 1765, was seen by Thomas Nugent on his visit to the royal palace at Neustrelitz in October 1766 and he noted:

> *A complete service of Chelsea porcelane, rich and beautiful in fancy beyond expression. I really never saw any Dresden porcelane near so fine; Her Majesty made a present of this choice collection to the Duke, her brother.*

The *Mecklenburg-Strelitz* service, now comprising some 137 pieces, plus five more replacements from the Konigliche Porzellan Manufactur, Berlin, was presented to Her Majesty Queen Elizabeth II in 1972 on her Silver Wedding Anniversary by James Clarke, a London dealer who had purchased it some years before, following its original acquisition by Sir Joseph Duveen in the 1920s. It is indeed very sumptuously decorated in the rococo style with exotic birds centrally located, with mazarin blue cartouches in the cavetto containing gilt butterflies and swags of flowers at the verge (Adams, *Chelsea Porcelain*, 1987). Had it borne armorial bearings it would undoubtedly be a very strong candidate for the earliest English armorial porcelain service. In fact, the earliest example of a Royal service in English porcelain which bears full armorial achievements must now be considered to be the *Duke of Clarence* service in Worcester porcelain of 1789 which bears the Royal coat-of-arms centrally and badges of the Order of the Thistle and Order of the Garter at the verge: this beautifully composed and decorated porcelain service is still admired for its technical brilliance and execution (Fig. 1.12). An earlier firmly dated Royal service, the *Prince of Wales* service, was commissioned from the Derby China Works in 1786 but bears no armorial achievements; this is recorded in the Derby archive and is still regarded as an exemplar of William Billingsley's rose painting as each piece contains an individual rose painted centrally in his characteristic style and enclosed within a circlet of gilt dots (Fig. 1.16). It has been mentioned above that the Liverpool porcelain manufactories specialised in Masonic emblems, although these are also now rather rare, but an early punchbowl of Christian's Liverpool factory, *ca.* 1770, decorated with scenes from Aesop's fables is known bearing the Masonic square and compasses. An even earlier example of a punch bowl from William Reid's factory, *ca.* 1756–58, does have family armorial bearings, namely an escutcheon with a stag trippant (dexter) and a black raven on a gold ground (sinister). The sinister arms have been identified as the arms of Corbett and the dexter arms have been associated with the ancient lords of Llywarch Holwbwrch, possibly assigned to the Lloyd family (Hillis, *Liverpool Porcelain, 1756–1804*, 2011), and judged to be extremely rare for Liverpool porcelains. Although chronologically pre-dating the armorial services discussed above this Liverpool armorial punchbowl exemplar does not really constitute an armorial service. The earlier English factories of Bow,

Fig. 1.16 Dessert plate from the *Prince of Wales* service commissioned by his Royal Highness Prince George, Prince of Wales and later King George IV in 1820, in 1786 from William Duesbury of the Derby China Works. Decorated with a salmon pink border, each piece bears a different individual rose depicted centrally situated with forget-me-nots and enclosed in a circlet of gilt dots enamelled by William Billingsley. Gilded by William Longden, whose gilder's number "8" appears on the underside adjacent to the footrim. *Private Collection*

Vauxhall and Limehouse, which were founded in the 1740s, in contrast, do not seem to have extant exemplars of armorial porcelains.

Detailed accounts and invoices from the College of Arms, the official heraldic authority for England, Wales, Ireland and the Commonwealth, have survived in the Josiah Wedgwood archive at Etruria, Staffordshire, for the depiction of coats-of-arms and crests supplied to him for enamelling on behalf of his titled clients on his commissioned armorial creamwares. Thereby, Wedgwood would be in receipt of diagrams of the officially approved crests, escutcheons and arms for his entitled clients and this would allay any argument later arising from a dissatisfied client about an incorrect heraldic portrayal on their armorial ceramic services. It is interesting that Josiah Wedgwood was apparently at first reluctant to undertake the provision of ceramic armorial services on commission from potential entitled clients and he commented on the 25th September, 1766, that;

> *Crests are a very bad thing for us potters to meddle with and I never take any orders for services so enamelled.*

Wedgwood argued that if items of crested armorial services were found to be faulty from the kiln after their second glost firing, subsequent to their enamelling and gilding having been undertaken, then they could not then be sold off as "seconds" elsewhere, thereby having to be written off as a total loss to the manufacturing business; however, he had to revise his strategy according to his client demand to match his competitor manufactories and he then started to accept commissions for armorial earthenwares and creamwares from the landed gentry and aristocracy. It is maintained that the Cambrian Pottery in Swansea closely monitored Josiah Wedgwood's success in marketing his ceramic creamwares and that indeed they launched their own Swansea versions of his creamwares with a very similar composition ceramic body, comprising a mixture of china clay, ball clays and feldspar, to which was added smalt to remove any vestige of an iron oxide contaminant in the clays which would have resulted in an unsightly brown colouration (Morton Nance, *The Pottery and Porcelain of Swansea and Nantgarw*, 1942). The Cambrian Pottery at Swansea even emulated Wedgwood's famous "*Frog*" service in creamware using a decoration of sepia landscapes and sepia borders of ivy leaves and berries in creamwares of their own; see later in this Chapter for a description of Wedgwood's *Frog* service, which was supplied to the Empress Catherine the Great of Russia (Ekaterina II) in 1773. Several illustrated examples are given in the excellent article by Gaye Blake-Roberts entitled "*Swansea and Wedgwood*" (Blake-Roberts 2003) of Swansea Cambrian Pottery creamwares that were closely modelled on Wedgwood's creamwares and his *Queen's Ware,* and several of these bear armorial devices.As well as family individuals, institutions and companies also commissioned heraldic armorial porcelains: we have already mentioned above that the HEIC placed a very large commission of armorial porcelain from Chamberlain's Worcester China Works, and they were certainly not unique in this regard. Masonic Lodges and the City of London Livery Companies also commissioned fine armorial porcelain services during the 18th and 19th Centuries, displaying their coats-of-arms and other heraldic devices and emblems for use by their members during their official dinners and banquets.

It has been claimed by some ceramics historians that Worcester is the single most collected English porcelain in the world and the *Ewers-Tyne Collection* at Cheekwood, Nashville, Tennessee, and the *Marshall Collection* at the Ashmolean Museum, Oxford, bear witness to this assertion with fine displays of Worcester armorial ceramics in their exhibition of almost three centuries of the finest and most sumptuous of porcelains from this manufactory, including several armorial examples (Sandon, *The Ewers-Tyne Collection of Worcester Porcelain at Cheekwood,* 2007; Sandon , *Flight & Barr Worcester Porcelain, 1783–1840*, 1978; *Armorial Worcester Porcelain of the First Period: Specimens from the Marshall Collection in the Ashmolean Museum and Other Sources*, 1964).

1.4 Named Porcelain Services

Armorial services should not be confused with the special "named services" made for the aristocracy, gentry and royalty which are also demonstratively some of the greatest accomplishments achieved in porcelain decoration but which differ from true armorial services in that they do not bear any heraldic achievement, namely a coat-of-arms, escutcheon, motto, coronet or heraldic crest of the armigerous person who commissioned the service. These named services are clearly too numerous to mention comprehensively here, but classic examples in English porcelain include the *Lord Ongley* (Fig. 1.17), *Prince of Wales* (Fig. 1.16) and *Earl Camden* (Fig. 1.18) services in Derby porcelain (Edwards, *Swansea and Nantgarw Porcelains: A Scientific Reappraisal*, 2017), the *Lady Mary Wortley Montagu* service in Worcester porcelain and the *Earl of Coventry (Blind Earl)* services in Chelsea and Worcester porcelain. Although the categorisation and classification of armorial services is seemingly straightforward to define on the basis of whether or not they display heraldic achievements, such as the depiction of the full coat-of-arms and/or a crest with or without an accompanying motto, a special classification resides with services that may not at first be considered as strictly armorial in that they do not display a coat-of-arms or a crest but nevertheless contain a simple monogram, cipher or initials which may be accompanied perhaps by a coronet or another genuine heraldic device. An example of the latter type of armorial service is the *Viscount and Lady Cremorne* service in Derby porcelain for which an extensive documentary correspondence exists in the Derby China Works archives between William

Fig. 1.17 Dessert plate from the *Lord Ongley* service in Derby porcelain which was commissioned by Lord Ongley, Robert Henley, the third Baron of Old Warden in Bedfordshire (1803–1877), probably around 1824 when he had reached the age of 21. Robert Bloor of the Derby China Works, who supplied the service, modelled its design on several Nantgarw China Works plates painted by James Plant that he had acquired from Robins and Randall's atelier in London. Although not an armorial service, the *Lord Ongley* service was one of the most expensive Derby services produced originally, but only seven pieces of this magnificent service now remain. Lord Ongley believed in entertaining and in holding large parties and he rapidly disposed of the family fortune, certainly not adhering to his family motto, "*Mihi Cura Futuri*", which translates as "*The Care of the Future is Mine*"! *Courtesy of Peter Frost-Pennington Esq., Muncaster Castle, Ravenglass, Cumbria*

Fig. 1.18 Dessert plate from the *Earl Camden* service, commissioned in 1795 from the Derby China Works by John Jeffreys Pratt, the second Earl Camden, created the Lord -Lieutenant of Ireland in 1795. This large service of over 100 pieces was decorated by William Billingsley and his characteristic rose painting can be seen; claimed by ceramic historians as representing the epitome of Billingsley's finest rose-painting. He was created Earl Camden in 1786 with a family seat at Wherwell House, Andover, Hampshire. Extensive correspondence exists in the Derby China Works archive between Lady Frances Camden and William Duesbury, in which she expressed her concern at the lateness in completion and in delivery of this service. *Private Collection*

Duesbury, the proprietor of the Derby China Works in the eighteenth Century, his London agent Joseph Lygo and Lady Philadelphia Cremorne relating to the commissioning of this particular service in late 1788 (Twitchett, *Derby Porcelain: An Illustrated Guide, 1748–1848*, 2002). The Derby porcelain *Cremorne* service consists of a simple randomly applied enamelled decoration of blue cornflowers and gilded foliage with a Viscount's coronet (displaying its statutory seven balls or "pearls", see further details in Chap. 2) above the initials *PHC* (for Philadelphia Hannah Cremorne) contained in a gilded foliate circlet in the reserve (Fig. 1.19). Lady Philadelphia Cremorne, granddaughter of William Penn, the founder of the state of Pennsylvania in the American colonies, was Lady-in-Waiting to Queen Charlotte, the Queen consort of King George III, in the late eighteenth Century. The small dessert comport shown in Fig. 1.17 is rather special in that it is one of only a pair made by the Derby China Works to the uniquely specific design of Lady Cremorne for her armorial service and was specially itemised in price in the invoice created for her by William Duesbury in 1789, which still exists in the Derby China Works archives, at £2 7 s and 6d. (in £sd, which today translates into £400, approximately $500 US).

Fig. 1.19 Dessert comport of unusual shape commissioned from the Derby China Works by Lord and Lady Cremorne: Thomas Dawson (1725–1813), Baron Dartrey of Dawson's Grove, County Monaghan, Ireland, was raised to the peerage as Viscount Cremorne in June 1785. In 1770 he married Philadelphia Hannah Freame, granddaughter of William Penn, founder of Pennsylvania, who was born in Philadelphia. This service, which can be dated to 1789 from archival correspondence, bears a viscountess coronet below which is the monogram *PHC* inside a circlet of gilt foliage. The decoration is a simple pattern of chantilly sprigs of blue cornflowers and gilded foliage. This service depicts perhaps the simplest kind of armorial on porcelain, namely a coronet and initials, but it nevertheless serves to attribute the commissioning of this service unequivocally. *Private Collection*

1.5 Pseudo-Armorial Services

A further type of service which strictly cannot come under the classification of an armorial service defined here as it does not portray arms, badges, a coronet or heraldic achievements but rather displays a device , emblem or a motif, which nevertheless could be seen to be indicative of its individual or family origin, its eventual location and its possible purpose, is exemplified by the famous *Frog* service in Wedgwood's creamware, which is of course strictly not porcelain as we are considering here. Nevertheless, this case study will be illustrative of several features which are shared with its porcelain analogues. On the first May 1759, Josiah Wedgwood founded his earthenware pottery at Burslem in Staffordshire and then successfully expanded his business to its new location at Etruria in 1766, which was built on the success of his fine glazed white earthenware production that had been achieved the year before. In the eighteenth Century, "china" was a generic term used for glazed ceramics which included porcelain at one end of the scale and earthenwares or pottery at the other, with different categories of creamware, celadons, faience, majolica and stoneware occupying the space in between (see Appendix I); this can be a potential trap for an unwary researcher and needs to be highlighted here. In 1767, Queen Charlotte, Queen consort of King George III, admired Wedgwood's creamware so much that she allowed him to call it *Queen's Ware* and she invested in several items of this especially admired earthenware pottery. This creamware body was lighter than normal earthenwares and also much more robust and was considered by some at that time to be equivalent in quality to many of the porcelains then being made elsewhere: being lighter in weight, it also was not subjected to the punitive national trade tariffs then operating in Europe based upon the weight of the ceramic items being exported and imported. This made the European ceramics manufacturers rapidly sit up and take notice as these creamwares were then effectively rendered considerably cheaper than contemporary porcelains to purchase and afforded a fine quality and robust ceramic that was clearly financially competitive with their other contemporary china and heavier faience and majolica earthenware analogues. The French manufactories in particular immediately responded to this challenge with a lighter version of their own heavier creamwares called *faience fine*. The supreme accolade accorded to Wedgwood's creamware, however, was the commission from another Queen, the Empress Catherine the Great of Russia (Ekaterina II), for a very large dinner and dessert service in *Queen's Ware*, which thereafter became known as the *Frog* service. The Empress Catherine wished to celebrate the victory of her Russian naval forces against the Ottoman Turks in the Russo-Turkish War (1768–1774) at the Battle of Chesma Bay, so she embarked upon the construction of the Chesma Palace, which was situated midway between her Winter Palace in St Petersburg and her Summer Palace at Tsarskoe Selo, on marshy ground known locally as the Kekerekeksinsky, which has been translated simply as "Frog Marsh" and known as *La Grenouillere* in the French language which was then officially used in the Russian Court. The Chesma Palace is now located on the Moskovskaya Prospekt heading out of St Petersburg towards Tsarskoe Selo. The Empress Catherine modelled her new Chesma Palace on Longford Castle in Great Britain, which had been originally built with gold and silver bullion recovered from a wrecked galleon of the Spanish Armada in 1588 and was then the seat of the Earl of Radnor, whose ancestor Laurens de Bouverie purchased the estate in the early eighteenth Century and landscaped it with the assistance of Lancelot "Capability" Brown. Empress Catherine the Great's order with Josiah Wedgwood, which was placed through her Russian Consul in London in 1773, was for a very large service of 944 pieces, in total some 50 place settings, comprising 680 pieces in a dinner service and 264 pieces in a separate dessert service. These services were to be hand painted in monochrome sepia, depicting in all 1222 different views of English stately homes and landmarks that were much admired by the anglophilic Empress Catherine, and with each piece bearing a green coloured enamelled motif of a frog, in deference to the site chosen for her new palace (and thereby so providing us also with an example of a *rebus*). At first, so Llewellyn Jewitt reports (Jewitt, *The Ceramic Art of Great Britain from the Prehistoric Times Down to the Present Day*, 1878), it seems that Josiah Wedgwood was extremely reluctant to "deface" his esteemed *Queen's Ware* with such a motif:

> *Wedgwood was very unwilling to disfigure the service with this reptile but he was told that it was not to be dispensed with.*

The landscape views were selected by Thomas Bentley, Josiah Wedgwood's business partner, and featured Wedgwood's own home, Etruria Hall, on a serving dish. This service was the most expensive ever made at Etruria and the Empress Catherine was invoiced for £2700, equivalent to approximately £400,000 today, i.e. approximately $500,000 (Edwards, *Porcelain Analysis and Its Role in the Forensic Attribution of Ceramic Specimens*, 2021). The service was made in Etruria and then sent to London to Wedgwood's decorating atelier in Little Cheyne Row, Chelsea, where his team of 30 painters decorated it with the requisite borders and scenes and then overglaze fired it. It is alleged that Wedgwood did not actually make a profit from this prestigious royal commission but he shrewdly put the service on show in London after its decoration and prior to its being shipped out to Russia, and it is said that he charged an entrance fee to the public for their admission and inspection of his noteworthy ceramic creation: the assembled finished service must have made a wonderful sight and undoubtedly provided Wedgwood with a phalanx of new aristocratic clients for his creamwares. Most of the *Frog* service that has survived into modern times is now held in the Hermitage Museum, St Petersburg, which now comprises some 300 pieces only, although some pieces do still occasionally surface at auction – the Wedgwood Museum in the Staffordshire Potteries now has the largest collection of *Frog* service items outside of St Petersburg. Recently, a serving platter from the *Frog* service made the sum of $46,000 at auction in 2019, a price clearly rivalling that of its contemporary fine porcelains in support of its importance, historical provenancing and ceramic quality. An example of the *Frog* service in Wedgwood's *Queen's Ware* is shown in Fig. 1.20, which illustrates a serving platter depicting Ditchley Park in Oxfordshire, the seat of the Earl of Litchfield, and is hand painted in sepia monochrome with a green enamelled frog motif at the verge. The border comprises scrolls with oak leaves and acorns in sepia, signifying that this platter is actually part of the Empress' dinner service. In Fig. 1.21 is shown the detail of the green enamelled frog in its "escutcheon" and in Fig. 1.22, a pair of covered vessels or porringers from the Empress' dessert service is shown with their characteristically distinctive and different border of intertwined ivy leaves in sepia. It should be noted that the enamelled frog motif which characterises these services is enclosed within an escutcheon or shield, known as a "wave" type escutcheon with the double curvature on its top line on the dexter and sinister halves (technically known as an *engrailment*), which gives the ceramic artefacts in the *Frog* service their "pseudo-armorial" context and a ceramics researcher could easily be mistaken for believing that the compound *Frog* dinner and dessert service was a genuine armorial service, which it clearly is not! The Empress Catherine's genuine heraldic coat-of-arms, in contrast, comprises a Romanov double eagle bearing a central escutcheon of St George on horseback slaying a dragon, with no supporters, the eagle holding a sceptre and an orb in its talons and being surmounted by the Imperial Russian crown, as it appears on other armorial services in porcelain commissioned by the Empress Catherine in the mid-to-late 1700s. Figure 1.23 shows the Empress Catherine's Imperial double-headed eagle on a Chinese porcelain Qianlong (1736–1795) service (known as the *Gatchina Palace* service (Howard, *Chinese Armorial Porcelain, Volume II*, p. 455, 2003). The Gatchina Palace was given to Count Grigory Orlov in 1765 by Catherine the Great in recognition of his assistance in deposing her husband Tsar Peter III some 3 years before in 1762: she also gave him a blue and gold porcelain service bearing his cipher. After Orlov's death,

Fig. 1.20 Serving platter from Josiah Wedgwood's Queen's Ware *Frog* service made for Empress Catherine the Great of Russia (Ekaterina II) in 1773 for use in her Chesma Palace – a very large, combined dinner and dessert service originally comprising 944 pieces with hand painted illustrations in sepia monochrome of individual English houses and estates, each containing a green enamelled frog motif contained within an escutcheon. This piece shows Ditchley Park in Oxfordshire, seat of the Earl of Litchfield, and is from the dinner service, which had a distinctive border of oak leaves and acorns. *City of Birmingham Museum of Art. Public domain*

Fig. 1.21 Detail of the green enamelled frog motif in an escutcheon from the *Frog* service; demonstrating that one could easily be misled into believing that this was an armorial service!

Fig. 1.22 Pair of porringers or covered bowls from the Queen's Ware *Frog* service of Empress Catherine the Great of Russia, showing the border of intertwined ivy leaves, indicating that these items were part of the dessert service component. ***Hermitage Museum, St Petersburg, Russia. Public domain***

Fig. 1.23 Armorial bearing of Empress Catherine the Great, dinner plate, Qianlong period (1736–1795) Chinese porcelain, destined for the Gatchina (Khotchino) Palace, St Petersburg, occupied by Count Orlov, a favourite of the Empress Catherine. ***Reproduced with the permission of Kingschina Ltd, Mayfair, London***

Catherine purchased the Gatchina Palace and it thereafter became a favourite home of the Imperial household. It is consistent chronologically that the Chinese armorial service bearing the Romanov arms illustrated in Fig. 1.23, therefore, would have been ordered by the Empress Catherine II for the Gatchina Palace sometime between

1765 and 1796 when she died, and when she was succeeded by Tsar Paul I. There are several *Gatchina Palace* services in existence bearing the Romanov arms, including a Wedgwood creamware service made for Tsar Alexander I, who succeeded Tsar Paul I in 1801. In 1823 the Imperial Porcelain Manufactory at St Petersburg was founded by the Romanovs and they made armorial porcelain services, at first exclusively for the royal palaces of the Tsars, including replacement pieces for those older services that had been damaged during use.Perhaps the classic example of an Empress Catherine II armorial porcelain service is that commissioned from the Sevres manufactory in France in soft paste porcelain in 1776 in a *bleu celeste* ground colour, as "*bleu celeste imitant la turquoise*", with cameos and flowers and this is indeed most lavishly gilded: the Sevres service depicts a stylised cipher *E II* picked out in floral wreaths and garlands, signifying Ekaterina II, beneath an Imperial crown (the only heraldic device) and was judged to be the most expensive service made by the Sevres porcelain manufactory at that time. After negotiation at Sevres with Prince Bariatinsky, Catherine's ambassador to Paris, the service cost 331,317 livres and was delivered in 1779. The service was destined as a gift for Prince Grigori Potemkin for use at his Taurida Palace in St Petersburg and comprised about 800 pieces initially, some 60 place settings, and included an *en suite* tea and coffee service comprising 48 pieces -- the service was re-purchased from Potemkin by Catherine for her own use in July 1782. The Empress Catherine stipulated that it must be made in soft paste porcelain, a departure for the Sevres factory which had moved onto the manufacture of hard paste porcelain in the early 1770s, but this was necessary as the hard paste body posed a problem for retention of the *bleu celeste* ground colour stipulated; this caused a severe processing problem for the Sevres manufactory and it is recorded that they could obtain only 800 pieces of the required quality from an initial 3000 pieces fired, representing a 73% kiln loss (Savill, *The Wallace Collection: Catalogue of Sevres Porcelain,* 1988; Savill 1982). A novel feature of this service was the inclusion of hard paste porcelain cameos which were placed strategically on the soft paste porcelain body of some pieces using a mixture of lead glass and *chymie* (a mixture of soap and animal glue) which was then fired in the kiln to promote adhesion. A fable surrounds this service, which has been linked with the saving of the Sevres manufactory from closure during the French Revolution by Catherine's settlement of the cost in a final payment in 1792, but it seems this is apocryphal as records show that she had already paid for her service by several instalments within 2 years of its delivery, i.e. by 1781, some 10 years before *The Terror* was launched in France (Preaud and Goodman 1995). Although most of the service still remains in the Hermitage Museum in St Petersburg it is claimed historically that 160 pieces of Catherine's Sevres service were looted from the fire which devastated the Hermitage in December 1837, when Tsar Nicholas I personally oversaw the saving of many exhibits; these looted pieces eventually found their way to auction rooms in London over the next few years, but many have now been restored to the Hermitage Museum.

The frog motif must assuredly now be re-classified to being that of a "rebus" despite its enclosure within what appears to be a genuine heraldic "wave" (or *engrailed*) escutcheon.

1.5.1 A Special Type of "Armorial" China

A special type of armorial ceramic is exemplified by Goss china, which was created in 1858 by William Henry Goss and was mass-produced cheaply for the souvenir market as commemoratives of family holidays in Victorian seaside resorts and spa towns; these items do not constitute armorial services or artefacts *per se* in terms of our coverage here, of course, but they certainly exhibit armorial bearings on miniature bone china objects and usually these display the coats-of-arms of towns and cities which are located near the resort (Fig. A1.1) However, they do not generally show the personal crests or arms of families, organisations or individuals, and hence they do not come into our categorically strict consideration of an "armorial" here (see Appendix I for a brief survey of Goss china presented in an armorial context). Although being an armorial in the generic sense, Goss china is also a pseudo-armorial in a secondary classification because it adopted a seemingly genuine heraldic crest as its trade-mark, namely a goshawk facing dexter and with wings extended sitting on a torse (see Fig. A1.2), which was designed to be a pun on the name Goss and, therefore, would really then be better described for our purposes here as a rebus. It is interesting that individuals from the Goss/Gosse family do have the right to bear arms as recorded in Fairbairn (Fairbairn, *Crests of the Families of Great Britain and Ireland and Their Kindred in Other Lands,* Volumes I and II, 1905) but it is not clear from the records whether these refer to the same branch of the family as that which instituted Goss china: the crest listed in Fairbairn is that of a falcon, wings expanded and inverted proper ducally gorged, or, and given therein as Plate 61/12 in Volume II. This true crest is a close match to but it is not identical with that adopted by Goss china as their trade-mark–so, the balance of opinion must revert to the statement of historical ceramicists that the Goss china crest is a fabricated one based on a goshawk rebus. Burke's *Armory* (Burke, *Burke's General Armory—The General Armory of England, Scotland, Ireland and Wales: Comprising a Registry of Armorial Bearings from the Earliest to the Present Time*, 1884) does differentiate between the two families listed therein named Goss and Gosse, the former having the individual crest indicated of a falcon with wings extended and described above and the latter, by reference to a Henry Gosse of Epsom, Surrey, a rather different crest of a pheon sable entwined by a branch of oak argent between two wings gold, guttee de sang: a pheon is a broad arrowhead with tangs having inner serrated edges, which may point either vertically upwards or downwards (see later, in the relevant discussion relating to its presence on a Nantgarw porcelain escutcheon in Chap. 4). Neither of these therefore can easily be assimilated into a genuine crest as depicted and used by the Goss family as a trade-mark on their china.

1.6 Swansea and Nantgarw

The theme of the current text is the comprehensive study of all known surviving armorial porcelain manufactured by the Swansea and Nantgarw China Works in the period between 1815 and 1820/1823, when porcelain production had by then ceased at both sites; closing sales had already taken place at Nantgarw and were scheduled for Swansea in the following years of the residual porcelain that was being decorated locally for purchase at each site. The active production of porcelain at Swansea occurred between 1815 and 1820 and at Nantgarw between 1817 and 1819/20, although residual porcelain stocks were decorated on site at each place after the formal closure of the porcelain manufactories for perhaps 2 or 3 years or more, at Swansea by the proprietors Timothy and John Bevington and at Nantgarw by William Weston Young and Thomas Pardoe (Edwards, *Porcelain Analysis and Its Role in the Forensic Attribution of Ceramic Specimens*, 2021). It is fair to say that even within the limited production runs of Swansea and Nantgarw porcelains during their operational period, armorial services are now considered to be rather rare compared with the other "named" and anonymous dinner, dessert, tea and coffee services that were produced and that to date the identification and attribution of the heraldic devices that are reproduced thereon has been fraught with difficulty (Gambon, *Porslen Abertawe a Nantgarw*, 2016). With regard to the Welsh armorial porcelains, Fergus Gambon (*Porslen Abertawe a Nantgarw,* 2016) has made the following comment recently in his book on Swansea and Nantgarw porcelains which summarises the current situation succinctly:

> *A number of Swansea crested services are recorded but most of the crests have not been identified ..... services with a full coat of arms are a great rarity at Swansea.*

And

> *Very little Nantgarw armorial porcelain is recorded.*

A survey of the existing historical literature produced on Welsh porcelains reveals the essential factual truth of these statements and it is found that only 12 books or monographs out of perhaps 30 or more published texts in total on Nantgarw and Swansea porcelains actually even mention the existence of armorial porcelain and in most of the cases cited these refer to specific duplicated, repetitive and well-known exemplars. The actual occurrence of items of individually crested heraldic armorial porcelain services from Nantgarw and Swansea is, therefore, finitely small, numbering perhaps a maximum of only ten or eleven distinctly different surviving examples known for each manufactory. It is important to approach this research exercise in a scientific and forensic manner as has been demonstrated successfully in other areas of ceramics research (Edwards, *18th and 19th Porcelain Analysis: A Forensic Provenancing Assessment,* 2020; Edwards, *Porcelain Analysis and Its Role in the Forensic Attribution of Ceramic Specimens,* 2021) where historical statements and the interpretation of events may themselves be open to forensic analysis. The comments of earlier historians, which in some cases were perhaps initially only expressed as opinions and have in the passage of time now been translated into irrefutable fact have occasionally been found to have distorted the reality of the situation. As in all research of this nature, the essence is to collate information from as many and diverse sources in the literature that are still accessible to the researcher: in some instances, this may have already been accomplished by previous historians but the recommended forensic scientific approach is to revert to the original documentation, if possible, and to verify or modify the conclusions or interpretations that have been made therefrom by earlier ceramics historians. It is appropriate here to cite the motto of the Royal Society of London, the world's oldest surviving and venerated scientific society, which is *"Nullius in Verba"*, translating literally as *"On the Word of no One"* or alternatively *"Take No One's Word for It"*, which was first proposed in 1663 with the granting of its arms by King Charles II. The motto refers to the emergence of experimental science and the establishment of facts in natural philosophy (before the later subdivision of natural science into chemistry and physics) from the arcane mysteries of alchemy and mediaeval mysticism, where everything was clouded in abstruse and often deliberately misleading or mysterious statements and "gobbledygook". In the Second Edition of Sir Isaac Newton's own classic and monumental work *Principia*, which he first published in 1687 (Newton, *Philosophiae Naturalis Principia Mathematica*, second Edition, published in 1713), he submitted an essay entitled *Scholium Generale*, which first established the basis of the new scientific procedure and he entreated colleagues to adopt the maxim "*Hypotheses Non Fingo*", which has been translated as *"Contrive no hypotheses that are not based upon experimental facts"*. This is precisely the maxim that has been adopted here: namely, to assiduously check everything at source wherever possible and examine all avenues before arriving at a measured and evaluated conclusion scientifically. By this means, several cases of mis-attribution that have been made in the literature relevant to armorial Welsh porcelains will become manifest in the adoption of this procedure; it seems that a great deal of guesswork or even simple assumptions have been made hitherto in attempting to interpret the origins of several of the relatively few heraldic symbols on Welsh porcelain and these should now be either confirmed, rejected or appropriately modified if possible. As a result, there will then be a firm basis established for the creation of a Welsh armorial database for the first time, which can be added to and expanded further with the future discovery of novel exemplars or through the examination of additional historical documentation relating to individual artefacts.It must be said that the philosophical shift away from abstract hypotheses and theories that had little basis in fact onto their mandatory experimental verification as vouchsafed by the Royal Society, and led by Sir Isaac Newton specifically, did not immediately have the universal support of some scientists and natural philosophers and many years after Sir Isaac Newton's impassioned plea to his colleagues for the establishment of a scientific basis for their experimental conclusions the respected German chemist Baron Justus Freiherr von Liebig (1803–1873) made the following rather disgruntled statement:

> *The loveliest theories are being overthrown by these damned experiments: it is no fun being a chemist anymore* (von Liebig, *Familiar Letters in Chemistry and its Relation to Commerce, Physiology and Agriculture,* 1843).

Nevertheless, it must be recognised in hindsight that Newton's philosophy did contribute materially towards the establishment of the basis of the modern scientific method, whereby the facts are discerned and used to advance, firstly, a weighted and measured hypothesis and then, secondly, by the observation of phenomena the proposition of a definitive theory to describe events. Von Liebig's assertion in the nineteenth Century that some rather fanciful theories, hypotheses or propositions were being overturned when subjected to a diligent checking and appraisal of the facts behind them reinforces the necessity and correctness for the adoption of this methodology, just as we propose to evaluate here the data for the correctness of attribution of the historical armorial assignments that have been made in the literature for Nantgarw and Swansea porcelains. It will probably come as a surprise to learn that several of the Welsh armorial porcelain artefacts that have been apparently "identified" in the historical literature hitherto do not stand up to their attribution in this current research project and, therefore, require a re-assessment of their assignment.

References

E. Adams, *Chelsea Porcelain* (Barrie and Jenkins, London, 1987)

Armorial Worcester Porcelain of the First Period: Specimens from the Marshall Collection in the Ashmolean Museum and Other Sources, Published by Percy Lund, Humphries & Co, and Illustrated by H. Norman Eastaugh, London, 1964

E. Barker, *"Burke and his Bristol Constituency", in Essays on Government* (Clarendon Press, Oxford, 1951)

Sir J. Bernard Burke, *Burke's General Armory—The General Armory of England, Scotland, Ireland and Wales: Comprising a Registry of Armorial Bearings from the Earliest to the Present Time* (Harrison & Sons, Pall Mall, London, 1884)

M. Beurdeley, *Porcelain of the East India Company* (Barrie and Ratcliffe, London, 1962)

G. Blake-Roberts, Swansea and Wedgwood, in *Welsh Ceramics in Context, Part I*, ed. by J. Gray, (The Royal Institution of South Wales, Swansea, 2003), pp. 75–98

P. Borschberg, The seizure of the *Santa Caterina* revisited: the Portuguese empire in Asia, VOC politics and the origins of the Dutch –Johore Alliance, 1602–1616. J. South-East Asian Stud. (January Issue), 5 (2002)

I. Bowen Backus, *Asia Materialized: Perceptions of China in Renaissance Florence*., Ph.D. Thesis (University of Chicago, 2014)

B. Brennan Ford, O.R. Impey, *Japanese Art from the Gerry Collection in the Metropolitan Museum of Art* (Metropolitan Museum of Art Publishing, New York, 1989)

T. Brook, *Confusions of Pleasures: Commerce and Culture in Ming China* (University of California Press, Berkeley, 1998)

J. Carswell, *Blue and White Chinese Porcelain Around the World* (British Museum Press, London, 2000)

M. Cassidy-Geiger, "Porcelain and prestige: princely gifts and *white gold* from Meissen", chapter 1, in *Fragile Diplomacy: Meissen Production for Foreign Courts, ca. 1710–1763*, ed. by M. Cassidy-Geiger, (Yale University Press, New Haven, Connecticut/London, 2007), pp. 2–23

P. Colomban, V. Milande, T.-A. Lu, Non-invasive on-site Raman study of blue decorated early soft paste porcelains. The use of arsenic-rich (European) ores compared with Huafalang Chinese porcelains. J. Eur. Ceram. Soc. **38**, 5228–5233 (2018)

P. Colomban, M. Gironda, D. Vangu, B. Kirmirzi, V. Cochet, The technology transfer from Europe to China in the 17th–18th centuries: non-invasive on-site XRF and Raman analysis of Chinese

Qing dynasty Enamelled masterpieces made using European ingredients/recipes. Materials **14** (2021). https://doi.org/10.3390/ma14237434

P. Conner, *The China Trade, 1600–1860* (Royal Pavilion Art Gallery and Museums, Brighton, 1986)

A. Cox, A. Cox, *Rockingham Porcelain* (Antique Collectors Club, Woodbridge, Suffolk, 2001)

E. Curtis, *Glass Exchange Between China and Europe, 1500–1800: Diplomatic, Mercantile and Technological Interactions* (Ashgate Publishing Ltd., Surrey, 2000), pp. 112–113

I. Davy (ed.), *Eat, Drink and be Merry: The British at Table, 1600–2000* (Philip Wilson, London, 2000)

R. Denyer, M. Denyer, H.G.M. Edwards, *The Pendock-Barry Derby Porcelain Service: A Reappraisal*, to be published, (2022)

H.G.M. Edwards, *Swansea and Nantgarw Porcelain: A Scientific Reappraisal* (Springer- Nature, Dordrecht, 2017)

H.G.M. Edwards, *18th and 19th Century Porcelain Analysis: A Forensic Provenancing Assessment* (Springer-Nature, Dordrecht, 2020)

H.G.M. Edwards, *Porcelain Analysis and its Role in the Forensic Attribution of Ceramic Specimens* (Springer-Nature, Dordrecht, 2021)

J. Fairbairn, *Crests of the Families of Great Britain and Ireland and Their Kindred in Other Lands*, vols. I and II, (T.C. and E.C. Jack, Edinburgh and London, 1905 and reprinted by New Orchard Publishing, London, 1986)

A. Farrington, *A Catalogue of the East India Ships' Journals and Logs, 1600–1834* (The British Library, London, 1999a)

A. Farrington, *A Biographical Index of the East India Company Maritime Service Officers, 1600–1834* (The British Library, London, 1999b)

A. Farrington, *Trading Places: The East India Company and Asia, 1600–1834* (British Library Publishing, London, 2002)

Sir C. Fawcett, The Striped Flag of the East India Company and its Connexion with the American *Stars and Stripes. The Mariner's Mirror,* **Vol. XXIII**, No. 4, October Issue, Cambridge University Press, 1937; and the *Journal for Nautical Research*, **23**, 449–476 (1937)

R. Finlay, The pilgrim art: the culture of porcelain in world history. J. World Hist. **9**, 141–187 (1998)

C. Froissart, *Collection d'un Amateur, Rouen 1680–1740: De la Premier Porcelaine de l'Age d'Or de la Faience* (Sotheby's, Paris, 2008)

F. Gambon, *Porslen Abertawe a Nantgarw (Swansea and Nantgarw Porcelain)* (Oriel Plas Glyn-y-Weddw, Llanbedrog/Abersoch/Gwynedd, 2016)

P. Glanville, H. Young (eds.), *Elegant Eating: Four Hundred Years of Dining in Style* (V&A Museum Publications, Kensington/London, 2002)

J. Graca, The Portuguese porcelain trade with China. Arts in Asia **7**, 45–47 (1977)

G. Grandjean, The porcelain of Rouen, in *Discovering the Secrets of Soft Paste Porcelain at the St Cloud Manufactory, ca. 1690–1696*, ed. by B. Rondot, (The Bard Graduate Center for the Studies in the Decorative Arts, New York, 1999)

G. Hanscombe, Armorial porcelain decorated in the London workshop of James Giles. Trans. English Ceramic Circle **20**(Part 1), 124 (2008)

M. Hillis, *Liverpool Porcelain, 1756–1804* (Maurice Hillis, Liverpool, 2011)

D.S. Howard, *Chinese Armorial Porcelain, vol. 1* (Faber & Faber, London, 1974)

D.S. Howard, *A Tale of Three Cities – Canton, Shanghai and Hong Kong: Three Centuries of Sino-British Trade in the Decorative Arts* (Sotheby's Publishing, London, 1997)

D.S. Howard, *Chinese Armorial Porcelain, vol. II* (Heirloom & Howard, Chippenham/ Wiltshire, 2003)

L. Jewitt, *The Ceramic Art of Great Britain from the Prehistoric Times Down to the Present Day, vols. I and II* (Virtue & Co. Ltd., Paternoster Row, London, 1878)

W.D. John, K. Coombes, G.J. Coombes, *The Nantgarw Porcelain Album* (The Ceramic Book Company, Newport, 1975)

R.A. Johnson, *Saint Croix, 1770–1776: The First Salute to the Stars and Stripes* (Author House Publishing, Bloomington, 2006)

J. Keay, *The Honourable Company: A History of the English East India Company* (Harper & Collins, London, 1993)
E. Keble Chatterton, A World for the Taking: The Ships of the Honourable East India Company, Fireship Press, Tucson, 2008
V. Keller, Accounting for invention:Guido Pancirolli's lost and found things and the development of desiderata. J. Hist. Ideas **73**, 223–245 (2012)
V. Keller, *Knowledge and The Public Interest, 1575–1725* (Cambridge University Press, Cambridge, 2015)
R. Kerr, N. Wood, *Science and Civilisation in China*, vol. 5, Part 12*: Ceramic Technology, S*eries editor (J. Needham, Cambridge University Press, Cambridge, 2004), pp. 364–368
G. Lagaly, Clay-organic interactions. Phil. Trans. R. Soc. A **31**, 315–332 (1984)
A. Lane, *Italian Porcelain* (Faber & Faber, London, 1954)
A. Lane, The Gaignieres-Fonthill vase – a Chinese porcelain of about 1300. Burlingt. Mag. **103**, 124–132 (1961)
C. Lestock Reid, *Commerce and Conquest: The Story of the Honourable East India Company* (Kennikat Press, Port Washington/New York, 1971)
E. Mako, J. Kristof, E. Horvath, V. Vagvolgyi, Kaolinitye-urea complexes obtained by Mechanochemical and aqueous suspension methods. Appl Clay Sci **83**(84), 24–31 (2013)
S.L. Marchand, *Porcelain: A History from the Heart of Europe* (Princeton University Press, Princeton, 2020)
J. McKeown, *English Ceramics; 250 Years Collecting at Rode Hall* (Philip Wilson Publishing/ Bloomsbury Press, London, 2006)
M. Medley, *Yuan Porcelain and Stoneware* (Faber & Faber, London, 1974)
M. Medley, *The Chinese Potter: A Practical History of Chinese Ceramics* (Phaidon Books, Oxford, 1989), pp. 176–191
C.A. Montalto de Jesus, *Historic Macao: International Traits in China, Old and New* (Kelly & Walsh Ltd., Hong Kong, 1902)
H.B. Morse, *The Chronicles of the East India Company Trading to China, 1635–1834, 5 volumes* (Clarendon Press, Oxford, 1926)
E. Morton Nance, *The Pottery and Porcelain of Swansea and Nantgarw* (Batsford, London, 1942)
I. Newton, *Philosophiae Naturalis Principia Mathematica*, 2nd edn. (Cambridge University Press, Cambridge, 1713)
J. Nicholson, *The History of Many Memorable Things in Use among the Ancients, but Now Lost and an Account of Many Excellent Discoveries Made by and Now in Use among the Moderns, 3 Volumes* (London, 1715)
G. Pancirolli, *Rerum Memorabilium, Iam Olim Deperditarium, Libro Duo,* 2 volumes, Machaelis Forster, Amberg, 1599 and 1602
S. Pierson, The movement of Chinese ceramics: Appropriation in global history. J. World Hist. **23**, 9–39 (2012)
S Pierson, "Chinese porcelain, the East India company and British cultural identity, 1600–1800", chapter 10, in *Picturing Commerce in and from the East Asian Maritime Circuits, 1550–1800,* Visual and Material Culture Series, 1300–1700, ed. T.H. Bentley, Amsterdam University Press, Amsterdam, pp. 275–292, 2019
M.A. Pinto de Matos, The Portuguese trade. Oriental Arts **XLV/1**(Spring Issue), 21–22 (1999)
M. Polo, *Il Milione,* cap. **CLVIII**, dell edizione a cuva di L.F. Benedetto, translated (1928)
T Preaud, J. Goodman, "The sevres porcelain service of Catherine II of Russia: the truth concerning payment", *Studies in the Decorative Arts*, University of Chicago and the Bard Graduate Center, **2/2**, pp. 48–54, 1995
W.R.H. Ramsay, E.G. Ramsay, *The Evolution and Compositional Development of English Porcelains from the 16th Century to Lund's Bristol c.1750 and Worcester c/1752—The Golden Chain* (Invercargill Press, Invercargill, 2017)
B. Redford, in *The Letters of Samuel Johnson, 5 Volumes*, ed. by B. Redford, vol. 3, (Princeton University Press, Princeton, 1992), pp. 70–71

H. Rissik Marshall, Armorial Worcester porcelains of the first period. Trans. English Ceram Circle **2**(Part 9), 199 (1946)

H. Rissik Marshall, *Coloured Worcester Porcelain of the First Period, 1751–1783* (Ceramic Book Co., Newport, 1951)

L. Rui, Portugal Embusca de China: imagens e Miragens (1498–1514). Ler Historia **9**, 33 (1990)

H. Sandon, *Flight and Barr Worcester Porcelain, 1783–1840* (Antique Collectors Club, Woodbridge, Suffolk, 1978)

J. Sandon, *The Ewers-Tyne Collection of Worcester Porcelain at Cheekwood* (Antique Collectors Club, Woodbridge, Suffolk, 2007)

R. Savill, "Cameo fever: six pieces from the Sevres porcelain dinner service made for Catherine II of Russia", Apollo Magazine, Volume CXVI, No. 249, November Issue, p.304, 1982

R. Savill, *The Wallace Collection: Catalogue of Sevres Porcelain* (The Trustees of the Wallace Collection, London, 1988)

W. Schulle, W. Goder, Die Erfinding des Europanischen Porzellans durch Bottger – Eine Systematische Schopferisch Entwicklung. Keram. Z. **34**, 598–600 (1982)

W. Schulle, B. Ulrich, Ergebnisse Gefugeanalytischer Untersuchungen an Bottgerporzellan. Silikatechnik **33**, 44–47 (1982)

S. Seifi, M.T. Diatta-Dieme, P. Blanchart, G.L. Necomte-Nana, D. Kobor, S. Petit, Kaolin intercalated by urea. Ceramic applications. Constr. Build. Mater. **113**, 579–585 (2016)

K. Smith, *The East India Company at Home, 1757–1857 –A Case Study in Armorial Porcelain, Case Study: The British Country House in an Imperial and Global Setting* (University of Warwick and University College, London, June 2014)

M.L. Solon, The Rouen porcelain. Burlingt. Mag. Connoisseurs **7**(26), 116–124 (1905)

J. Twitchett, *Derby Porcelain: An Illustrated Guide, 1748–1848* (Antique Collectors Club, Woodbridge, Suffolk, 2002)

G. Vasari, *Le Vite de piu eccelenti pittori, scultori e architettori*, Lorenzo Torrentino, Florence, 1568, translated into English by E.L. Seeley, 1908

T. Vollker, *Porcelain and the Dutch East India Company* (E.J. Brill, Leiden, 1954)

J.F. von Liebig, *Familiar Letters in Chemistry and Its Relation to Commerce, Physiology, and Agriculture (1843)* (Createspace Independent Publishing Platform/Booksurge Publications, Scotts Valley, 2015)

G. Walton, The era of Louis XIV: a turning-point in the history of diplomatic gifts, in *Versailles: French Court Style and its Influence*, ed. by H. C. Collison, (University of Toronto Press, Toronto, 1992)

G. Walton, Diplomacy and ambassadorial gifts in the 16th and 17th centuries, in *Gifts to the Tsars, 1500–1700; Treasures from the Kremlin*, ed. by B. Shifman, G. Walton, (H.N. Abrams, New York, 2001)

A. Weiss, Ein Geheimnis des Chinesischen Porzellans. Angew. Chem. **75**, 755–762 (1963)

A. Wild, *The East India Co.: Trade and Conquest from 1600* (Lyons, New York, 2000)

X. Xiandiong, Europe-China-Europe: the transmission of the craft of painted enamels in the 17th and 18th centuries, in *Goods from the East, 1600–1800, Trading Eurasia*, ed. by M. Berg, (Basingstoke, Palgrave Macmillan, 2015), pp. 92–106

G. Yanyi, The raw materials for making porcelain and the characteristics of porcelain wares in northern and southern China in ancient times. Archaeometry **29**, 3–19 (1987)

Chapter 2
Heraldic Achievements

Be not merely a man of letters, but let literature be an honourable augmentation to your arms not just fill the escutcheon.

Samuel Taylor Coleridge (1772–1834), Biographia Litteraria, 1817.

First appearance deceives many.

Publius Ovidius Naso (Ovid) (43 BCE-17 CE).

All truths are easy to understand once they are discovered. The point is to discover them.

Galileo Galilei (1564–1642).

There are as many characters in men as there are shapes in nature.

Publius Ovidius Naso (Ovid) (43 BCE-17 CE).

Abstract A discussion of heraldic achievements and devices, including the composition of a coat-of-arms, comprising an escutcheon, supporters, crest, torse, helm, motto and lambrequins. A listing of the animals that are commonly found as supporters and on crests and their symbolic meaning. Reading of the coat of arms to deduce the rank of the arms bearer and a description and meaning of heraldic terminology. The importance of cross-checking carefully the various components such as the escutcheon, crest and motto to verify their matching: some pitfalls in the assignment of a crest or arms which are claimed by several families. Survey of the types of armorials in Swansea and Nantgarw porcelain, *viz.* full coat-of-arms, partial coat of arms, escutcheon only, escutcheon and crest, crest only, crest and motto, and examples of mistakes that could ensue from a mis-interpretation of the detail, particularly regarding similar crests. Distribution of the different types of armorial bearing amongst Swansea and Nantgarw porcelains.

H. G. M. Edwards, *Welsh Armorial Porcelain*,
https://doi.org/10.1007/978-3-030-97439-8_2

Keywords Heraldic achievements · Heraldic symbolism · Heraldic colours and terms · Composition of a coat-of-arms · Pitfalls in interpretation of crests · Swansea and Nantgarw armorial porcelain types

2.1 The Origins of Heraldry and Coats-of-Arms

It is useful to embark here upon a brief survey of the historic origins of heraldry, the symbolism of the armorial devices used therein and the heraldic composition of a coat-of-arms from which the essential features of armorial bearings on porcelain can then be understood and appreciated. This has been summarised by the French artist, poet, novelist and dramatist, Jean Cocteau (1889–1963), as follows:

> *Every poem is a coat of arms. It must be deciphered. How much blood, how many tears in exchange for these axes, these muzzles, these unicorns, these torches, these martlets, these seedlings of stars and those fields of blue?*
> *Jean Cocteau, Le Sang du Poete (The Blood of a Poet), Orphic Trilogy, 1930.*

The operative idea in this statement relevant to the current text relates to the decipherment of the coat-of-arms, which demands an understanding of the basis of the heraldic devices used in the coat-of-arms and that is precisely what is required to enable the correct assignment or attribution to an individual of an armorial bearing or crest that is depicted on a ceramic artefact to be made. In the thirteenth century and the early years of the fourteenth century there existed several lists and compilations of heraldic devices and achievements which arose from the description of the participation of knights in jousting tournaments and in the orders of battle: these are too numerous to list comprehensively here but several examples of these lists with their date years and citation of the numbers of arms displayed are Glover's Roll (*ca.* 1240, 53 coats-of arms presented), Walford's Roll (1275, 185 coats-of-arms presented), the St George's Roll (1285, 677 coats-of-arms presented), the Lord Marshal's Roll (1295, 565 coats-of-arms presented), the Falkirk Roll (1298, 115 coats-of-arms presented), the Siege of Caerlaverock Roll (1300, 110 coats-of-arms presented) and the Banneret's Roll (1312, 1110 coats-of-arms presented). Even in these early years, the relatively large numbers of coats-of-arms cited are noteworthy. One of the first comprehensive accounts of the principles of heraldry and its derivations which attempted to trace its origins historically is that of Sir John Ferne who, writing in the mid-sixteenth century, claimed that heraldic arms could be traced back to the Trojan War:

> *The bearing of arms is no less ancient than the battle of Troy*

which he followed with "*The first sovereign that ever gave a coat of arms to his soldiers was Alexander the Great*". This means that Ferne would date the origins of heraldry to at least the Graeco-Trojan conflict of 1260–1186 BCE and then more personally somewhat later in time to Alexander III of Macedon around 356–323 BCE. Most historians now accept that Ferne's book (Ferne, *The Blazon of*

Gentrie. Divided into Two Parts. The First Named the Glorie of Generositie. The Second Named Lacyes Nobilitie. Comprehending Discourses of Armes and of Gentry. Wherein is Treated the Beginning, Parts and Degrees of Gentlenesse, With Her Lawes: of the Bearing and Blazon of Cote-Armers: Of the Lawes of Armes and Combats, 1586), although a classic of its type and being perceived to be better than some earlier texts and others of its genre, is written in a rather tedious style in the form of a dialogue between a knight and his esquire and sometimes is rather diffuse in its statements of apparent fact and legend. This was certainly highlighted over two centuries later by another early heraldic writer, James Dallaway (*Inquiries into the Origin and Progress of the Science of Heraldry in England with Explanatory Observations on Armorial Ensigns,* 1793), who preferred to assign the origin of heraldic arms to the Romans, through their adoption of eagles for the standards of their legions, stating:

> *Eagles, the ensign of the great Roman Empire, were adopted very early as being one of the hieroglyphs of royalty.*

However, most historians and heraldic experts would maintain that the Roman eagle on its standard, although forming a rallying point and a focus for their troops in battle, was not strictly a coat-of-arms in that it does not contain symbols or indicators of personal familial descent as does a genuine coat-of-arms. A more modern version of the Roman eagle might be the swastika symbol held by a spread eagle as the motif adopted by the National Socialist (Nazi) Party in Germany in the 1930s and 1940s. Both of the heraldic texts of Ferne and Dallaway, despite their drawbacks and unsubstantiated chronological assumptions about the origins of heraldry, do nevertheless contain a considerable amount of background information regarding the adoption and the early usage of heraldic devices and symbols that have come down through the ages and will be of help in our understanding of the composition of the elements that comprise the coat-of-arms used by the armigerous individuals and their families that we are investigating here . Two more early texts devoted to heraldic principles are those of William Wyrley (1592) and Sir William Dugdale (1682), both of whom were Rouge Croix Pursuivant Heralds at the College of Arms in London, the latter becoming Garter Principal King of Arms.Several statements in the early armorial literature, such as those selected below, form a good basis for understanding the origin, purpose and growth of heraldry and its social standing in the context of the bearing and the granting of a coat-of-arms. These statements are reproduced from the very early heraldic works described above and elsewhere to provide a background for our future discussions of the composition of coats-of-arms and their family origins:

> *The tournaments held with such magnificence towards the end of the 11th century were an introduction to the more general use and assumption of arms" – Susana Dawson Dobson (1793).*
>
> *There are twelve dignities of regality, whereof six be Noble and the others Princely. The six Noble be these: Gentlemen, Esquire, Banneret, Knight, Baron or Lord and Viscount. The six degrees of regality which are Princely (because they may wear crowns) be these: Earl, Marquis, Duke, Prince, King and Emperor – Sir John Ferne (1586).*

So great is the honour given to a Gentleman that if another do usurp or challenge to bear the same marks of nobleness he bears, the coat-armour, there will be held a Combat and single fight between these two persons – Sir John Ferne (1586).

When as a Gentleman of coat-armour hath married an heir to a Gentleman of coat-armour and hath issue by her, that issue as heir, bears the arms of his father and of his mother in his Shield quarterly, and it is called Coat quartered plain or rather, a Shield quartered plain – Sir John Ferne (1586).

By the wife the husband may be enriched but not ennobled – a husband of no coat-armour is ignobilem – Sir John Ferne (1586).

The first Norman arms were rude and consisting of few figures, but as feudal influence increased armorial distinctions were more necessary and by the conquest of France by King Edward III the armories were extended and the bearers proportionately increased – James Dallaway (1793).

Quartering of arms is evidence of the maternal descent of the son who then becomes the heir general – James Dallaway (1793).

In a long Line of illustrated descent each member has a peculiar distinction. They bear their paternal ensigns, attaching others by impalement or quartering. They are connected thereby with some other family, either by marriage or descent – James Dallaway (1793).

Not only enamelling of armorial bearings upon plate was in use as early as the beginning of the 13th Century but they were engraved likewise upon different pieces. A blank escutcheon is argent, dotted is or, perpendicular lines is gules, horizontal lines is azure, crossed lines is sable, dexter diagonal lines is vert, and sinister diagonal lines is purpre – James Dallaway (1793) (Author: this "heraldic tincture" representation on non-chromatic engraved or printed arms is known as "tricking" and its relevance in our later discussions will be apparent).

King Richard II (Author: reigned 1377–1399) *was the first to use supporters to his arms* – Susana Dawson Dobson (1793) (Author: this is perhaps rather controversial, as other authors claim that King Edward III (reigned 1327–1377) was the first monarch to use a lion gardant and a falcon as the supporters to his arms, whereas King Richard II later used a pair of stags collared and chained as his supporters).

A distillation of similar comments from the same sources reveals the following important summary points concerning the use of animals or heraldic devices depicted on coats-of-arms:

- The lion has always been considered as the emblem of royalty and appears as the first, as well as the most frequent, beast to appear on armorial bearings.
- Talbots and greyhounds are of a high armorial antiquity as they were necessary for hunting pursuits.
- Pheons or arrowheads, derived from the Saxon *floene*, are more commonly represented than whole arrows; they are made of fine tempered metal, barbed and scalloped on the inside to increase the difficulty of their extraction from the wound. Occasionally, the pheon is said to be *imbrued,* which means that blood spots are also depicted issuing from the arrowhead point.
- Chequy (*eschiquette*) is the oldest geometric figure found in heraldic coats-of-arms.
- The *escutcheon* is the major feature of the coat-of-arms which bears definitive information about the arms-bearer. There are many different ways of presenting an escutcheon to display heraldic features in emblazonment and the main ones are as follows: per fess, the shield is divided in two horizontally; per pale, the

shield is divided in two vertically; per bend, the shield is divided in two diagonally; per chevron, the shield is divided by a chevron or inverted "vee"; per pile, the shield is divided in two with an inverted triangle, its base merging with the top of the shield; per saltire, a cross quartering the shield diagonally; per quarterly, a cross quartering the shield horizontally and vertically; per gyronny, the shield is divided into eighths radiating from the centre; barry, horizontal bars of two colours (tinctures); paly, vertical bars of two colours (tinctures); bend dexter, a bar extending from the top left to the bottom right on the shield; bend sinister, a bar extending from the top right to the bottom left on the shield; bordure, a coloured edge around the shield; quarter, the top left quarter (to an observer, otherwise known as the dexter quarter) of the shield .

- The shape of the escutcheon (Latin *scutum*, a shield; Norman *escuchon*) was directly relevant to the type of shield borne by knights in battle. It is indicative that the most likely adoption of heraldic achievements must have arisen during the Norman era since the first escutcheon that bore heraldic arms was of the rounded top and elongated inverted teardrop shape tapering to a point, which can still be seen on the effigies of tombs of Norman knights in English cathedrals-this is often referred to as a "kite" shield. This soon became more triangular and then squared-off at the top, becoming the shorter version that is now more commonly depicted in coats-of-arms, known as a "heater" shield, which was first carried by cavalry and mounted knights in battle. From Tudor times, a greater variation in the escutcheon shape was developed and modifications such as a *bouche* (which represented a small annulet cut out from the top dexter part of the shield to accommodate the lance during jousting tournaments) and an *engrailment* (comprising two concave portions to the dexter and sinister top halves of the shield and coming to a point in the middle) were made.
- Generally, a lady who had been accorded arms in her own right would display these on a lozenge-shaped escutcheon, which was never accompanied by a helm or its crest since she would not be anticipated to use a shield or to participate in a battle.
- Clergymen, for a similar reason, would usually also bear their arms on a cartouche or oval-shaped escutcheon, but this was not always so.
- These points relating to the shapes of the escutcheon apply to British heraldry and, when considering the shapes of escutcheons which have evolved from the heraldic coats-of-arms in other countries, it is found that between 20 and 30 different types of escutcheon are known internationally, which reflects the variation in shapes of the shield used by their knights in combat.
- There are many types of crosses depicted in heraldry and their differentiation is important: most forms date from the earliest Norman portrayals and the names adopted are a mixture of Latin and French. A detailed survey is given by H. Stanford London (*Coat of Arms*, 1958), which especially describes the technical details of the cross paty (pattee), cross patonce, cross formy and their evolution in heraldry.

In summary, therefore, it can be proposed that the concepts of heraldry and the granting of arms probably originated in the eleventh/twelfth century, with the Norman escutcheons first appearing in English heraldry and developing with the parallel growth in variation of the shield and crest designs appearing in later mediaeval times, probably in the thirteenth and fourteenth centuries, as a consequence of the staging of elaborate jousting tournaments involving knights and men-at-arms, when the heraldic additions such as supporters also then came into being. It will now be appropriate to consider the heraldic components which together comprise the full coat-of-arms that are borne by an armigerous individual.

2.2 Components of Armorial Bearings

In British heraldry, the coat-of-arms is emblazoned onto an escutcheon (shield) which is associated with two supporters, named sinister (left) and dexter (right) as viewed by the shield bearer and, therefore, seen in mirror image when viewed by an observer from the front of the shield (when sinister now becomes "right" and dexter is "left" as viewed from the front of the shield). An order or badge of honour can be affixed in the lower compartment beneath the escutcheon, where the motto may also sit. Above the escutcheon with its blazon may sit a crown or a coronet, with perhaps a chapeau (a soft cap), each of which marks the rank or noble standing of the bearer of the arms; the coronet indicates the status and rank in the peerage of the arms bearer through the presence, number and arrangement of small balls on spikes (known as "pearls") and strawberry leaves around the coronet (a circlet for nobles). In the upper compartment of the coat-of-arms is found the helmet (helm) or galero. For the regal status of an arms bearer the coronet is replaced with a jewelled crown. The helm comprises several types of open-faced or closed visor, again indicating the rank of nobility of the shield bearer – usually a knight or a baronet will have a closed visor on the helm, whereas higher ranks in the peerage will have a barred or an open visor. The helm may also adopt a frontal or a side profile image, whether open or closed, but a regal status is always accompanied by a front-facing open visor. Sometimes the helm is accompanied by a prominent *bavier* (known as a "beaver"), or protective chinpiece. The helm is always accompanied by strips of cloth mantling (lambrequins), which continued as the linen surcoat worn on top of a knight's suit of armour and is surmounted by a twisted silk torse or cloth wreath, which may be twin-coloured and which functions as a supporter of the crest on the helm itself. The torse is usually curved in a convex way to match the profile of the top of the helm. Above the crest a slogan may also appear which represents the ancient battle cry (see later). In the Roman Catholic Church, clerics entitled to bear arms replace the helm with a galero; for example, this may be a red wide-brimmed cardinal's hat bearing 15 tassels at each side and carrying a suspended gold patriarchal cross centrally on a chain. Anglican Church clerics who are entitled to bear arms use a galero which is typically a black wide-brimmed hat with tassels in various colours and Anglican arms-bearing bishops can alternatively use a mitre in place of a helm

(Scott-Giles, *Boutell's Heraldry*, 1954; Woodcock, *The Oxford Guide to Heraldry*, 1988).

The number and arrangement of the strawberry leaves and balls on spikes, or "pearls", which alternate on a noble's coronet situated above the shield or escutcheon gives the rank in the peerage of the entitled bearer of the arms displayed. Hence, a Duke's coronet (and that of a Duchess) does not have "pearls" but will have eight strawberry leaves in total, of which five will be seen when viewed in two dimensions from the front, comprising three fully facing the observer and another two which are viewed at each side in profile. Similarly, the coronet of a Marquess (and Marchioness) will have four strawberry leaves and four "pearls" or balls, of which three strawberry leaves and two balls will be visible from the front. An Earl (and Countess) will have eight strawberry leaves and eight balls (an example of the arms and crest of the Earl of Gosford which exhibits these features is shown in Fig. 2.1), of which four strawberry leaves and five balls will be visible from the front. A Viscount (and Viscountess) will have 16 balls of which seven will be visible from the front, and a Baron (and Baroness) will have six balls of which four will be visible from the front. Hence, the monogram and crest of the Viscountess Philadelphia Cremorne, as depicted on the dessert comport from her armorial Derby China Works *Cremorne* porcelain service shown earlier in Fig. 1.19, has a coronet with seven balls ("pearls") clearly visible from the front, indicative of the rank of a Viscount or Viscountess (Twitchett, *Derby Porcelain, 1748–1848: An Illustrated Guide*, 2002).

The heraldic arrangements of other countries of these leaves and pearls on their coronets differ somewhat between each other and from those in English heraldry described above: for example, we have seen already in Chap. 1 that the coronet of a French Comte, Jacques Asselin de Villequiers, which was portrayed on a piece of armorial Rouen porcelain dated to 1695, possessed nine balls or "pearls" when viewed from the front – whereas that of a French Marquis would have had three strawberry leaves and 13 balls displayed (several of which would have been symmetrically arranged on a single spike in trios of three pearls) as seen from the front and a French Duc would have five strawberry leaves and five balls on his coronet when viewed similarly. Italian nobility utilise rather different escutcheon shapes and arrangements of the pearls on the coronet, which are not supported on spikes.

A coronet is normally depicted for all ranks of the British peerage from Baron to Duke, whereas royalty use a jewelled crown in their arms, which has a different shape and composition to that of a coronet (Scott-Giles, *Boutell's Heraldry*, 1954; Fox-Davies, *A Complete Guide to Heraldry: with Illustrations by G. Johnston, Herald Painter to the Lyon Court*, 1909). It is interesting that in one of the earliest books on heraldry (Ferne, *The Blazon of Gentrie. Divided into Two Parts. The First Named the Glorie of Generositie. The Second Named Lacyes Nobilitie. Comprehending Discourses of Armes and of Gentry. Wherein is Treated the Beginning, Parts and Degrees of Gentlenesse, With Her Lawes: of the Bearing and Blazon of Cote-Armers: Of the Lawes of Armes and Combats*, 1586) the rank of Viscount specifically was not considered suitable for the wearing of a coronet and was downgraded from Ferne's upper *Princely* category to the lower grade of *Nobility* (see the quotation in Sect. 2.1 above).

Fig. 2.1 Example of an heraldic coat-of-arms and crest; a bookplate of the second Earl of Gosford, Archibald Acheson (1776–1849), of Gosford Castle, Markethill, Armagh, Ireland, who was for some time Lieutenant -Governor of Canada and Governor-General of British North America in the early nineteenth Century, becoming the second Earl in 1807. A noted Swansea porcelain service, superbly decorated in a London atelier, was commissioned for Gosford Castle between 1815 and 1817. The *Gosford Castle* service is identical in decoration and gilding to the *Marquess of Exeter* service made for Brownlow Cecil (1795–1867), the second Marquess of Exeter and Lord Burghley, of Burghley House, Stamford, Lincolnshire, except that each piece of the *Marquess of Exeter* service is marked SWANSEA in red stencil and none of the *Gosford Castle* service is so marked. Neither service is an armorial. *Private Collection*

In heraldry, there are but two metals recognised, namely *or* and *argent*, these being gold and silver. There are five colours depicted, namely *azure* (az., blue), *gules* (gu., red), *vert* (green), *sable* (sa., black) and *purpure* (purple), with the colours yellow and white being additionally represented by *or* and *argent* (ar.), respectively. There are four furs used in heraldry: namely, *ermine*, a white background (field) with black dots, *ermines*, a black field with white dots, *erminois*, a gold field with black dots, and *pean* a black field with gold dots.

The mediaeval symbolism of some of the animals and devices used on the heraldic crest, in the supporters of the arms or in the escutcheon itself which are of potential relevance here and in our future discussions can be summarised with a selection as follows:

> *Boar* (sanglier): fiercely combative, courageous; *Otter*: denotes one who is intelligent and industrious; *Lion*: the noblest of the wild beasts, depicts strength and valour; *Leopard*: signifies one who is valiant and who braves danger with courage; *Dolphin*: charity; *Eagle*: fortitude and magnanimity; *Stag*: will fight if provoked; *Sprouting tree stump*, rebirth and growth; *Bird*: eloquence ; *Escallop shell*: a successful leader and a protector of providence; *Paschal lamb*: possesses faith, innocence and is resolute in spirit; *Martlet*: continuous effort and virtue; *Dove*: exemplifying gentleness and peace; *Lamb:* symbol of Christianity; *Hound*: leadership; *Pelican*: self-sacrifice and piety, *Phoenix*: resurrection; *Falcon/Hawk*: signifies someone of action; *Frog:* harmony and peace; *Griffin*: mythical beast combination of an eagle and a lion, valour; *Cock*: virility and bravery; *Chough*: death; *Wildcat*: vigilance and courage; *Butterfly:* peace; *Bull*: valour; *Bee:* industrious; *Bat*: watchfulness; *Antelope:* peace and harmony: *Raven*: divine providence; *Snake:* wisdom and medicine ; *Swan*: the royal bird, harmony; *Wyvern*: mythical beast , combination of a dragon and serpent, courage; *Imperial eagle:* two-headed, signifying fortitude and magnanimity. In addition, several animate and inanimate objects signify a particular symbolism , such as an *Arm:* industry; *Spear*: readiness for battle, a knight in service; *Tower*: strength and protection; *Axe*: military service and duty; *Bezant*: banking; *Book*: learning; *Dagger*; honour in battle; *Fetterlock*: a symbol of authority; *Gauntlet*: military service; *Hand*: faith; *Portcullis*: protection; *Mullet:* a star with 5,6,8 or 12 points, signifying military service; *Pheon*: dexterity and nimbleness of wit.

Originally, the crest would have been worn on top of the helm on the field of battle, secured by the torse wreath and the cloth mantling on the helm (lambrequin), as reflected in the older and mediaeval coats-of-arms, but more recent coats-of-arms awarded by the College of Arms exhibit in comparison perhaps less practical crests for this purpose, such as castles, towers and ships. For instance, Admiral Lord Horatio Nelson's coat-of-arms has two crests, one of which depicts the stern of a Spanish battleship that he captured at the Battle of Cape St Vincent in 1797, namely the 112-gun *San Josef*. The second crest is a *chelengk,* an Ottoman Empire honour awarded to Nelson after his defeat of the Napoleonic French navy at the Battle of the Nile, at Aboukir Bay in August 1798. Nelson's motto was "*Palmam qui meruit ferat*", which translates as "*Let Him Who Has Earned It Bear the Palm*", and a palm tree also features above his escutcheon along with a three-masted man-o'-war ship and a shore battery. Nelson's coat-of-arms and these heraldic devices can all be seen on the armorial plate of the *Horatia* service in Chamberlain's Worcester porcelain shown later in Fig. 7.1, and his escutcheon is of the type described above as "engrailed". The earliest coats-of-arms and crests tend to be rather simple in concept and design compared with the later ones; for example, the earliest version of the Barry arms that is depicted centrally in the *Barry-Barry (Pendock-Barry)* Derby China Works porcelain service, a dinner plate from which is shown in Fig. 1.13 (Edwards, *Porcelain Analysis and its Role in the Attribution of Unknown Specimens,*

2021; Denyer et al., *The Pendock-Barry Derby Porcelain Service: A Reappraisal,* 2022). This comprises just ten alternating red and white horizontal bars on the shield, dating back to its Norman origins, with a crest of a wolf's head erased sable. The later arms and escutcheon of Pendock-Barry of Roclaveston Manor, Nottinghamshire, a descendant of the Barry family who laid claim to the Barry arms in the early nineteenth century, are more complicated and his crest depicts an embattlemented tower charged with red roses which has the motto "*A Rege et Victoria*", which translates as "*From the King and Conquest*". Incidentally, the presence of the pink/red roses depicted on this later Pendock-Barry crest could well have inspired the inclusion of the symbolic wreath of roses enamelled around the verge of the armorial porcelain plate of the *Pendock-Barry (Barry-Barry)* service shown in Fig. 1.13 and painted so superbly by William Billingsley, the esteemed master painter of the rose on porcelain.

One of the most ancient lineages in English heraldry is that of Percy, descended from William de Percy, who accompanied William the Conqueror from Normandy in his invasion and conquest of Britain in 1066. Josceline of Louvain, brother of Queen Adeliza, the second wife of King Henry I, and son of Godfrey, Duke of Brabant, married Agnes de Percy and established the Percy Dynasty with William, Lord Percy of Alnwick, Northumberland, and its several cadet branches. The family crest is a *lion rampant or*, a favourite theme in heraldry and a very simple one, which will also feature on a hitherto unattributed Swansea porcelain crested service and some potential assignments made to its individual and family of origin will be discussed later.

Reference to the factory workbooks and lists of services completed by the proprietors of porcelain manufactories can be invaluable for the chronological placement of an armorial service, especially when the letters or notes of communication still exist between the factory proprietors and their clients, as is the case for William Duesbury's Derby China Works and his London agent, Joseph Lygo, from whom many services were commissioned in Derby porcelain by the aristocracy (Twitchett, *Derby Porcelain: An Illustrated Guide, 1748–1848,* 2002; Edwards, *Porcelain Analysis and its Role in the Attribution of Unknown Specimens,* 2021). An example of the usefulness of such correspondence is provided by the letters between Lady Philadelphia Cremorne and William Duesbury relating to the commissioning and delivery of her armorial service in 1788, another is the record of the purchase of an armorial tea service (Fig. 1.14) by Bishop John Egerton, the Bishop of Durham, in 1784, the sale of which was logged in the Covent Garden showroom of the Derby China Works in London; it is interesting that the Bishop's tea service displays his engrailed escutcheon and his crest but not his mitre or galero. However, the manufactory documentation is often found to be incomplete, and it then becomes a much more daunting task to trace the provenance of an armorial service completed by the factory, which involves a diligent and meticulous search of supportive archival records where these are available. Unfortunately for ceramics historians, these detailed accounts for many porcelain manufactories no longer exist and the prime source for armorial research then becomes the armory literature. The primary process of identification of a coat-of-arms or an heraldic crest involves the consultation of several reference texts in the armorial literature, the major ones being: Burke, *A Genealogical and Heraldic Dictionary of the Peerage and Baronetage of the United Kingdom*, 1826; Burke, *Burke's General Armory – The General Armory of England, Scotland, Ireland and Wales: Comprising a Registry of Armorial Bearings from the Earliest to the Present Time*, 1884; Burke, *A Visitation of the Seats and Arms of the Noblemen and Gentlemen of Great Britain and Ireland,* 1895; Cockayne, *The Complete Peerage of England, Scotland, Ireland, Great Britain and the United Kingdom Extant, Extinct or Dormant,* 1910; Fairbairn, *Crests of the Families of Great Britain and Ireland and Their Kindred in Other Lands,* 1905; Fox-Davies, *Armorial Families: A Complete Peerage, Baronetage and Knightage, and a Directory of Some Gentlemen Coat -Armour and being the First Attempt to Show Which Arms are in Use at the Moment are Borne by Some Legal Authority*, 1895; Fox-Davies, *A Complete Guide to Heraldry: with Illustrations by G. Johnston, Herald Painter to the Lyon Court,* 1909. Additionally, the consultation of Elvin's classic work (*A Handbook of Mottoes Borne by the Nobility, Gentry, Cities, Public Companies Etc.*, 1860) is especially useful in this regard. In particular, Fairbairn (Fairbairn, *Crests of the Families of Great Britain and Ireland and Their Kindred in Other Lands,* 1905) is found to be useful for its listing of mottoes (approximately 7000 of which are given in Volume 1 with their family association) and for its pictorial representation of crests, which occupy 314 pages of plates in Volume 2 with 15 crests per page.

The consultation of these reference works is a highly recommended if not a mandatory exercise for the purposes of the assignment of the correct attribution of a crest, a motto or a coat-of-arms displayed on porcelain or other ceramics, however much support that may already exist for the assignment historically. Even if the manufactory records indicate clearly the source of the commissioning of an armorial service, it is good forensic practice to check this against the official listings of the arms and crests in the literature described above: an excellent exemplar of the need to cross- check a coat-of-arms with documents and literature in this way is provided by the case study of the *Barry-Barry (Pendock-Barry)* armorial service in Derby porcelain, (Fig. 1.13) which has been fully described elsewhere, where small differences in the escutcheon displayed were critically important for the determination of the correct genealogy and eventual chronological placement of the commissioning of this important armorial service (Edwards, *Porcelain Analysis and its Role in the Attribution of Unknown Specimens,* 2021; Denyer et al., *The Pendock-Barry Derby Porcelain Service: A Reappraisal*, 2022). As a result, much of the conjecture and indecision that has occurred in the literature hitherto about the attribution of the *Barry-Barry* porcelain service chronologically has now been addressed and corrected through a careful and critical analysis of the heraldic armorial achievements involved in the Pendock-Barry and Barry-Barry arms.

2.2.1 The British Peerage

In descending order of rank in the hereditary British peerage there are Duke (Duchess), Marquess (Marchioness), Earl (Countess), Viscount (Viscountess) and Baron (Baroness). An additional grade was created in the mid-twentieth century with Life Peers, who could sit in the House of Lords along with a selection of nominated hereditary peers: a Life Peer sits below a Baron or Baroness in rank.

An individual peer can hold several peerage titles of different rank which may have been created and inherited over several centuries. The precedence of a peer among others of his own rank is dependent upon the antiquity of the peerage in question: basically, the older the title the more senior the peer in that rank. The lowest rank of Life Peer, according to the Life Peerage Act of 1958, was created to elevate to the peerage for their lifetime only citizen commoners of the realm in recognition of a particular service that they had performed. Theoretically there is no limitation imposed on the number of Life Peers created and Prime Minister Tony Blair with the assent of Queen Elizabeth II created more than 500 of these. All peers are listed in one of the five peerages; the Peerage of England, created by the Kings and Queens of England prior to the Act of Union of 1707, the Peerage of Scotland, created by the Kings and Queens of Scotland before the Act of Union of 1707, the Peerage of Ireland, created by the Kingdom of Ireland before the Act of Union of 1801, the Peerage of Great Britain, created by the Kings of Great Britain between 1707 and 1801, and the Peerage of the United Kingdom for titles created since 1801.

The Royal Dukes (sons of the Sovereign, and conferred upon Royal Princes upon their marriage, such as Edinburgh, York, Kent, Cambridge, Sussex and Gloucester) take precedence before other Dukes in the peerage, independent of their date of foundation. Other Royal Dukes are the Sovereign (Duke of Lancaster) and the Prince of Wales (Duke of Cornwall). There are 24 Dukes currently in the peerage, with the senior non-Royal Duke in the English Peerage being the Duke of Norfolk (1397). The senior Dukes in the Scottish and Irish peerages are the Duke of Hamilton (1643) and Duke of Leinster (1766), respectively. Of the 34 Marquesses, the senior in the English Peerage is the Marquess of Winchester (1551) and in the Scottish Peerage, the Marquess of Huntley (1599). Of the 191 Earls, which are preceded by the Royal Earl of Wessex, the senior in the English Peerage is the Earl of Shrewsbury (1442) followed by the Earl of Crawford (1398) in the Irish Peerage. There are also 4 Countesses in this rank by their own right. The 115 Viscounts have as their senior in the English peerage, Viscount Hertford (1550), Viscount Falkland (1620) in the Scottish Peerage and Viscount Gormanstoun (1478) in the Irish Peerage. Finally, there are 426 hereditary Barons and 9 Baronesses in their own right, which have as precedence Lord de Ros (1264) in the English Peerage and Lord Kingsale (1273) in the Irish Peerage. It should be noted that, although carrying a hereditary honour and being a member of the aristocracy, a Baronet is not included in the peerage. It goes without saying though that the largest number of armigerous families are neither in the peerage nor entitled as are Knights and Baronets. Excluding Life Peerages, the number of arms-bearers in the peerage can be estimated as 800; the relative number of arms-bearers must exceed 7000 from the mottoes listed in Fairbairn, most of which are replicated by several families. In the 611 pages of crest-bearing families listed by Fairbairn (Fairbairn, *Crests of the Families of Great Britain and Ireland and Their Kindred in Other Lands,* Volume I, 1905) with an average of 75 families listed in the three columns on each page, it is calculated that there are some 45,000 arms-bearing families in this database, which gives an idea of the dimension of difficulty facing an armorial researcher who has no primary evidential clues on which to commence starting a search for the attribution of a crest.

2.3 Swansea and Nantgarw Armorial Porcelains in the Literature

Before we can start to evaluate even the limited range of extant Swansea and Nantgarw armorial porcelains, we should first examine the types of armorial service that could potentially exist in our database for the Swansea and Nantgarw China Works. This will provide the basis for the establishment of a scientific methodology of investigation whereby the holistic approach that has been demonstrated to be so successful in assisting in the identification of porcelain specimens that were of an unknown factory attribution or provenance can also be utilised here to define some standard procedures for operation that will influence and facilitate the process of the correct identification of armorial porcelains (Edwards, *Porcelain Analysis and Its Role in the Forensic Attribution of Ceramic Specimens,* 2021). The types of Welsh porcelain service that are known to bear an armorial decoration can be briefly summarised as potentially belonging to one of the following categories:

- Plain porcelain services that bear a crest, which is usually gilded, either in the reserve or centrally located on the artefact.
- Plain porcelain services that bear a crest and an associated motto.
- Services that bear a coat-of-arms, which may be fully or only partially depicted heraldically, including the supporters, a motto, helm and a crest, which is usually centrally and prominently placed on the porcelain field of flatwares or on the curved sides of comports and tureens. An unusual placement of the crest on an armorial porcelain plate has been observed in the *Pitt* service of 1705, where the crest is neither located centrally nor in the reserve but on the cavetto of the plate and oddly situated beneath the escutcheon (Fig. 1.9), which is the reverse of common heraldic practice. The torse upon which the crest sits is also incorrectly depicted with a concave profile rather than convex, which would not sit properly on the helm. These coats-of-arms might be expected to be executed in polychrome enamels and will be accompanied by some gilding, perhaps in the crest itself and additionally on the lambrequins, motto and the supporters. The coat-of-arms may alternatively only be a partial one, possibly featuring an escutcheon only, and may also omit one or more of the other heraldic achievements compris-

ing the supporters, crest and motto, or perhaps some abbreviated combination of these may be found placed around the basic heraldic escutcheon.

- Services that contain a crest or a coat-of arms, with or without a motto, but also accompanied by an associated decoration, which may be simply gilded or enamelled, or perhaps possess more intricate geometric gilding patterns and may comprise, in addition, scattered flowers, floral bouquets and other decorative geometric features and motifs. This can be especially useful for chronological purposes if a particular factory painter can be identified from the accompanying decoration or patterns on an armorial artefact, for example, Henry Morris, William Pollard or David Evans at Swansea and William Billingsley, Thomas Pardoe or William Weston Young at Nantgarw.

We are now able to propose a standard methodological procedure for the identification of the family bearer of the crest or coat-of-arms, which can avoid some pitfalls for reaching an unwary and less prescribed assignment which could lead one down an incorrect pathway or course of action and potentially result in a misattribution of the arms, or perhaps even a complete failure to attribute the heraldic devices to a particular individual. For instance, it will be seen that a crest on its own can be insufficient to interpret unequivocally and assign to a family arms-bearer without undertaking much diligent searching through the heraldic literature. Even then, there may be several commonalities to be found with similar crests used by other noble arms-bearing families which could potentially result in the attainment of an incorrect conclusion if small nuances are not noted, such as the "nowing" of a rampant lion's tail or the type of cross borne on a flag or the finial carried on the pennon of a Paschal lamb crest. Fairbairn (*Crests of the Families of Great Britain and Ireland and Their Kindred in Other Lands,* 1905) provides a comprehensive listing of the crests, many of which are illustrated, and of the family individuals who have used them, and this can indicate a potentially valuable lead for the armorial researcher. The attitude of the head of a bird of prey which may be depicted with its wings spread and talons drawn (heraldically "armed"), such as an eagle whose head is facing to the front (affronted) or looking in the sinister or dexter direction, is also critical in this respect. The situation pertaining to the crest of a stork and its assignment to the correct Pitt family has already been discussed in Chap. 1, where the detail of the stance of the stork and the precise position of its legs, and especially whether the dexter leg was raised or not and if it was resting on or holding another object such as an anchor, was seen to be instrumental in the decision for its correct attribution to the individual arms-bearer. A crest with an associated motto does assist here in the narrowing down of the potential field, whilst bearing in mind that the motto may itself be an acquired one and perhaps may not have been officially registered to the arms-bearer by the College of Arms. Some armories, such as Fairbairn (*Crests of the Families of Great Britain and Ireland and Their Kindred in Other Lands,* 1905), provide lists of the families who share mottoes (usually stated in their English, Latin, French or Welsh languages) and this is extremely useful for an initial assessment of likely contenders in an investigation of the adoption of crests to be made. Elvin (Elvin, *A Handbook of Mottoes Borne by the Nobility,*

Gentry, Cities, Public Companies Etc., 1860) also provides a list of mottoes that have been commonly used in heraldic achievements without displaying the crests. The depiction of the full coat-of-arms with its crest, motto, helm, with or without a coronet, and its supporters undoubtedly affords the best chance of success for the researcher to deduce its unequivocal identification and for deriving the correct family and thence the individual arms-bearer attribution.

The complete listing of extant armorial services on Swansea and Nantgarw porcelains has been extracted from the existing ceramics literature of these two manufactories and is given in Table 2.1; here, the 12 texts or articles in the literature which mention armorial porcelains when discussing the operations of the Swansea and Nantgarw manufactories are listed along with details of the armorial decoration, crests or coats-of-arms depicted therein. Several important points need to be made from the list given in Table 2.1:

- Of the 35 listed crests or coats-of-arms mentioned in the existing literature on Welsh porcelains only 19 can be identified here as being distinctly different from each other and many references and citations in the historical literature texts, therefore, relate to identical crests or coats-of-arms that have already been described by other authors and historians. The full references, from which the given illustration plate and the page numbers where appropriate have been extracted and reproduced in the Table, have been provided in the footnote to Table 2.1. The actual numbers of armorial porcelains cited in these texts can be summarised as follows for the Swansea and Nantgarw manufactories, according to the authors, their texts and the year of their citation: Gambon (*Porslen Abertawe a Nantgarw (Swansea and Nantgarw Porcelain)*, 6 Swansea +1 Nantgarw), 2016; Jones and Joseph (*Swansea Porcelain: Shapes and Decoration*, 3 Swansea), 1988; John (*Swansea Porcelain*, 5 Swansea), 1958; John (*Nantgarw Porcelain*, 2 Nantgarw), 1948; John et al. (*Nantgarw Porcelain Album,* 1 Nantgarw), 1975; John (*William Billingsley*, 1 Nantgarw), 1968; Meager (*Swansea and Nantgarw Potteries*: *Catalogue of the Collection of Welsh Pottery and Porcelain on Exhibition at the Glynn Vivian Art Gallery, Swansea,*1 Swansea), 1949; Fairclough (*Welsh Ceramics in Context,* Volume II, 1 Swansea), 2005; Morton Nance (*The Pottery and Porcelain of Swansea and Nantgarw*, 1 Swansea +2 Nantgarw), 1942; Phillips (Heraldic Swansea Emblems, *Nantgarw & Swansea Newsletter*, 9 Swansea), 2006; *Gartre'n Ol* (*Gartre'n Ol /Coming Home Exhibition Catalogue – An Exhibition of the Finest Nantgarw Porcelain,Nantgarw China Works Museum, Tyla Gwyn, Nantgarw,* 1 Nantgarw), 2019; Edwards (*Nantgarw and Swansea Porcelains: An Analytical Perspective,* 1 Nantgarw), 2018. This gives the total distribution of the 26 Swansea and 9 Nantgarw armorials recorded and referenced in the literature, many examples of which, as stated above, are repetitive citations and are not, therefore, distinctive individually from each other.
- Of the 26 Swansea armorials cited in the literature hitherto, 16 have been classified as "Unidentified", with 10 being labelled as being "Identified" by their authors; the 9 Nantgarw armorials cited in the literature fare significantly better

Table 2.1 Swansea and Nantgarw Armorial Porcelains: recorded service exemplars

Author	Book/Article	Factory	Page /Ref.	Type	Depiction	Comment
Gambon	Porslen Abertawe a Nantgarw	Swansea	25 /S19	Crest	Gilt dolphin	Unidentified
			25/S20	Crest	Gilt demi-leopard rampant	Unidentified
			25/S21	Crest	Gilt scallop shell	Unidentified
			25/S22	Crest	Gilt bird on tree stump	Unidentified
			27/S26	Crest	Gilt boar's head	Unidentified
			33/S40	[a]p Coat-of-arms	Polychrome +crest	Clarke of Hereford/Parkinson
		Nantgarw	63/N25	p Coat of arms	Polychrome +crest	Wyndham
Jones & Joseph	Swansea Porcelain	Swansea	119/4	Crest	Gilt demi-leopard rampant	Unidentified
			121/3	p Coat-of-arms	Monochrome	Unidentified[c]
			213/1	Crest	Gilt boar's head	Unidentified
John	Swansea Porcelain	Swansea	Plate33	Crest	Gilt dolphin	Garnett-Orme
			Plate35	p Coat-of-arms	Polychrome +crest	Clarke of Hereford/Parkinson
			Plate78C	p Coat-of-arms	Polychrome +crest	Clarke of Hereford/Parkinson
			Plate 84C	Crest	Gilt demi-leopard rampant	Unidentified
			Plate 84C	Crest	Gilt boar's head	Unidentified
John	Nantgarw Porcelain	Nantgarw	Plate34B	Crest+motto	Gilt + Polychrome	Ramsay
			Plate 54B	p Coat-of-arms	Polychrome /Pardoe	Unidentified
John *et al.*	Nantgarw Porcelain Album	Nantgarw	Plate56	Coat-of-arms	Polychrome	Viscount Weymouth
John	William Billingsley	Nantgarw	Plate89C	Crest	Gilt otter speared	Homfray of Penllyne
Meager	Swansea Nantgarw Potteries	Swansea	38/111	Crest	Gilt boar's head	Sir Watcyn Williams-Wynn
Fairclough	Welsh Ceramics	Swansea	200	Crest	Boar +holly tree	Thomas Lloyd + motto
Morton Nance	Swansea & Nantgarw	Swansea	335	p Coat-of arms	Polychrome	Abel Gower
	Porcelains	Nantgarw	Plate CLX	Crest	Gilt otter speared	Homfray of Penllyne
			Plate CLXIIIB	Crest + motto	Polychrome + Gilt	Ramsay
Phillips	Heraldic Swansea	Swansea	18/45	Crest	Leopard rampant	Unidentified

Author	Book/Article	Factory	Page /Ref.	Type	Depiction	Comment
			18/46	Crest	Spread eagle	Venn
			18/47	Crest + motto	Paschal lamb	Unidentified
			18/48	Crest	Gilt demi-leopard rampant	Unidentified
			18/48	Crest	Bird on tree stump	Unidentified
			18/48	Crest	Scallop shell	Unidentified
			18/49	Crest	Boar's Head	Unidentified
			18/50	p Coat-of-arms	Polychrome – crest	Clarke of Hereford/Parkinson
			18/51	Crest + motto	Boar in holly tree	Thomas Lloyd + motto
Nantgarw CW[b]	Gartre'n Ol	Nantgarw	Page 35	Crest + motto	Polychrome + Gilt	Ramsay
Edwards	Nantgarw & Swansea Porcelains	Nantgarw	Figure 1.10	Crest	Gilt demi-lion rampant	Baron Phipps

Edwards (2018), Fairclough (2005), Gambon (2016), *Gartre'n Ol* (2019), John (1948, 1958, 1968, 1975), Jones and Joseph (1988), Meager (1949), Morton Nance (1942), Phillips (2006)

A. Renton, National Museum of Wales Amgueddfa Cymru, personal communication to the author, July 2021

[a]p Coat-of-arms: only a partial coat of arms is depicted. Clarke of Hereford/Parkinson shows the escutcheon, crest and motto but no supporters, whereas the unidentified coat-of-arms shows only the escutcheon without the crest and supporters

[b]Nantgarw China Works Museum Trust: *Gartre'n Ol /Coming Home Exhibition Catalogue – An Exhibition of the Finest Nantgarw Porcelain, Nantgarw China Works Museum, Tyla Gwyn, Nantgarw,* 2019

[c]Athough the current situation from the literature records gives this partial coat-of-arms as unidentified, recent evidence from a personal communication to the author (Renton, 2021) assigns this escutcheon and arms to Lockwood of Dawes Hall, Essex, impaling Manners-Sutton, which is more fully described in Chap. 5, Sects. 5.9 and 5.9.1

[d]Out of the 35 citations of armorial porcelains covered in these texts, comprising 26 Swansea and 9 Nantgarw exemplars, there are only 19 discrete exemplars, including the dubious Garter Star motif, namely, 7 (8 including the Garter Star motif) for Nantgarw and 11 for Swansea

in this respect with only 1 classed as "Unidentified" and 8 labelled as "Identified". It will transpire that several of those which have been labelled as "Identified" in the Swansea list, in fact, have also been found to have been incorrectly attributed as a result of the current research, which somewhat worsens the apparent analytical numerical situation. Hence, some 64% of Swansea armorials are accepted thus far as being "Unidentified" and only 11% of Nantgarw armorials similarly, not accounting for those artefacts that have been incorrectly labelled according to the current work as having been "Identified" in the literature. In fact, no fewer than 8 of those Swansea armorials citations mentioned hitherto as "Identified" now have some degree of doubt cast upon their original assignment but, in contrast, all but two of the 8 Nantgarw armorials labelled as "Identified" in the literature have been confirmed as having been correctly attributed through the current research exercise.

- Of the 35 crests and coats-of-arms on Welsh porcelain cited in the literature, it appears that these can be reduced to only 11 distinctly different exemplars on Swansea porcelain and 7 (possibly 8, if one includes the "Garter Star" motif, see below) on Nantgarw porcelain.
- Crested armorial pieces on Swansea porcelain far outnumber those which display the full or partial coats-of-arms: there are only two partial coats-of-arms represented on the 11 extant Swansea porcelain exemplars, representing some 18% of the total, and 9 crests, representing 82% of the total armorials.
- Armorial pieces on Nantgarw porcelain are clearly less commonly encountered than those on Swansea porcelain, with only 3 coats-of-arms, namely one full and two partial coats-of-arms, being represented on the 7 exemplars, some 42% of the total, and 4 crests representing 56% of the total. The "Garter Star" motif has been excluded from this analysis, as will be explained further below.
- Only one armorial exemplar depicted on Swansea and Nantgarw porcelain collectively has the full armorial coat-of-arms displayed, comprising the escutcheon, supporters, coronet, lambrequins, crest, torse and motto, which has been identified (John et al., *The Nantgarw Porcelain Album*, 1975) as those of Lord Henry Frederick Thynne, Viscount Weymouth, on Nantgarw porcelain (Fig. 2.2). Lord Henry Frederick Thynne (1797–1837), who later briefly became the third Marquess of Bath (as cited and illustrated by John et al., *Nantgarw Porcelain Album,* Plate 56, 1975) in 1837, but his tenure of the title following his accession to the Marquessate was very brief as he died later in the same year. It is noted that Lord Henry Frederick Thynne would have attained the age of 21 in 1818, which may have been a possible reason for the commissioning of this Nantgarw service and assigning it to him and this would be a very tempting speculation. However, genealogically it has been discovered that Lord Henry Frederick Thynne did not succeed to the title of Viscount Weymouth until January 1837 upon the death of his elder brother, some 17 years after Nantgarw ceased production, so the service has been incorrectly attributed historically to Lord Henry Frederick and should now be re-labelled as belonging to Lord Thomas Thynne, who was Viscount Weymouth between 1796 and 1837. There are some other serious anomalies with this service regarding the heraldic portrayal of these arms and these will be fully discussed in Chap. 4.

Fig. 2.2 Dinner plate from the Nantgarw China Works, *ca.* 1817–1820, impressed NANT-GARW C.W., bearing the full coat-of-arms of Viscount Weymouth, Lord Henry Frederick Thynne (1797–1837), who became the third Marquess of Bath in 1837 just months before he died. The escutcheon is quartered: first and fourth, a barry of ten or and sable; second and third, or a lion rampant tail nowed gu. Crest: on a wreath of gu. and or, a reindeer statant or collared sable. Supporters: dexter, a reindeer or gorged with a plain collar sable; sinister, a lion, tail nowed and erect, gules. Motto: *"J'ay bonne cause"* which translates as *"I have good reason"*. Family seat, Longleat House, Wiltshire. Illustrated in Plate 56 John et al., *Nantgarw Porcelain Album*, 1975. *Private Collection*

The last column in Table 2.1 gives the current attribution of the armorial displayed as "Unidentified" or "Identified", where one has been made or proposed in the literature; this now needs to be evaluated carefully for each case cited. It will transpire that in several cases the arms depicted on the porcelain have been imperfectly assigned to a particular individual family arms-bearer by ceramic historians in the literature when the original heraldic records are cross-checked in the armory books against the assignments that have been made in the lists enumerated above. However, for the sake of historical and literary accuracy the assignment of the arms as made by the original author in the literature cited is reproduced here in Table 2.1 and the evidence for the correctness or otherwise of these attributions will then be discussed fully and individually in detail appropriately in the following Chapters, namely,

Chap. 4 for the Nantgarw porcelain armorials and Chap. 5 for the Swansea porcelain armorials. These assignments will then be confirmed or amended accordingly, if such a proposal is deemed to be necessary. In the holistic approach adopted for the scientific evaluation of a porcelain artefact (Edwards, *Porcelain Analysis and Its Role in the Forensic Attribution of Ceramic Specimens,* 2021), the three components comprising the scientific analytical data derived from the body, glaze and the pigments used in the decoration, the stylistic appearance of the porcelain considered from an expert connoisseurship point of view and the use of historical documentation and the appropriate extant records, are all examined and determined synergistically and independently to then arrive at a weighted judgemental decision for the attribution of a source manufactory. For the armorial porcelains being considered here from the Swansea and Nantgarw factories, the question arising of the source factory attribution from the stylistic concerns does not usually arise as many of the artefacts or pieces have been correctly marked with their factory of origin, either by the use of an impressed mark or by an applied stencil mark, so a greater emphasis and reliance is then being placed solely upon the supportive documentary or historical evidence for the provision of novel information. Therefore, this must be considered very carefully as it provides a unique pathway to determine the origin perhaps not of the porcelain itself as is normally required in this area but rather of the family to whom belongs the individual coat-of-arms or crest depicted upon it. The stylistic considerations that are so important for the assignment of the more anonymous porcelain artefacts from the Swansea and Nantgarw manufactories, however, should not be dismissed completely as they will come to the fore again when we are seeking additional information about the gilding and enamelling decoration which sometimes accompany the heraldic crest or escutcheon and which can provide some additional information, especially regarding its assignment to a local enamelling workshop or to a London atelier decoration and sourcing. Perhaps the most surprising result to emerge from the collection of data which comprise Table 2.1 is that of the 35 items mentioned in the 12 literature texts that cite Nantgarw and Swansea armorial porcelains, no fewer than 15 crests and 2 partial coats-of-arms mentioned cumulatively in the literature (representing some 50% of the total involved in the survey) have hitherto remained unidentified. Furthermore, it transpires, in addition, that several of those that have been identified thus far in the literature have either been reported incorrectly or have been mis-assigned, so demanding that an essential complete re-assessment and forensic evaluation now be made of all assignments of the Welsh porcelain armorial bearings that have been accomplished hitherto. Although it has been said that generally there is no dispute arising from the identification of the manufactory, that is Swansea and Nantgarw, where the armorial porcelains were made, in one particular case a punch bowl has recently been catalogued as Swansea porcelain having hitherto been thought to have been an earthenware product of the Cambrian Pottery (see Chap. 3). However, in the absence of chemical analytical data for this artefact, the composition of the porcelain body remains unknown and being of a very poor translucency it cannot be assigned to one of the three Swansea bodies and it could possibly be an imported artefact which was decorated at the Swansea site. Further research, ideally involving an analytical scientific component, is now required to contribute to the assessment of this exemplar and a decision as to its authenticity as a porcelain body made in the Swansea China Works.

We need to enquire about the root cause of such a high incidence of failure to identify a crest or a coat-of-arms in our database of armorial exemplars from the Nantgarw and Swansea China Works, when clearly there are comprehensive lists of these in existence as cited here earlier in the armories literature. The major difficulties experienced in the tracing of the source of an armorial heraldic coat-of-arms, achievement or a crest from the armories literature are as follows:

- Heraldic lists of coats-of-arms and crests are generally arranged alphabetically in armories according to the surname of the arms bearer and ancillary information is also provided about their location, seat, ancestors, the date of granting of the arms and any modifications to the original crest or coat-of-arms that may have arisen from marriage, inheritance or the succession and the adoption of other arms or related family associations. A further complication is that titled members of the aristocracy and peerage appear in the armory listings under their family names: for example, the Marquess of Bute and Lord Dumfries appears under Crichton-Stuart, the Marquess of Anglesey under Paget, the Marquess of Bath under Thynne, the Duke of Northumberland under Percy, the Duke of Beaufort under Somerset and Earl Camden under Jeffreys.
- Hence, starting with a crest on a porcelain artefact there is normally no indication given as to the identity of the individual or to the family to whom it belongs and generally some further information is essential to progress the research and provide a lead-in to familial names for cross-checking. Armories also do not usually list the crest by type and then afford a means of checking out the entitlement of the family by name; for example, the families who adopt a boar's head, couped and erased, as their crest, which would make the task of assignment so much easier as it narrows down the list of possible arms bearers. It is true that Fairbairn (*Crests of the Families of Great Britain and Ireland and Their Kindred in Other Lands,* Volumes I and II, 1905) does offer a comprehensive list (in Volume I) and illustrated examples of selected crests (in Volume II) accompanied by the names of the families who share them, but an assiduous search is still necessary to elucidate the assignment of a particular crest. Likewise, in Fairbairn (Volume I) a list of mottoes and the families who have adopted their usage is given, occupying some 95 pages of text in total with some 60–70 entries per page, giving a cumulative list of between 6000–7000 heraldic mottoes available for research.
- It will be seen later that in several cases concerning the partial coats-of-arms that are depicted on the Cambrian Pottery, Swansea, earthenwares and creamwares the name of the person involved is associated with the arms, or possibly there is an accompanying monogram, cipher or initials, and the reason for its commissioning may also be stated, along with a date and the painter who executed the armorial work, such as "*Thomas Pardoe, 1804*". It is exceptionally fortunate for a researcher to discover such a useful documentary appendage to the coat-of-arms or crest on these Cambrian Pottery, Swansea earthenwares – given that the name of the entitled bearer of the arms is provided thereon and often with a date

Fig. 2.3 Nantgarw porcelain oval dessert dish, *ca.* 1820–1823, with moulded border and a monogram in cursive script, *TW*, for Thomas Williams, also known as Gwilym Morgannwg (William(s) of Glamorgan) in his role as Bardic poet and Eisteddfodwr. Painted centrally with a view of Pontypridd Bridge, which was built over the River Taff by William Edwards in 1756 and for a time was the largest single span stone bridge structure in the world. *Copyright and reproduced with permission from the National Museum of Wales, Amgueddfa Cymru, Cardiff*

of its commission – but, unfortunately, this is not encountered at all for their Nantgarw and Swansea porcelain armorial analogues. Very rarely indeed does one find the presence of even a monogram, cipher or initials on Nantgarw or Swansea porcelains. A noted exception is the Nantgarw moulded oval dish painted with a scene of Pontypridd Bridge by Thomas Pardoe and illustrated in Fig. 2.3, bearing the initials *TW* at the verge, which have been correlated with Thomas Williams of Pontypool, the Welsh Eisteddfod Bard, who was also known by his Bardic name of Gwilym Morgannwg (William(s) of Glamorgan). Such initialled exemplars are usually found as solitary pieces and it is a little difficult to judge whether they were originally parts of a service or perhaps created as unique cabinet artefacts – the *Cremorne* Derby porcelain initialled dessert comport illustrated earlier (Fig. 1.19) is known to be part of a service as other artefacts such as plates, ice-pails and soup dishes are known to bear the same decoration with the monogram and coronet. Similarly, there are suggestions that the Nantgarw oval dish discussed above is part of a dinner-dessert service. The situation is much clearer when monogrammed or ciphered pieces such as large ewers, porter mugs and punch bowls are found, although special items such as ice pails, muffin dishes, comport centrepieces and egg cups would normally have been included along with breakfast, dessert and dinner services. Special cases could be made for cabaret sets or solitaire sets, for example, which would have a tray, teapot, lid and stand, covered sucrier, slop bowl, cream jug, tea cup and saucer and perhaps a coffee pot, coffee cup and a small plate as well. Other items of porcelain, such as writing sets or chamber sticks, could possibly bear an armorial device or a monogram, cipher or initials but to the author's knowledge none are known for Swansea or Nantgarw porcelains. In some cases, such as the armorial *Viscount Weymouth* service or the *Wyndham* service, only solitary examples seem to have survived from the literature even though they would have originally been part of much larger services as the percentage survival of even some of the larger porcelain services may not have been great. It should be appreciated too that some manufactories did prepare sample specimen plates for proposed prestigious services to prospective clients, which were not eventually selected by the client, and these truly then afford unique examples of armorial porcelains. An incidence of this practice is provided by the sumptuous Royal Rockingham armorial porcelain service commissioned by King William IV and Queen Adelaide in 1832, for which some dozen or more special one-off armorial plates with different designs and colour palettes were made for the personal approval of King William IV and Queen Adelaide, who selected just one example from these for their service commission to be undertaken and manufactured (Cox & Cox, *Rockingham Porcelain, 1745–1842*, 2001). Several of these unique armorial Royal Rockingham Royal service specimen plates are still in existence and can be seen in Cox & Cox (*Rockingham Porcelain, 1745–1842*, 2001).

- The multiplicity of names associated with crests and coats-of-arms and the intermarriage between families creates a further geographical parameter for consideration in the assignment of the individual named arms-bearer, which it is necessary to consider in our assessment. For example, the attribution of a dolphin *naiant* crest on a Swansea armorial porcelain service, one of the very few Swansea crests in the literature to have been identified thus far, has now been found to have actually been incorrectly attributed in the literature to an arms-bearing Irish family, namely Garnett-Orme, of Abbeytown, Co. Mayo, Ireland, which upon careful checking of the actual crest as it has been depicted on the relevant Swansea porcelain artefact is seen not to be correct when compared with the crest that is assigned to that family arms bearer as it appears in the armory. In fact, the correct assignment here is to be accorded to an English branch of the same family, Orme, with a family seat in Northampton, whose crest in the armory now matches precisely that shown on the Swansea porcelain service! This case will be discussed more fully in Chap. 5 when the individual Swansea porcelain armorials are being considered. The importance of this procedure particularly in the assignment of a Swansea armorial crest and coat-of-arms, where a relatively large number of crests remain unassigned in the existing ceramics literature, will soon become apparent. This means that for the same family name, the coat-of-arms or the crest may differ subtly, or sometimes significantly, for different branches of the same family and further examples of this will be provided in due course. In the case alluded to here, the dolphin *naiant* crest of the Irish Garnett-Orme family arms-bearer should have a poleaxe spearing the leaping dolphin, which the Swansea armorial porcelain crest clearly does not possess, so the inevitable conclusion must be that the historical attribution in this case to Garnett-Orme of co. Mayo must be incorrect.

- In some cases, the same heraldic blazon (escutcheon or shield) will have entirely different crests associated with it which are designed to accentuate the link with another branch or ancestry of the titled family, perhaps through marriage and descendancy. Some family individual coats-of-arms officially even have two crests (as does the coat-of-arms of Lord Nelson referred to earlier and shown in Fig. 7.1) as will be illustrated below. For example, Wyndham Lewis, the influential MP for Maidstone, a partner in the Dowlais Ironworks and a nominated supporter of the Nantgarw China Works as documented by William Weston Young, had two crests associated with his escutcheon and arms, the first being that of Lewis and the second that of Price/Pryce, a family descendant with whom he proudly claimed an ancestry (see later). Another was Admiral Lord Horatio Nelson, who also had two crests, firstly, that of the stern of the Spanish battleship *San Jose* which he captured at the Battle of Cape St Vincent in 1797 (Nelson was Commodore at this time and in command of the *Captain,* a 74-gun battleship) and secondly, an Ottoman Empire *chelengk* honour that was awarded for his victory at the Battle of the Nile in 1798. The link to another family branch or ancestry, as mentioned above for Wyndham Lewis, can be developed further here: he commissioned a beautiful Nantgarw porcelain service in a particularly fine *rose du Barry* ground colour in simulation of Sevres porcelain called the *Greenmeadow* service, after his estate in Llantrisant. This service is not an armorial one but research into Lewis' coat-of-arms in the heraldic literature reveals that he displayed two crests, one for Lewis and the other for Price/Pryce, whom he counted in his distant ancestry. His coat-of-arms, therefore, displays these two crests, but only one of these would normally be used for a simple armorial crested porcelain service. It is the second of these crests which appears on a hitherto unidentified Swansea armorial artefact associated with his motto in Welsh, namely "*Duw Ar Fy Rhan*" ("God for my Protection"), which incidentally provides only the second exemplar of Welsh porcelain which bears a Welsh inscription! Normally, the armories will definitively indicate the possibility of the entitlement of the bearer to use more than one crest to demonstrate a familial ancestral link elsewhere to another family. Several titled families have a motto in Welsh but this may not be provided with their partial coat-of-arms or crest as depicted on an armorial artefact. For example, The Duke of Beaufort and Sir Watcyn Williams-Wynn both have mottoes in Welsh and in Latin associated with their arms but these do not appear in the incidence of their arms adoption on Welsh ceramics. As a result, only two Welsh mottoes have been found to be associated with the 18 armorial Welsh porcelain exemplars identified here (i.e., 11% of known examples); in all, on the different Welsh armorials studied here, three mottoes are provided in Latin, two in French and two in Welsh, so giving a 40% representation of mottoes on the known existing armorial exemplars from Swansea and Nantgarw porcelains.

- A particular difficulty can arise for the depicted partial coats-of-arms, such as an escutcheon which does not display a crest, as the combination of the two heraldically can provide a mutual support for the correct assignment of an attribution and remove any possibility of confusion arising from the same escutcheon bearing a different crest for a different branch of the same family. Even the association of a crest and escutcheon does not necessarily provide an unequivocal identification in all cases as instances occur where different branches of the family have adopted both the same crest and escutcheon, but then perhaps not the same motto. The relevance of this in an armorial service has already been demonstrated in Chap. 1 with the attribution of the armorial *Pitt* service of Chinese porcelain to Pitt of Ewerne Stapleton in co. Dorset and not to individuals in the other branches of the Pitt family (Figs. 1.10 and 1.11) who have the same escutcheon but who have adopted different crests and/or mottoes to accompany it. Despite this, some authors have attributed the incorrect motto to Pitt of Ewerne Stapleton!
- Crests can be easily mistaken for each other if drawn rather carelessly and can be assigned to individuals of different families with only small differences apparent in their visual depiction. This can be a trap for the unwary and it seems that some misconception has arisen from this source in the past; for example, a Swansea armorial porcelain crest assigned to a particular individual in the literature is found to be factually incorrect in the current investigation because the crest depicted for that individual contains an additional heraldic symbol which then indicates that it should be attributed elsewhere for its correct attribution. Commonly encountered in this category are animals which are portrayed holding objects such as branches or crowns in their talons or paws, or which have neck collars or coronets displayed (when they are then said to be *gorged* heraldically). The case of Pitt of Ewerne Stapleton alluded to earlier provides an example of the basic stork crest but other Pitt individuals have some modifications occurring to their crests wherein the stork is portrayed with its dexter claw raised or resting upon another heraldic device such as an anchor (as has been demonstrated in Figs. 1.10 and 1.11).
- If the coat-of-arms is incomplete or contains an escutcheon which has possibly been over-simplified by the enameller, gilder or decorator then it can prove to be extremely difficult to assign unequivocally to an individual bearer of those arms. A case of this sort will arise later for a Nantgarw armorial porcelain dessert plate with a simply drawn and gilded escutcheon, which possesses no additional armorial devices to aid its attribution but which nevertheless still fortunately has enough information contained within it to effect the potential identification of the individual concerned.
- Crests which appear on armorial porcelain are primarily executed in gilt and this can also create a problem where the crest differentiation between individuals of arms-bearing families in an armory is based upon a colour recognition (known as an heraldic *tincture*) or the highlighting of a particular feature on the crest in colour. A case in point here, which will be a topic for discussion later, is the escallop shell crest, which for a particular individual in an arms-bearing family,

Fig. 2.4 Crest of a dolphin *hauriant* from Fairbairn's list of crests. Fairbairn, *Crests of the Families of Great Britain and Ireland and Their Kindred in Other Lands,* Volumes I and II, T.C. and E.C. Jack, Edinburgh and London, 1905 and New Orchard Publishing, London, 1986

cited here as belonging to Graham-Clarke of Newcastle-upon-Tyne, should be quartered in *gules* and *or* heraldically, i.e. portrayed in alternate quarters of red and gold, so how does one define a crest that is represented on porcelain as a gilt escallop shell which has quarterings indicated geometrically in gilt but with an undefined colour or tinctorial basis for that quartering? Similarly, a plain gilt dolphin *naiant* on a Swansea armorial porcelain artefact which appears as a crest on a helm in the armories is recorded in the armorial lists as appearing as sable (i.e. black) with its highlighted fins and teeth in gold *(or)* – yet, as a crest on the particular Swansea porcelain artefact under investigation it is depicted in gilt only, but it is otherwise an exact representation of the twin coloured crest version (see later, Chap. 5 and Fig. 5.9). In this context, the description of the attitude adopted by the dolphin on the crest is also vitally important: in this instance, the crest on the armorial plate depicts a dolphin *naiant* horizontally skimming the waves with its body arched, whereas an alternative version showing a dolphin *hauriant* is pictured leaping vertically from the water, as reproduced here in Fig. 2.4.

- A noted exception to our discussion of gilded crests here is the carefully drawn and enamelled crest of Thomas Lloyd of Bronwydd on Swansea porcelain, as discussed later and shown in Chap. 5 and illustrated in Figs. 5.3 and 5.4, which is executed in polychrome enamels and has the appropriate distinctive motto in Welsh accompanying it, namely *I Dduw B'Or Diolch* ("Thanks be to God"), which renders its identification and matching to the individual Thomas Lloyd unequivocal. Generally, however, most crests on Welsh armorial porcelain do not have a polychrome decoration and are thus simply gilded.

2.3.1 *Distinction Between Monogram, Cipher and Initials*

In several instances in the above text the phraseology of monogram, cipher and initials has been used: these are not the same, although all involve combinations of letters, and it is important at this point to explain the distinction between them as they have been used rather interchangeably and fairly loosely to describe their presence on china in the past. A *monogram* is a combination of two or more letters in which one letter forms part of another and cannot be separated from it: the monogram, therefore, represents an artistic design in which the letters are interdependent – the removal of one letter component then causes the design to fall apart. A *cipher,* on the other hand, is composed of an interlacing of the letters whereby the removal of one letter does not affect the others and they still exist in their original design form (an example is the *VOC* logo of the Dutch East India Company shown in Fig. 1.6). *Initials*, in contrast, are simply either a single letter or combinations of two or more letters which are separate spatially and these also may have a design component (an example is the *Cremorne* service in Fig. 1.19 which displays the initials *PHC*). Historically, there has been much confusion in the use of these terms, which prompted Albert Turbayne to write his classic text (*Monograms and Ciphers: Designed and Drawn by Albert Angus Turbayne and the Other Members of the Carlton Studio,* 1905) in which he illustrates examples of artistically designed monograms and ciphers; some monograms and ciphers are so artistically drawn that it is sometimes difficult to extract from them the correct sequence of initials.

2.4 Examples of the Heraldic Descriptors for Coats-of-Arms and Crests

It is appropriate next to consider the meaning of the heraldic terms that will occur in our description and discussion of armorial achievements and an abbreviated list of many of these has been made in Table 2.2 which will assist the reader in the appreciation of some of the heraldic descriptors that will now follow in this text. To illustrate some of the points made above and to illuminate the care that needs to be taken in the interpretation of the crests and partial coats-of-arms that are to be found on Welsh porcelains, some exemplars are considered and the corresponding entries as they appear for them as selected from *Burkes Armory,* for example, will now be transcribed here and will be followed by a summary of the salient points that emerge which need to be considered by the researcher of an armorial porcelain artefact. Selected entries for individuals of arms-bearing families that have already been "identified" on Swansea and Nantgarw porcelains, or which have been proposed as assignments to the armorial features displayed thereon, are given below in the accepted heraldic language and abbreviations as these appear in the armory literature, especially the books of Burke and Fairbairn (*Burke's General Armory – The General Armory of England, Scotland,Ireland and Wales: Comprising a Registry of*

Table 2.2 Heraldic terms, their abbreviations and meaning

Heraldic term	Abbreviation	Meaning
Abasement		Lowering of dignity or station
Accollee		Side by side
Accompanied		Between
Accorne		Horned
Accrued		Fully grown
Achievement		Complete display of arms: Shield, helmet, mantling, torse, crest, motto.
Addorsed		Turned back-to-back
Affrontee		Full-face to the front
Allerion		Eagle without beak or feet
Allusive		Arms based on name
Ambulant		Walking
Anche		Curved
Annulet		Small ring, circlet
Apaumy		Open handed, showing palm
Argent	ar.	Silver, white
Armed		Beast exhibiting claws, teeth, talons
Aspectant		Animals facing each other
Assurgent		Rising out of the sea
Armed		Horn, beak, talon, usually in colour
Azure	az.	Blue
Banner		Flag of knight
Barbed		Bearded
Barry		Transverse bars of contrasting colours on shield
Base		Lowest part of shield
Beaver		Chinguard of helmet
Bend		A diagonal from dexter to sinister on the shield
Blazon		Shield and coat-of-arms in colours
Bordure		Border around shield distinguishing different branches of same family
Brased		Interlaced
Brouchant		Overlying
Caboched		Head of beast, facing, without neck
Canton		Corner of shield
Charge		Figures in escutcheon
Chapeau		Cap used in place of torse
Checky		Alternate squares of different colours
Chevron		An inverted vee, ^
Chevronet		A small chevron
Chimera		Mythical beast: Griffin, dragon, gryphon, cockatrice, harpy, wyvern ....
Collared		Wearing a collar

(continued)

Table 2.2 (continued)

Heraldic term	Abbreviation	Meaning
Combatant		Fighting
Compony		Squares of alternate colours on shield
Couchant		Lying down with head raised
Couped		Head, limbs cut off at neck
Courant		Running
Coward		Animal with tail between its legs
Crenelle		Embattled
Demi		Top half: e.g. demi-lion, demi -leopard
Dexter		Right side of, as viewed by bearer of shield
Diadem		Pointed circle of gold
Difference		Added to escutcheon to differentiate families
Displayed		Bird with wings expanded and legs spread
Dormant		Sleeping
Eaglet		Several eagles in an escutcheon
Embrued		Bloodied, especially weapons
En arriere		Shown from the back
Ensigned		Ornamented
Erased		Cut off, leaving jagged edge
Erect		Horizontal parts, displayed vertically
Ermine		White fur with black spots
Ermines		White spots on white fur
Escallop		Seashell
Estoile		Star with rays
Fess		Third part of shield between Centre and Top
Fess point		Centre of escutcheon
Field		Surface of escutcheon
Fleur-de-lys		Lily
Gambe		Leg
Garb		Wheatsheaf
Gaze		Intent look
Gauntlet		Gloved hand Armour
Gemel		Double, twinned
Gorged		Collar around the neck of an animal,
Gules	gu.	Red
Guttee de sang		Drops of blood
Haurient		Fish, dolphin, vertical with head pointing upwards
Helm		Visored helmet
Imbued		Spotted with blood
Impaled		Two arms on one shield
Inescutcheon		Small escutcheon in larger shield
Issuant		Rays

(continued)

Table 2.2 (continued)

Heraldic term	Abbreviation	Meaning
Lambrequin		Mantling, cloth covering for Armour
Martlet		A swallow
Mullet		Five-pointed star, roundel of spur
Or		Gold, yellow
Pale		Vertical band on a shield
Pheon		Broad arrow, tang with converging blades and serrated inner edges
Pineapple		Pinecone
Proper	ppr.	In standard tincture but stylised
Rampant		Rearing beast with paws in attack position
Sable	sa.	Black
Salient /saliant		Leaping or springing beast, hind legs on the ground
Saltire		St Andrew's cross, diagonal cross
Segreant		Swimming
Sejant		Sitting down on haunches with forepaws on the ground
Sinister		Left side of, as viewed by bearer of shield
Tierce		a third division
Torse		Twisted silk cloth wreath of two colours between helmet and crest
Vert		Green
Vulned		Ensanglante, animal wounded and bleeding

Armorial Bearings from the Earliest to the Present Time, 1884; Fairbairn, *Crests of the Families of Great Britain and Ireland and Their Kindred in Other Lands,* Volumes I and II, 1905):

Lewis Lewis (Green Meadow, co. Glamorgan). Quarterly, 1st, sa. a lion rampant ar.; 2nd, sa. a chevron between three spear heads ar. embrued gu.; 3rd, sa. a chevron between three fleurs-de-lys or; 4th, or on a quarter gu. two lions passant gardant of the first. *Crests* – 1st LEWIS: a lion sejant ar. 2nd PRICE: A Paschal lamb glorified or, bearing a pennon of St George. Mottoes – for Lewis: "*Patriae Fidus*" meaning "*Faithful to the Fatherland*" and "*Ofnar na ofno angau*", a Welsh motto which dates to the ancient Princes of Powys and Lords of Ceredigion and their meeting with King Edgar in 959 CE in Caerlleon Gawr (Chester) and which translates as "*Let him who fears not death be feared*" (Williams, *The Carmarthen Journal*, 1919). The Welsh motto adopted by Lewis, "*Ofnar na ofno angau*", is also shared with Watkin, but the crests are very different, being a lion sejant and a wolf's head erased with three arrows. The motto of Pryce is "*Duw ar Fy Rhan*", "*God is my Protection*", with the crest of the Paschal lamb.

Lewis (Gilfach, co. Carmarthen). Quarterly, 1st and 4th, az. a stag trippant ar. unguled and attired and bearing between his horns an imperial crown or; 2nd and 3rd, az. a chevron between three eagles heads erased or. *Crest* – a stag and an eagle's head, as in the arms.

Lewis (Llanarchayon, co. Cardigan). Gu. on a mount in fesse vert three towers triple towered ar. between three scaling ladders or. *Crest* – out of a mutual coronet gu. a demi-wolf saliant ar. Motto – "*Libertas*" meaning "*Liberty*".

Lewis (Gwynfe, co. Carmarthen).1st and 4th gu. a griffin segreant or for Lewis of South Wales; 2nd and 3rd, az. three nag's heads as for Lloyd. *Crest* – a demi-griffin segreant couped or.

Lewis (Clynfiew, Pembroke). Gu. three serpents nowed in triangle ar. within a bordure engr. or. *Crest* – a nag's head couped, bridled ppr. Motto – "*Be wise as a serpent, harmless as a dove.*

Williams-Wynn

Wynn (Wynnstay, Denbigh, Bt.). Quarterly, 1st and 4th, arms of Owen Gwynedd, Prince of North Wales (d. December 1169), vert three eagles displayed in fess or for Wynn; 2nd and 3rd, those of Cardrod Hardd, ar. two foxes counter salient in saltire, the dexter surmounted of the sinister gules for Williams. *Crest*- an eagle displayed or.

Clarke

Clarke (co. Hereford). Gu. two bars ar. in chief three escallops or. *Crest* – An escallop quarterly gu. and or.

Clarke (Ipswich, Suffolk). Or two bars az. Overall a nag's head erased ar. in chief three escallops gu. *Crest* – A nag's head erased sa.

Clarke (Ipswich, Suffolk). Ermine on a bend engr. sa. three conger eel's heads erased ar. collared with a bar gemel gu. *Crest* – A conger eel's head erased and erect gu. collared with a bar gemel or.

Clarke (co. Northampton). Per pale gu. and or, a bend and border counterchanged. *Crest* – Out of a mural coronet ar. a cubit arm in armour holding a scimitar hilted or.

Clarke (Norfolk). Or, on a bend engr. az. a mullet ar. *Crest* – An eagle's leg gu. joined to a wing or.

Clarke (of London). Sa. on a bend engr. ar. three lozenges of the first. *Crest* – A talbot's head gorged with a fesse az. charged with three lozenges gold.

Clarke (co. Hereford). Ar. a chevron between three lions rampant or. *Crest* – A lion rampant vert holding a pen ar.

Parkinson

Parkinson (Kinnersley Castle, co. Hereford). Gu. on a chevron between three ostrich feathers erect ar. as many mullets sa. *Crest*- a cubit arm erect vested erminois cuff ar. holding in the hand ppr. and ostrich feather also erect gu.

From armory entries such as these, one can make some general observations, such as:

- It is seen that the family name itself does not generate a unique coat-of-arms and crest: this is very much a personal heraldic achievement, which should in theory enable the identification of an individual to be made unequivocally if sufficient information is provided in the armorial representation which is depicted on the porcelain artefact.
- Family entitlement to the bearing of a coat-of-arms is dependent upon the lineage and descendancy of the individuals and this gives rise to changes in the arms with the geographical location of different branches of the arms-bearing

family. Hence, in *Burke's Armory* the family name is generally followed by the family seat or its locality: this can result in many iterations of a coat-of-arms and crest occurring for different branches of the family, some of which are significantly different from each other and others which are similar and will seem to be of relatively minor importance to the deduction of the attribution of the arms.

- In some cases, the arms and crest entries in the armory are unique to just one individual, for example Sir Watcyn Williams-Wynn of Wynnstay (see above), whereas others may have a multiplicity of family entries for similar or only slightly different arms, for example Lewis has 34 entries, Clarke (including Clark) has 82 entries and Parkinson has 5 entries for individuals in *Burke's Armory*. Whereas the crests for Clark(e) have numerous varied representations in several forms including a falcon, swan, demi-huntsman, lark, serpent, eagle, demi-griffin, mullet, cross, lion, bear, talbot's head, dragon's head, stag, goat, scimitar, arrow, fox, fouled anchor, pheon, sword, battleaxe, dolphin, nag's head, peacock, elephant's head, tiger's head, seahorse, greyhound, conger eel and fleur-de-lys, only two, namely Clarke of co. Hereford and Graham-Clarke of Frocester Manor have an escallop shell quartered in gules and or.
- Whereas several important *named services* are recorded as having been commissioned from the Swansea and Nantgarw manufactories, such as the *Burdett-Coutts, Mackintosh, Lord Jersey, Vernon, Wilde, Lysaght, Duke of Gloucester, Duke of Cambridge, Marquess of Anglesey, Marquess of Bute, Spence-Thomas, Sir John Williams, John Vivian, Duncombe, Lady Seaton, Edward Edmunds, Gosford Castle, Marquess of Exeter* and *Wyndham Lewis' "Greenmeadow"* services, there is a noted and parallel absence of these names associated with the family arms or crests, albeit these being of a rather limited number that have been recorded to date, that have been identified or assigned already in Table 2.1 for their armorial bearings. Likewise, other major landowning families in South Wales that were particularly close geographically to these manufactories have not appeared to feature individuals in any identification process of the armorials that have been assigned thus far: these would include, for example, the Mansel-Talbots of Margam (cadet branch of the Earl of Shrewsbury), the Turbervills of Ewenny Priory, the Mansels of Penrice Castle, the Phillips of Picton Castle, the Nicholls of Merthyrmawr, the Turbervilles of Coity Castle, the Stradlings of Cowbridge and the Corys of Dyffryn.
- A particular problem arises in the assignment of a crest which is bare of any other information, such as a motto, which could assist in its correct identification as has been highlighted earlier: even so, it is occasionally found on detailed checking of the arms that the motto is not an accepted one for the designated family bearer of those arms and for that particular crest – and this can then create another problem in the attribution process. In the Scottish heraldic arms lists (which are drawn up by the Court of the Lord Lyon, which has regulated Scottish heraldry since 1532) the motto is officially credited along with the arms but this is not so in English heraldic arms lists (as compiled by the College of Arms for England, Wales and Ireland) where the motto seems to have been acquired by usage assimilation rather than by an official grant. The correct matching of a

motto with the crest, supporters and escutcheon is a powerful confirmation of the correct attribution having been achieved – but as will be seen later, this does not always occur in practice and then difficulties must arise around the certainty of the attribution or one must peruse the armory more assiduously to verify if a better match can be achieved between the various heraldic components.

2.4.1 Welsh Heraldry and Mottoes

Because the topic of this book is armorial Welsh porcelain, it was reasoned that local Welsh arms-bearing families may have been associated with the commissioning of armorial porcelains from the Nantgarw and Swansea manufactories and, therefore, their crests could be represented on the arms that have thus far remained unidentified. Of course, it must also be recognised that English arms-bearing families who owned land in Wales would also be expected to be enumerated in this discussion. The prime source of information about ancient Welsh arms-bearing families is the text by Michael Siddons (Wales' Herald of Arms Extraordinary, 1994–2010), namely *The Development of Welsh Heraldry*, published in four volumes, the first three between 1991 and 1993 and the last volume in 2007 (Siddons, 1991, 2007); this comprehensive scholarly text has been described as moving in "strange and uncharted territory" and providing the finest account of the development of heraldry in Wales. Basically, Siddons has identified three categories of arms-bearing persons in Wales from about 1100: the *advenae,* or Norman lords following the Conquest, the Welsh aristocracy and the non-tribal Welsh families. Although the original and older Welsh coats-of-arms clearly indicated a family descendancy from the princes and lords of Wales, such as Nefydd Hardd (Nefydd the Handsome or Fair, 1180–1200) of Caernarvon, whose lineage could be traced back to Cunedda and Owen Gwynedd, the arms were generally quite simple in construction. Those of Nefydd Hardd, for instance, were argent a chevron sable between three spear heads of the second imbrued, points upwards (which we will come across again painted on a piece of Nantgarw armorial china!). Nefydd Hardd was Lord of Nant Conway and the credited founder of the VIth Noble Tribe (Jones, Coat of Arms, *The Arms of the XV Noble Tribes of North Wales*, 1958). Siddons also compiled a list of Welsh family pedigrees up to about 1700 in his *Welsh Pedigree Rolls*, published in 1996.

Slightly later, arms were revealed for Llywelyn mab Iorwerth of Gwynedd (Llywelyn the Great) (1173–1240) and his grandson, Llywelyn ap Gruffydd (Llywelyn the Last) who was killed by a lancer at Cilmeri near Builth Wells in 1282, so ending the ancient Welsh line of Princes. Ithel ap Bleddyn (1300) had a coat of arms showing two lions rampant addorsed with tails entwined with a crest of a coronetted maiden's head (affording the earliest known example of a Welsh family crest). Howel ap Meredydd (1295) had a paly of six on a fess with three mullets. It seems, however, that the earliest coats-of-arms in Wales were those of the Norman *advenae*; in contrast, it appears that the earliest contemporary Welsh heraldic devices were more of the *impresa* type, which comprised a pictorial figurative representation accompanied by a motto, which retained their popularity well into the fourteenth century. After this, the Welsh coats-of-arms became less "conventional" and matched better their English analogues.

Mottoes have been popular throughout heraldic history and are believed to have been developed from the original rallying cries used by forces in battle: in Scotland, warring clans rallied to their leader's call with a "*sluagh-gairn*", from which was derived a *slogan,* the earliest example of which is believed to be the Roman legionary war cry of "*Dexter et sinister*" when rallying around their eagle, meaning "*To the Right and to the Left*". King Edward III at the Battle of Crecy in 1346 used the war cry "*Dieu et Mon Droit*" *(God and My Right)*, which was then adopted as the Royal motto. Mottoes, unlike the arms, are not required to be registered by the College of Arms and so are not protected against use by other families or arms-bearers (Neubecker et al., *Heraldry: Sources, Symbols and Meaning*, 1989). Sometimes, a motto may be used as an heraldic pun: such an instance is provided by the Roche family motto, "*Mon Dieu est Ma Roche*" *(My God is my Rock)*! The use of different languages for the mottoes of Welsh arms-bearing families is of interest and relevance too: by the seventeenth century the four languages used for mottoes by Welsh arms-bearing families had the following occurrence for the 206 families surveyed – English 15: 7%, Welsh 47: 23%, French 30: 15% and Latin 114: 55%. A 207th family could not be transcribed into the database because it had a motto which could neither be deciphered nor translated! The preponderance of Latin is said to reflect the general standing of literacy amongst the arms-bearers, but certainly French and Welsh are not without their significant presence. One in four mottoes is in the Welsh language: interestingly, when we come to survey the results of our armorial research into Welsh porcelains it is found that of our 18 identified and distinct armorials on porcelain, there are 7 which bear a motto with the following linguistic distribution: 3 Latin, 2 French and 2 Welsh. This means that 40% of our Welsh porcelain armorials in the database bear a motto and of these 2/7 are Welsh, i.e. 28%; in view of the much smaller number of mottoes surveyed in the armorials, this realistically means that we are in the same boundary region as the occurrence of Welsh language mottoes in the family heraldic survey and this cannot be coincidental. Several Welsh armigerous families do cite mottoes that are bi-lingual, i.e. they have mottoes in two of the four languages mentioned, namely English, French, Welsh and Latin. An example is the motto of Goodman of Ruthin which is given as "*Da gloriam Deo*" (Latin) and "*Gogoniant i Dduw*" (Welsh), meaning "*Give glory to God*" – this same Latin motto also appears on the arms (granted in 1577) of The Worshipful Company of Dyers in the City of London (Siddons, *Welsh Pedigree Rolls*, 1996). Although, again, recognising that the numbers in the dataset are small it appears that the prevalence of mottoes on Nantgarw porcelain is significantly greater (approximately twice) than those which occur on Swansea porcelain: Nantgarw has four mottoes (2 crests plus 2 coats-of-arms) from 7 armorial exemplars (56%) whereas Swansea has only three mottoes (2 crests plus 1 coat-of-arms) from 11 armorial exemplars (27%).

Some mottoes for Welsh families have a meaning that is lost in time: for instance, Sir Richard Wynn of Gwydir Castle, Ruthin, Denbighshire, had the motto "*Non mihi glis servus nec hospes hirundo*", which has the translation of "*My servant is not a dormouse nor my host a leech*". This message seems to reflect an articulated response to a long lost and particular historical situation.

2.5 What Constitutes an Armorial Service?

It may seem a trivial task to define an armorial porcelain service for the purpose of this study but it is actually a fundamental concept of any scientific analytical study that one must first understand and define the basis of the establishment of the exemplar and specimen database upon which the research is to be focused and thereby impose some real, definitive and tangible limitations for its coverage and selection from a wider portfolio of potentially inclusive specimens. Simply, an armorial service can be defined here as a porcelain or more widely, a ceramic, artefact that has displayed upon it one or more heraldic emblems and this can range from a simple crest or coronet, with or without its torse, to the full coat-of-arms depicting the escutcheon, badge/honour, coronet, helm, lambrequins/ mantlings, torse, crest and motto. The service component of the definition, of course, implies that there would have been originally multiple artefacts which bore the arms or the crest but with inevitable loss or breakage incurred of many items with time over the period of 200 years or more since their manufacture these may now be reduced to just single or maybe a few extant items. In other cases, we might expect that an armorial ceramic artefact may be of a unique construction, as perhaps exemplified by a punch bowl, a cabinet piece, a jug, vase or a mug, but it appears that, apart from possibly the Swansea Llwchwr punchbowl, these do not feature in our database of porcelain exemplars at all. As will be seen later, the armorial porcelains and other ceramics under consideration here contain a range of heraldic achievements and their synergistic combination is critical for the correct attribution of the heraldic bearings to a particular individual or family. Of equal importance is a statement as to what does or does not come into this classification and perhaps what outliers may be considered for inclusion in our armorial database. The following cases and examples, therefore, strictly do not come within the scope of armorial china, although some earlier texts in the literature do tend to avoid this distinction in a somewhat arbitrary way and thereby have a seemingly wider coverage of factory output but with a consequent distortion arising for the amount of armorial porcelain or ceramic earthenwares that were made at the manufactory. This can, of course, result in giving an observer a false impression of the extent of the numerical importance of armorials in the overall factory production when compared with their other contemporary "anonymous" artefacts. Basically, the following general categories of ceramic artefact are well recognised for what they are but they strictly do not come within our classification of armorial porcelains for the generation of our database of exemplars:

- Commemorative inscriptions which do not contain any heraldic device: although these are of great interest historically, especially for cultural heritage preservation and for social science historians who need to appraise and evaluate the public sentiments and reactions of people to important local, national and international events of the time. These commemorative pieces include captions and often relatively lengthy inscriptions, which reflect the thoughts and beliefs that were current amongst the populace in the eighteenth and nineteenth centuries. A good example of this is the anti-Buonaparte feelings expressed in 1812 on a Cambrian Pottery, Swansea, earthenware jug which celebrated the military victories of the Marquess of Wellington (just before he was raised to his Dukedom in 1813) over the Napoleonic forces in the Iberian Peninsula from 1808 to 1812. This jug decoration comprised a medallion of Britannia seated with a shield, spear and a lion with her feet on a fallen French tricolor standard and supported by a Winged Victory figure bearing the palm of peace and the laurel wreath of victory. The inside rim of the jug has an enamelled decoration of a border of leaves and berries with crossed Union flags and the names of the battles in which Wellington and his Spanish and Portuguese allies defeated the Napoleonic French occupational forces in Spain and Portugal: namely, Albuera, Almeida, Rodrigo (author: Ciudad Rodrigo), Fuentes (author: Fuentes d'Onor), Badajos, Vimiero, Talavera and Salamanca. Later examples of ceramic artefacts on this same theme also contain reference to the battle of Vittoria, which was fought in 1813. However, there is no evidence of any heraldic emblem or coat-of-arms belonging to either Sir Arthur Wellesley or to the Duke of Wellington, so technically such pieces would not therefore be encompassed in our basic definition for their inclusion as an armorial artefact for our database, aside from the fact that they do not qualify as porcelain anyway! Another Cambrian Pottery, Swansea, earthenware jug delivers a thinly veiled diatribe upon the feelings of a defeated Napoleon in 1814: a caricature of the figure of Napoleon is shown surmounted by the phrase "BONAPARTE DETHRONED, April 1st 1814" surrounded by figures uttering sentences such as "What says Conscience now, eh Boney?" and "Down with Boney", "Galloping Boney, Wolloping Boney, Where are You Now". Napoleon Buonaparte himself is portrayed exclaiming, "Oh, Cursed Ambition, What Hast Thou Brought Me to Now", as he is faced with the figure of a horned red devil with a hook, who is graphically intent upon pulling him down into a deep pit. On the other side of this jug is depicted a scene of the single span Pontypridd Bridge, built by William Edwards in 1756, which at the time of its construction was the longest single span stone bridge in the world. Again, this ceramic artefact cannot be categorised as an armorial for the purposes of our discussion here, interesting as it is historically, as it does not exhibit any heraldic device, even though it undoubtedly provides an example of an important social indicator and revelation of the public sentiments of the period.
- Political support for local MPs or Acts of Parliament or Bills. A classic example of such a political inscription is the one created by the Cambrian Pottery, Swansea, on a jug for John Owen of Orielton, Pembroke, which reads:

Elected for the County of Pembroke, Friday October 30th 1812 after a contest of 11 days against the Honourable John Frederick Campbell Esqr., and a Trio of Peers and created Baronet on Tuesday, November 3rd 1812. Owen 1529/Mr Campbell 1344/Majority 185.

- Whilst this inscription again is clearly a commemorative one and does not in itself qualify the ceramic piece as an armorial item, on the other side of the same jug the coat-of-arms of Sir John Owen also appear – and the jug therefore seems to serve two purposes: firstly, the political celebration of John Owen's victory in the Parliamentary election for the county of Pembroke and, secondly, his elevation to the peerage a few days later. The coat-of-arms featured for Sir John Owen comprise a quartered escutcheon with two "wild men" (*sic.* as described in Fairbairn, *Crests of the Families of Great Britain and Ireland and Their Kindred in Other Lands,* Volumes I and II, 1905) supporters carrying clubs, a lion rampant crest and the motto: *"Honestas optimia politia"*, which means "*Honesty is the Best Policy"*. Although clearly a commemorative piece, this jug is also an armorial in terms of our definition but of course it will not enter our specific database for analysis here as it is not derived from the Swansea China Works and does not qualify as a porcelain service in our classification. It will be apposite to return to the arms and crest of Sir John Owen depicted later when research into several of the unidentified crests on Swansea porcelain are being investigated.
- This is not the only example of an inscription on Cambrian Pottery, Swansea, earthenwares which also contain some sort of heraldic device and which therefore confers upon them some sort of "hybrid" category between commemorative and armorial ceramics but which still nevertheless renders them admissible as *bona fide* exemplars of armorial ceramic artefacts within our broader remit, which may serve to inform our research into the armorial porcelains of Swansea and Nantgarw. Another series of Cambrian Pottery, Swansea, creamwares and earthenwares in this "hybrid" category is exemplified by the *Stannary* dinner services which were made exclusively for Cornish tin mining customers in the early nineteenth century: a particular example depicts a tin mine (or, alternatively, in some cases, St Michael's Mount, Cornwall), fish and copper ingots, with an inscription "Success to the Trade and Co. of Cornwall" and "Fish/Tin/Copper ONE AND ALL". The latter, a Cornish motto, is accompanied by the heraldic escutcheon of Cornwall, a shield containing 15 bezants in an inverted triangle of five horizontal rows, numbering 5-4-3-2-1 from the top and continuing vertically downwards: a bezant has already been described as an heraldic device which is derived from the Byzantine gold coin, *bizantius aureus,* which appears heraldically as a gold circle or roundel on the arms portrayed. It is designated as a symbol of banking, commerce and finance in heraldic achievements. The Cornwall fishing industry is separately commemorated through the UNITY series of plates of a blue and white transfer willow pattern design on Cambrian Pottery earthenware plates and dishes containing a central cartouche depicting a pilchard over the letters UNITY but with no accompanying armorial bearings. These were supplied to the nine fishery warehouses of Newlyn in Cornwall by the Cambrian Pottery and would have been used in their annual festive celebrations of the annual fishing harvest.

- Inscriptions relating to a marriage. For example, an early tea canister of William Coles' first period Swansea creamware from the Cambrian Pottery presented to Jane Ridgway upon her marriage to George Ridgway on July 28th 1777, with the accompanying legend:

Steal not this thing
For fear of shame for
Onit is the one's name
Jane Ridgway her canister.

A second example is an inscription on the marriage of Thomas Cornish to Mary Northern, June 30th, 1818, on the base of a teapot of a Cambrian Pottery tea service which was also decorated with shells and seaweed. Another earthenware supper service commemorates the marriage of an Illtyd Nicholl of Cowbridge to a Miss Bond of Usk in October 1807.

- Announcements of the birth of a child. For example, Jane Davies, born September 11th, 1799 – Jane's birth was registered at St Mary's Church in Swansea, known locally as "The Potters' Church", so perhaps her father and/or mother worked at the Cambrian Pottery?
- Personal gift inscriptions in appreciation of assistance rendered: an excellent example is the *Cymro* stoneware Cambrian Pottery jug inscribed "From Mary Ann Morris to her Kind Teacher, Mrs Howells, 1847. A Tribute of Affection", one of a pair of such jugs – the other being presented to the Revd. W. Howells of Bethany Chapel, the husband of Margaret Howells.
- Inn or trade inscriptions: there are numerous examples of these from the Cambrian Pottery including, Elizabeth Rees, St Ives, Swansea; David Evans, the Angel Inn, Carmarthen; John Daniel, Cooper, Carmarthen; Success to the Clock and Watch Trade, Cardiff, 1808, John Thackwell, clockmaker; B. Hawkins, The Ship Swan, London.
- A Rebus – which is a pictogram representing an idea, a person or an item which is a play or pun on words, but which in itself is not necessarily heraldic in concept but is designed to simulate one: hence, Richard Holdship, a partner in the Worcester China Works in the eighteenth century used an "anchor" (*viz.*, a device for holding a ship, so by inference therefore it is a "holdship") as a rebus mark on his Worcester porcelain. Similarly, the emblem of Swansea City Association Football Club in the nineteenth century was a swan swimming on water; surprisingly, the official crest of the town and borough of Swansea itself did not have a swan on water at this time but rather an osprey with wings spread, at least until 1985, when this was replaced and a swan now appears in turreted embattlements as the new crest of Swansea City. Despite this, a Masonic punch bowl from the Cambrian Pottery, Swansea, inscribed "Indefatigable Lodge No. 333 W. Essery RW Master AL5806", corresponding to the year 1807, has a design inside the bowl with a rebus for Swansea, namely a swan on water! The frog motif enamelled on Empress Catherine the Great's (Ekaterina II) Wedgwood creamware service described earlier in Chap. 1, the *Frog* service, is another example of a rebus, in this case for the marshy site (Frog Marsh) of the palace where it would

have been intended for use near St Petersburg, but this time it is enclosed additionally in an engrailed escutcheon which could be easily inferred, therefore, to be a genuine armorial bearing!

- Monograms, ciphers and initials: these do not constitute an armorial bearing unless accompanied by a crest or by another similar heraldic device. The dessert service ordered by Viscount and Viscountess Cremorne from the Derby China Works in 1788 and mentioned earlier (from which a specially designed comport dish is shown in Fig. 1.19) bore the initials *PHC*, for Philadelphia Hannah Cremorne, Lady-in-Waiting to Queen Charlotte, inside a cartouche and surmounted by a Viscount's coronet, an heraldic device which clearly makes it part of an armorial service. Mr. Kyrle Fletcher possessed a Cambrian Pottery, Swansea, creamware cabaret service decorated with chinoiserie and panels of fruit and flowers bearing the initials *EH,* which was presented to Admiral Lord Horatio Nelson and Mrs. Emma Hamilton on their visiting Swansea in 1802 (Morton Nance, *The Pottery and Porcelain of Swansea and Nantgarw*, 1942). Despite the documented historical association of this particular piece, the absence of heraldic achievements negates this from being formally classified as an armorial ceramic artefact, unlike the Derby *Cremorne* service mentioned above. Later, we shall describe a genuine Lord Nelson armorial porcelain breakfast service that the Admiral ordered from Robert Chamberlain's Worcester China Works in 1804, which was called the *Horatia* service, and which accompanied Nelson aboard *HMS Victory* at the Battle of Trafalgar in October, 1805. This Chamberlain service is a genuine armorial service as it depicts the full coat-of-arms of Nelson, including the supporters, his crest and his motto, as shown in Fig. 7.1. Nelson's coat of arms, awarded to him in 1800 after the Battle of the Nile, comprise a gold escutcheon with a cross flory sable and on a bend three bombs fired proper; the escutcheon is accompanied by a garter with the motto *"Tria Juncta in Uno"*, meaning *"Three Joined into One*", and surmounted by a disabled ship, a date palm and a ruined shore battery. There are two crests shown, the sinister one being the stern of the Spanish battleship *San Josc* captured by Nelson at Cape St Vincent, bearing a statement "*Faith and Works*", the whole under a Viscount's coronet with its seven gold balls on spikes (or "pearls") on the circlet. The dexter crest is a representation of a Turkish *chelengk*, a diamond-studded *aigrette* with foliage and petals and upwards-facing 13 radial rays (each one representing a French battleship that was sunk at the Battle of the Nile), which was awarded by Sultan Selim III of the Ottoman Empire for Nelson's outstanding bravery in the Battle of the Nile in 1798; a plate from this Chamberlain's Worcester armorial service made for Lord Nelson is shown in Fig. 7.1. Nelson was very proud of his *chelengk* and is frequently depicted wearing it in his naval officer's tricorn hat. Clearly, with this wealth of heraldic information provided on the armorial porcelain it is relatively straightforward for a researcher to identify the coat-of-arms of Admiral Lord Horatio Nelson and to match it chronologically with the dating of his awards as displayed on his crests.
- Monograms, ciphers, initials and motifs are rare on Swansea and Nantgarw porcelain artefacts: one of the best-known examples of rare initialled Welsh porcelain items is the Nantgarw oval dish, now residing in the National Museum of

Wales Collection in Cardiff since it was purchased in 1895. This depicts a painting by Thomas Pardoe of the Pontypridd Bridge over the River Taff built by William Edwards in 1756 which at that time was the longest single-span stone bridge structure in the world: the plate is illustrated in Fig. 2.3 and bears the initials *TW* in cursive script and assigned to Thomas Williams (1778–1835), the Bardic poet and Eisteddfodwr from Pontypool, who adopted the Bardic name Gwilym Morgannwg (William(s) of Glamorgan). It is to be noted that despite the high quality of the porcelain and the painting, the monogram and edging are picked out in a chocolate brown pigment: this was a characteristic of the Pardoe and Young period when they were attempting to minimise the cost of commissions by not using any gilding for the decoration of remnant Nantgarw porcelain in 1821–22.

- Finally, motifs which are characteristic of a particular organisation but which are not perhaps qualifying *per se* as primary heraldic symbols may be commissioned for special ceramic services: these are generally included in a survey of "armorials" but strictly lie outside our definition as prescribed here. Freemasonry grew from the guild of stonemasons and architects in the fifteenth century. An example of this category of ceramic artefact is a Masonic item which displays the symbolic emblem of a square and compasses (Fig. 2.5), which represents "*The square to square our actions. The compasses to circumscribe and keep us in bounds with all mankind*". The letter G occasionally appears with this emblem, as shown here, and is said to stand for Geometry, "*one of the noblest of sciences*". It is apposite here to illustrate an example of what is generally believed to be the last porcelain commission personally undertaken by William Billingsley at the Nantgarw China Works in December 1819, namely, the decoration of a set of six Masonic mugs in Nantgarw porcelain with gilt symbolic emblems for a Thomas Johnes of Cardiff, one of which is shown in Morton Nance (*The Pottery and Porcelain of Swansea and Nantgarw*, 1942, Plate CLXC). An example of one of this set in Nantgarw porcelain from the Welsh ceramics collection at the National Museum of Wales, Amgueddfa Cymru, is illustrated here in Fig. 2.6., showing the square and compasses, and in Fig. 2.7 showing a sunburst motif with rays on the other side. Although this particular set of mugs is now believed lost as a set,

Fig. 2.5 Masonic emblem symbolising the square and compasses of early stonemasons and adopted into Freemasonry: the G is believed to represent Geometry, "one of the noblest of the sciences"

Fig. 2.6 A Masonic beaker displaying the square and compasses symbol, Nantgarw porcelain, attributed to William Billingsley's decoration; an order/invoice is extant which was placed for six beakers of this type placed with the Nantgarw China Works in December 1819, probably the last personal work carried out by Billingsley before he departed Nantgarw for Coalport in early 1820. Illustrated in Ernest Morton Nance (*The Pottery and Porcelain of Swansea and Nantgarw*, Batsford, London 1942: Plate CLXC). *Copyright and reproduced with permission from the National Museum of Wales, Amgueddfa Cymru, Cardiff*

Fig. 2.7 The other side of the Masonic beaker shown in Fig. 2.6., showing a sunburst with rays containing a face. *Copyright and reproduced with permission from the National Museum of Wales, Amgueddfa Cymru, Cardiff*

nevertheless, one does occasionally turn up at auction. The invoice in Billingsley's handwriting still exists in The National Museum of Wales, Amgueddfa Cymru, and states:

Nantgarw Porcelane Manufactory
Bot of Wm. Beely 'December 4th
Ballanced at Nantgarw in December 1819
Rec'd the above thomas Jones.

Here, it is appropriate to mention a special porcelain service which has been described in the ceramics literature as an "*unusual English armorial service*". An order for an armorial porcelain service was placed by His Highness Willajah Nabob Azam Jah with Chamberlain's Worcester manufactory through their agents in Madras, Griffiths, Cooke & Co., in 1820 for a large compound porcelain service comprising integral breakfast and dinner-dessert services of over 1000 individual pieces in total; these services are unusual in that the "coat-of-arms" or heraldic device depicted thereon is not immediately recognisable as such but has nevertheless been considered by many writers to be armorial in its appearance. The dinner-dessert service comprised a cobalt blue ground border and central illustrations from *Curtis' Botanical Magazine* with the Arabic inscription in the form of an armorial "escutcheon" included in a vignette in the reserve. Founded in 1787 at Kew Gardens and still extant today, it is the world's longest running botanical periodical. An example of this particularly fine botanical service is shown in Fig. 2.8, where the

Fig. 2.8 Large oval dish from the dinner-dessert service ordered from Chamberlain's Worcester manufactory in 1820, for Azam Jah, Nawab of the Carnatic (reigned 1819–1825). This service has a dark cobalt blue ground border and this particular dish displays a central botanical plant taken from *Curtis' Botanical Magazine* of 1819/1820 (Volume XLVII, Plate 282), inscribed on the reverse *"Shewy Arctotes". Reproduced with permission from the Collection of Indar and Aruna Pasricha, Kensington, London*

Fig. 2.9 Detail of the "armorial" feature resembling an escutcheon on the *Carnatic* breakfast service ordered from Chamberlain's Worcester manufactory in 1820, for Azam Jah, Nawab of the Carnatic (reigned 1819–1825) bearing the Arabic script, *"Amir al-hind nawab Azam Jah bahadur"* above a date of *1236 Hijri* (Hegira). *Reproduced with permission from the Collection of Indar and Aruna Pasricha, Kensington, London*

escutcheon containing the Arabic script is also seen: the oval dish also bears the picture of a plant labelled in script on the reverse as "*Shewy Arctotes*", which is taken from Plate 282, Volume 47, of *Curtis' Botanical Magazine* (1819/1820) .The breakfast service in contrast has a deep pink border and the same inscription is contained in a cartouche that is now located centrally and which is surrounded by a wreath of garden flowers. The "escutcheon" actually depicts the name and title of Azam Jah, Nawab of the Carnatic (reigned 1819–1825) in Arabic script and reads "*Amir al-hind nawab azam jah bahadur*" above a date of *1236 Hijri*, which equates to 1820 in the Western calendar (Fig. 2.9). Known as the *Carnatic* service, it took 3 years to complete at the Chamberlain's Worcester manufactory and was eventually shipped out to India in 1823 for the Nawab Azam Jah on the HEIC East Indiaman, *Windsor Castle*, under the command of a Captain Lee. It is indeed a moot point as to whether this constitutes an armorial service on the basis of our definition proposed here, since it does not contain any heraldic bearings, devices or achievements that are recognisable as such and really is more akin to several of the personal inscriptions found on the Cambrian Pottery, Swansea, earthenwares that have been referred to earlier, some of which are described in more detail later in this book and which have been strictly excluded from the classification of being armorial "china". Nevertheless, porcelain dealers, museum curators and collectors do still refer to this Chamberlain's Worcester *Carnatic* service as "armorial" porcelain even though it is devoid of any true heraldic devices or a coat-of-arms because the inscription sits in a cartouche which could be loosely regarded as an escutcheon but is strictly not so heraldically as it does not contain any symbols relating to familial descendancy.

An interesting parallel to the topic of this book is the appearance of several armorial achievements or devices on Swansea Pottery creamwares and earthenwares, basically covering both the Cambrian Pottery and the Glamorgan Pottery, which have been described in previous works as "Swansea China", and some of these have already been mentioned above. Whilst technically being outside the remit of the current text, which addresses Swansea and Nantgarw porcelains exclusively, the author realised that the presence of some well-defined and genuine armorials on Swansea pottery might potentially assist in the identification of some crests and coats-of-arms on the analogous porcelains as they generally relate to commissions from local arms-bearing entitled individuals – and this theme will be developed further later. A rather unusual armorial representation on a Cambrian Pottery plate from about 1814 has occasioned much debate and has been assigned to a foreign coat-of-arms and designated probably Italian or French because of its curious heraldic construction: the author discusses this in more detail later and has attributed it to Mrs. Hester Thrale Piozzi (Cannadine 2004). It should be noted that the heraldic devices on the Swansea Cambrian Pottery artefacts are normally accompanied by statements and monograms which materially assist in their attribution and identification as they contain the name of the person who commissioned the artefact, the date of its commissioning and often the reason for their doing so! This is undoubtedly a boon to researchers in armorial ceramics as it provides a definitive historical starting point for their armory research.

References

J.B. Burke, *A Genealogical and Heraldic Dictionary of the Peerage and Baronetage of the United Kingdom* (Privately Published, London, 1826)

Sir J. Bernard Burke, *Burke's General Armory – The General Armory of England, Scotland, Ireland and Wales: Comprising a Registry of Armorial Bearings from the Earliest to the Present Time* (Harrison, London, 1884)

Sir J. Bernard Burke, *A Visitation of the Seats and Arms of the Noblemen and Gentlemen of Great Britain and Ireland*, vol I-II (Hurst & Blackett Publishers, London, 1895)

Sir D. Cannadine, Thraliana: The diary of Mrs Hester lynch Thrale (later Mrs Piozzi), 1776-1809, in *The Oxford Dictionary of National Biography*, ed. by General Editor, (Oxford University Press, Oxford, 2004)

G.E. Cockayne, in *The Complete Peerage of England, Scotland, Ireland, Great Britain and the United Kingdom Extant, Extinct or Dormant*, ed. by V. Gibbs, (St. Catherine's Press, London, 1910)

S.T. Coleridge, *Biographiae Litteraria: Biographical Sketches of my Literary Life and Opinions*, Volumes I and II, Rest Ferrer, Paternoster Row, London/Kirk & Mercien, New York, 1817

A. Cox, A. Cox, *Rockingham Porcelain, 1745–1842* (Antique Collectors Club, Woodbridge, 2001)

Curtis' Botanical Magazine, or *The Flower-Garden Displayed*, vol XLVII, ed, J. Sims (Sherwood, Nealey and Jones, Paternoster Row, London, Plate 282, 1819/1820)

J. Dallaway, *Inquiries into the Origin and Progress of the Science of Heraldry in England with Explanatory Observations on Armorial Ensigns* (R. Raikes, Gloucester, for T. Cadell in the London and for B. and J. White, London, 1793)

S. Dawson Dobson, *Historical Anecdotes of Heraldry and Chivalry: Tending to Shew the Origin of Many English and Foreign Coats of Arms, Circumstances and Customs* (Holl and Brandish, Worcester, 1795)

R. Denyer, M. Denyer, H.G.M. Edwards, *The Pendock-Barry Derby Porcelain Service: A Reappraisal*, to be published (2022)

Sir William Dugdale, *The Ancient (Antient) Usage in Bearing of Such Ensigns of Honours as are Commonly Call'd Arms with a Catalogue of the Present Nobility of England to Which is Added a Catalogue of the Present Nobility of Scotland and Ireland & C.,* Theater for Moses Pitt/Richard Davis, Oxford, and sold by Samuel Smith, St Paul's Churchyard, London (1682)
H.G.M. Edwards, *Nantgarw and Swansea Porcelains: An Analytical Perspective* (Springer, Dordrecht, 2018)
H.G.M. Edwards, *Porcelain Analysis and Its Role in the Forensic Attribution of Ceramic Specimens* (Springer, Dordrecht, 2021)
C.N. Elvin, *A Handbook of Mottoes Borne by the Nobility, Gentry, Cities, Public Companies Etc.* (Bell & Daldy, London, 1860. Reprinted with Supplement and an Index by R. Pinches, Heraldry Today, London, Genealogical Publishers, Baltimore, 1971)
J. Fairbairn, *Crests of the Families of Great Britain and Ireland and Their Kindred in Other Lands,* vols I and II (T.C. and E.C. Jack, London, 1905, New Orchard Publishing, London, 1986)
O. Fairclough, The use of set pattern decoration at the Swansea China works, in *Welsh Ceramics in Context*, ed. by J. Gray, vol. II, (The Royal Institution of South Wales, Swansea, 2005), pp. 192–208
Sir J. Ferne, *The Blazon of Gentrie. Divided into Two Parts. The First Named the Glorie of Generositie, The Second Named Lacyes Nobllitie.Comprehending Discourses of Armes and of Gentry. Wherein is Treated the Beginning, Parts and Degrees of Gentlenesse, With Her Lawes: of the Bearing and Blazon of Cote-Armers: Of the Lawes of Armes and Combats,* John Windet for Toby Cooke at the Tiger's Head in St Paul's Churchyard, London, 1586
A.C. Fox-Davies, *Armorial Families: A Complete Peerage, Baronetage and Knightage, and a Directory of some Gentlemen Coat -Armour and Being the First Attempt to Show Which Arms Are in Use at the Moment Are Borne by Some Legal Authority*, vols I and II (T.C. and E.C. Jack, Grange Publishing Works, London, 1895)
A.C. Fox-Davies, *A Complete Guide to Heraldry: With Illustrations by G. Johnston, Herald Painter to the Lyon Court* (T.C. & E.C. Jack, London, 1909)
F. Gambon, *Porslen Abertawe a Nantgarw (Swansea and Nantgarw Porcelain)*, Oriel Plas Glyn-y-Weddw, Llanbedrog, Abersoch, Gwynedd, 2016
Gartre'n Ol*: Coming Home – An Exhibition of the Finest Nantgarw Porcelain, Nantgarw China Works Museum, Tyla Gwyn, Nantgarw,* 2019
W.D. John, *Nantgarw Porcelain* (Ceramic Book, Newport, 1948)
W.D. John, *Swansea Porcelain* (Ceramic Book, Newport, 1958)
W.D. John, *William Billingsley* (Ceramic Book, Newport, 1968)
W.D. John, K. Coombes, G.J. Coombes, *Nantgarw Porcelain Album* (Ceramic Book, Newport, 1975)
A.E. Jones, S.L. Joseph, *Swansea Porcelain: Shapes and Decoration* (D. Brown, Cowbridge, 1988)
F. Jones, *The Arms of the XV Noble Tribes of North Wales, Coat of Arms, The Heraldry Society,* 36, October Issue, 1958
K.S. Meager, *Swansea and Nantgarw Potteries: Catalogue of the Collection of Welsh Pottery and Porcelain on Exhibition at the Glynn Vivian Art Gallery* (Swansea Corporation, Swansea, 1949)
E. Morton Nance, *The Pottery and Porcelain of Swansea and Nantgarw* (B.T. Batsford, London, 1942)
O. Neubecker, J.P. Brooke-Little, R. Tobler, *Heraldry: Sources, Symbols and Meaning* (Tiger Books International, London, 1989)
D. Phillips, Heraldic symbols on Swansea. *Nantgarw & Swansea Newsletter*, p. 18 (2006)
C. Scott-Giles, *Boutell's Heraldry* (Frederick Warne, London, 1954)
M.P. Siddons, *Welsh Pedigree Rolls* (National Library of Wales, Aberystwyth, 1996)
M.P. Siddons, *The Development of Welsh Heraldry,* vols I, II and III, National Library of Wales, Aberystwyth, 1991, vol IV, National Library of Wales, Aberystwyth, 2007
H. Stanford London, Crosses Paty, Patonce and Formy. *Coat of Arms, The Heraldry Society*, 33/34, January and April issues (1958)
A.A. Turbayne, *Monograms and Ciphers: Designed and Drawn by Albert Angus Turbayne and the Other Members of the Carlton Studio* (The Caxton Publishing, London, 1905)

J. Twitchett, *Derby Porcelain: An Illustrated Guide, 1748–1848* (The Antique Collectors Club, Woodbridge, 2002)

R.E. Williams, Cymru Fu, *The Carmarthen Journal and South Wales Weekly Advertiser*, p. 8, 12 Sept 1919

T. Woodcock, *The Oxford Guide to Heraldry* (Oxford University Press, Oxford/New York, 1988)

W. Wyrley, *The True Use of Arms*, 1592, reprinted by J.G Bell, London, 1853

Chapter 3
History of China Production at Swansea and Nantgarw

> *It must not be forgotten that porcelain was manufactured by mere empirical processes of art and without the guidance of correct scientific principles.*
> *Baron Justus Freiherr von Liebig.*
> *Familiar Letters on Chemistry and Its Relation to Commerce, Physiology and Agriculture, 1843.*
> *An error is more dangerous in proportion to the degree of truth which it contains.*
>
> *Ancient proverb.*

Abstract A history of porcelain production at Swansea and Nantgarw, with a brief survey of the main personalities involved, namely William Billingsley, Samuel Walker, William Weston Young, Lewis Weston Dillwyn and Thomas Pardoe, over the broad production period between 1815 and 1823. The Cambrian Pottery at Swansea on an adjacent site to the Swansea China Works, also produced armorial earthenwares and creamwares earlier and later in their history, which could provide a cross-link to the armorial porcelain being made at Swansea at a similar time as both ventures were owned by Dillwyn for a critical period. The Cambrian Pottery Swansea earthenware armorial punchbowls and a literature survey of other Cambrian Pottery armorials. The role of William Weston Young at the Nantgarw China Works.

Keywords Cambrian Pottery Swansea · Armorial earthenwares · Armorial punchbowls · Lewis Weston Dillwyn · William Weston Young

It is appropriate first to summarise here the history of the Nantgarw China Works and that of its neighbour at Swansea to facilitate the placement of their armorial porcelains in the chronological context of their porcelain production. In common with several other porcelain manufactories that were founded in the late eighteenth and early nineteenth Centuries, Swansea already had a thriving earthenware ceramics production at the Cambrian Pottery from 1764, some 50 years prior to the foundation of the Swansea China Works and the manufacture there of fine porcelains. In addition, the Glamorgan Pottery was established at Swansea some 50 years later in 1812, the same year that the production of porcelain at the Swansea China Works

H. G. M. Edwards, *Welsh Armorial Porcelain*,
https://doi.org/10.1007/978-3-030-97439-8_3

was first attempted unsuccessfully by Lewis Weston Dillwyn. In contrast, this was not the case at Nantgarw, where William Billingsley and Samuel Walker, with financial support from William Weston Young, established the Nantgarw China Works on a site where there was no previous history of ceramics production.

3.1 The Nantgarw China Works

Although Swansea had a ceramics industry prior to that at Nantgarw, fine porcelain was made first at Nantgarw early in 1814 and hence that shall be considered first here. The Nantgarw China Works was the ultimate vision of William Billingsley, who was born in Derby in 1759; he was apprenticed to William Duesbury at the Derby China Works in 1776 and rose to become the head of the enamelling workshop and chief decorator in succession to his mentor, Edward Withers (Edwards, *Nantgarw Porcelain: The Pursuit of Perfection*, 2017a; Edwards, *18th and 19th Century Porcelain Analysis: A Forensic Provenancing Assessment*, 2020). He rapidly developed a national reputation for the quality of his ceramic enamelling, especially for his beautiful rose painting, and for the adoption of a "washing-out" brushwork technique which gave his roses a lifelike and three-dimensional naturalistic quality: this was a major innovation in its day as we can ascertain from a comparison of Billingsley's lifelike depiction of a single rose on individual pieces of the *Prince of Wales* service on Derby porcelain as shown on the dessert plate in Fig. 1.16, which was executed in 1786. In comparison, the rose painting of his mentor at Derby, Edward Withers, is shown on the shell dish with an *oeuil-de-perdrix* ground in Appendix 2 Fig. A2.3, which was painted just a few years earlier, *ca.* 1780, and also on Derby porcelain. Billingsley's porcelain rose painting certainly belies the statement and opinion vouched by Eric Wilson that "*The porcelain rose is not as pretty as the one that decays*" (Wilson, *Against Happiness: In Praise of Melancholy*, 2008). The quality of Billingsley's rose painting at Derby earned him the title there of "*The Rose Painter*", no mean accolade when one looks at the phalanx of quality porcelain enamellers that were in the Derby China Works decorating workshop at that time. He was also noted for his realistic portrayal of the accompanying foliage, which included a brown tingeing to the edges of the leaves as if they had been left in water too long, and for his painting of full-blown pink roses and rosebuds viewed from the underside. Examples of his rose painting at Derby are seen in Figs. 1.16 and 1.18, these being the *Prince of Wales* service commissioned by HRH Prince George in 1786 and the *Earl Camden* service commissioned by Judge Jeffreys, Earl of Camden, in 1790, respectively. A later and superb example of Billingsley's rose painting is seen on the *Barry-Barry* (*Pendock-Barry*) service (Fig. 1.13) which shows his characteristic rose painting skills perfectly (Edwards, *Porcelain Analysis and Its Role in the Forensic Attribution of Ceramic Specimens,* 2021). The wreath of pink roses around the verge of the *Barry-Barry* dinner plate shown in Fig. 1.13 contains 26 roses, plus nine rosebuds, of which seven roses are depicted from their underside. In contrast, the *Earl Camden* plate (Fig. 1.18) has a significantly different composition which comprises 16 roses plus 41 rosebuds and with 4 roses shown from their underside, which gives a numerically different statistic. Billingsley's ambition was to create the world's finest translucent porcelain as a canvas for his exquisite flower painting and to this end he left Derby in 1795 and helped to establish the Pinxton China Works at the invitation of the founder and proprietor, John Coke. He was still restless there, however, and he left Pinxton in 1799, proceeding to Mansfield, where he set up a porcelain decorating business before moving on to Brampton-in-Torksey in about 1804, and thence, after several diversions in early 1808 to seek work briefly elsewhere, to the Royal Worcester China Works of Barr, Flight and Barr in 1808. At Brampton-in-Torksey he met Samuel Walker from a local farming family, who later became his son-in-law upon marrying Sarah Billingsley at Worcester in 1810. Walker had the requisite engineering expertise in kiln construction and in kiln temperature control and the firing of ceramics. At Worcester, encouraged personally by Martin Barr, the proprietor of the Royal Worcester China Works, Billingsley and Walker experimented with novel body compositions, kiln design and the synthesis of porcelain and they eventually achieved a beautiful and novel porcelain body. After the death of Martin Barr, Messrs. Flight, Barr & Barr took over the management at the Royal Worcester China Works from 1813 under the senior partnership of Joseph Flight, along with his junior partners George and Martin Barr. The new management did not wish to embark upon the production of a novel porcelain body and instead confirmed their intention of retaining their existing hybrid china body which was appreciated by their clientele, so Billingsley and Walker, with Sarah and Lavinia Billingsley, departed Worcester for Nantgarw. At this quiet Welsh village, situated adjacent to the newly constructed Glamorgan Canal (1793) and with excellent access to high quality anthracite coal from the Rhondda Valleys and a deep-water port at Cardiff, 7 miles away, they produced the first examples of Nantgarw china of an exceptionally high quality in 1814. Although failing to secure government funding in September 1814, their submitted specimens of the first Nantgarw porcelain excited much interest among the Committee and despite strong support given from the influential Sir John Nicholl of Merthyrmawr for the continuation of their production at Nantgarw, Billingsley and Walker were forced to close down operations there and they then joined Lewis Weston Dillwyn at the Swansea China Works later in 1814. Upon recruiting Billingsley and Walker to join him at Swansea for the manufacture of porcelain, Lewis Dillwyn was delighted: he wrote to his father-in-law, John Llewellyn of Penllergaer, in October 1814 saying "*I never have ever done a better day's work in my life*". Early in 1815 they started to make porcelain in the Swansea China Works with new kilns constructed by Samuel Walker. During 1815–1817, Dillwyn undertook some empirical trial experiments in the variation of his porcelain body paste composition with Walker as his kiln manager and therefrom came Dillwyn's finest duck-egg translucent body, which was held in great esteem by an appreciative clientele. Billingsley obviously approved of this very fine quality, highly translucent china body, with its characteristic blue-green translucency, and he executed some of his finest ceramic painting on it whilst he was based at Swansea. In an attempt to improve the robustness of the Swansea porcelain and to minimise his kiln losses which are believed to have approached 75%, Dillwyn decided that he would significantly decrease the phosphatic bone ash component of his duck-egg porcelain and he substituted this in experimental trials with a steatitic magnesian component, resulting in his "trident porcelain". Although achieving his objective in being much more robust, unfortunately, this new body was significantly less translucent and deemed to be a poor substitute for the delicate duck-egg variety of porcelain which his clients had admired so much and proved much less popular with his discerning clientele. Just 2 years later, Dillwyn decided to lease the Swansea China Works to Timothy and John Bevington, who wound up the operation over the next few years by selling off the stocks of residual porcelain, both of trident and duck-egg composition, which were locally decorated by the resident Swansea China Works artists. In the interim, William Billingsley had parted company with Lewis Dillwyn in late 1816 or early 1817; some ceramic historians have ascribed his departure to his disapproval of Dillwyn's venture into the manufacture of the new, robust but visibly inferior trident porcelain body, and Billingsley then returned to Nantgarw, where he was joined later in 1817 by Samuel Walker. Together with William Weston Young, who invested £1100 of his own money and a further £1000 subscribed from the "Ten True Men of Glamorgan" (Morton Nance, *The Pottery and Porcelain of Swansea and Nantgarw*, 1942), which is discussed further in 5, the Nantgarw China Works went into production later in 1817. The Nantgarw porcelain was immediately successful with an approving London clientele. Almost all of the Nantgarw porcelain output was negotiated for sale in his London showrooms in 47, Oxford Street, by John Mortlock, who seemingly could not get enough of this wonderful new porcelain to satisfy the client demand in the capital. It sold in London at a very high premium which was comparable with prices achieved for the much-esteemed Sevres china, then still considered to be the epitome of the finest European porcelain even whilst operating under its emergence from a sales embargo instituted during the Napoleonic Wars. Unfortunately for Billingsley, Walker and Young, the porcelain losses in kiln firing at Nantgarw approached 90% and the Nantgarw China Works was unable to meet the continuing and rising demands of its London clients through John Mortlock's agency for supplies of Nantgarw porcelain in the white for decoration in the London ateliers of John Sims and Robins & Randall. It is believed that the Nantgarw china supplied to John Mortlock would have been glazed on site at Nantgarw and the enamelling and gilding, therefore, would have been applied overglaze in the London ateliers. The locally decorated china at this stage would have been ordered directly through the manufactory and, therefore, would have been but a very small percentage of the output in comparison with the London decorated china commissions where most of the output was directed. Perhaps some local pieces would have had small visible defects or blemishes which rendered them unsuitable for sending to John Mortlock, who demanded perfection in fulfilment of his order to maintain his premium sale prices rivalling those of the finest Sevres porcelain. It is claimed that a sales premium of between 300% and 500% was typical for Mortlock's sale of Nantgarw porcelain from his showrooms in London. Eventually, the horrendously large kiln wastage of 90% made the business operation unviable commercially at Nantgarw, when realistically only one in every ten pieces of porcelain after firing was suitable for sale in London and the Nantgarw China Works closed, near to bankruptcy, in 1820. Billingsley and Walker had by that time departed Nantgarw to join John Rose at his Coalport China Works, it has been suggested soon after April 1820 (Morton Nance, *The Pottery and Porcelain of Swansea and Nantgarw*, 1942). William Weston Young, the third partner in the Nantgarw China Works enterprise in 1817 along with Billingsley and Walker, then took on the personal responsibility to try to revive the fortunes of the Nantgarw China Works and he persuaded Thomas Pardoe to leave his porcelain decorating business in Bristol to assist him in this venture at Nantgarw. Young and Pardoe had known each other from the early 1800s as they had both been employed as young ceramic decorators by Lewis Weston Dillwyn at the Cambrian Pottery in Swansea. They had lodged together, were close friends, and had collaborated together in the decoration of several earthenware and creamware services (Young, *The Diaries of William Weston Young, 1776–1847 (1802–1843)*, West Glamorgan Records Service, Swansea). Pardoe joined Young at Nantgarw in January 1821 and together they decorated and glazed stocks of remnant porcelain from the Billingsley/Walker era for sale at auction locally, whilst also undertaking commissions received directly from clients. The first Nantgarw auction sale was held in October 1820 before the arrival of Pardoe at the Nantgarw China Works; following on from the last auction sale of enamelled Nantgarw china in October 1822, Young became disillusioned with the operation and management of the Nantgarw China Works and he departed Nantgarw early in 1823, leaving Thomas Pardoe and his son, William Henry Pardoe, who was then just 20 years old, to decorate the stock of porcelain that remained there in the white. There is evidence from recent analytical chemical data from Nantgarw porcelain shards excavated from the waste pit discovered at the Nantgarw China Works site (Colomban et al. 2020) that Young did experiment with producing a Nantgarw hard paste porcelain at the Nantgarw China Works during 1821 and 1822 (Edwards, *Porcelain to Silica Bricks: The Extreme Ceramics of William Weston Young, 1776–1847*, 2019; Edwards, *Porcelain Analysis and Its Role in the Forensic Attribution of Ceramic Specimens*, 2021). However, Young's porcelain could never approach the composition and the quality of the Billingsley/Walker Nantgarw porcelain body as he had no knowledge of the Nantgarw porcelain phosphatic raw material composition, particularly the component derived from calcined and finely milled bone ash in which therefore his new porcelain would have been seriously deficient. It can be assumed that any surviving examples of Young's hard paste synthetic efforts in this direction would not now be easily recognised as "Nantgarw" porcelain by connoisseurs and experts who are more used to the highly translucent phosphatic standard that represented the true Nantgarw porcelain (Colomban et al. 2020).

In 2019, a special exhibition of the finest Nantgarw porcelain (*Gartre'n Ol: Coming Home – An Exhibition of the Finest Nantgarw Porcelain, Nantgarw China Works Museum, Tyla Gwyn, Nantgarw*, 2019) was held at Tyla Gwyn, the house where William Billingsley and his family were resident for their tenure at the site of the Nantgarw China Works and which now forms the focal point of the Nantgarw China Works Museum with a permanent display of fine Nantgarw porcelain. At this exhibition, to celebrate 200 years of porcelain production at Nantgarw special samples of the recently reproduced Nantgarw china of the highest translucency which are made in accordance with the ancient Billingsley/Walker formulation but using modern high technology kiln temperature control (Edwards, *Porcelain Analysis and Its Role in the Forensic Attribution of Ceramic Specimens*, 2021) were also shown. As with the antique porcelain exemplars, this reproduction truly deserves the accolade of *Aur Gwyn o Gymru*, the *White Gold of Wales*. This phrase is reminiscent of Keating's work in which he described the search for porcelain synthesis at Meissen by Augustus the Strong, Elector of Saxony, in the opening decade of the eighteenth century, materially assisted by Ehrenfriede von Tschirnhaus and the alchemist Johann Bottger, which finally resulted in the production of a commercially viable hard paste porcelain in Europe to rival that of the much-admired Chinese imports (Keating, *The Hunt for White Gold*, 2011).

3.2 The Cambrian Pottery and the Swansea China Works

3.2.1 The Cambrian Pottery

The Cambrian Pottery was founded by William Coles of Cadoxton, Neath, on the 19th September, 1764, when he was granted a 41 years lease on the old copper works site at the Hafod, Swansea (currently the site of the First Great Western, GWR and British Rail Swansea mainline railway station), on the understanding that he would demolish the copper works and erect new pottery buildings there. The Swansea Corporation in the eighteenth century was keen to dispel remnants of the old copper smelting factory there which had generated so much atmospheric pollution and they anticipated a move to make Swansea a healthier resort to attract business and visitors to its prime site on Swansea Bay – for many years hitherto Swansea had been called "Copperopolis" and noted for its sulfurous and unhealthy atmosphere which arose from the smelting of the copper sulfide ores to produce copper. The first recorded production of earthenwares at the Cambrian Pottery was recorded in 1767 and the oldest exemplar from this first period of production is a salt-glazed flask inscribed to Morgan John and dated March 28th, 1768 (Gray, *Welsh Ceramics in Context, Part I*, 2003). William Coles operated the Cambrian Pottery until he died in 1778, then a succession of managers ended when the entrepreneurial George Haynes took over in the mid-1780s, until William Dillwyn assumed control through his purchase of the manufactory in 1802. William Dillwyn, who had left Philadelphia in 1777 during the Revolutionary War, bought Higham Lodge, Walthamstow, Essex, and settled there with his wife Sarah (nee Weston); Lewis Weston Dillwyn was born in Higham Lodge in 1778. William Dillwyn took over the Cambrian Pottery which he purchased from the Duke of Beaufort, Lord of the Manor, for the sum of £12,500 (equivalent today to £1.25 M). William Dillwyn intended that his son should take over the management at the Cambrian Pottery but Lewis Dillwyn was more interested as a scientist in the flora of the Vale of Neath and its hinterland and he was elected to the Royal Society for his research on the botanical *Confervae* (Dillwyn, *British Conferva or Coloured Figures and Description of British Plants Referred by Botanists to the Genus Conferva*, 1809), the publication of which he accomplished with the aid of William Weston Young, an accomplished draughtsman and artist, who was formally employed as a decorator in the Cambrian Pottery workshop but who accompanied him on his local expeditions and sketched the botanical specimens (Edwards, *Porcelain to Silica Bricks: The Extreme Ceramics of William Weston Young, 1776–1847*, 2019). Hence, George Haynes was retained as manager of the Cambrian Pottery until 1810 when Lewis Dillwyn finally stepped in as manager at his father's behest. In the meantime, Lewis Weston Dillwyn had married Mary Adams, the natural daughter of Colonel John Llewelyn of Penllergaer, Llangyfelach, in 1807; their eldest son, John Dillwyn was born in 1810 and would eventually inherit his grandfather's estate of Penllergaer and Ynysygerwn. Lewis Weston Dillwyn desired to manufacture fine porcelain at Swansea and he first attempted to do this unsuccessfully between 1812 and 1814 by hiring two people from the Coalport China Works, namely Messrs. Briggs and Burns, which proved to be an unsuccessful venture on his part and no porcelain was manufactured at Swansea at this time. It has been reported that a type of stoneware was made but certainly no porcelain resulted from these early efforts. It needed Dillwyn's engagement of William Billingsley and Samuel Walker from their financially compromised Nantgarw China Works Phase I enterprise in late 1814, resulting from their unsuccessful attempt to secure a grant for its continuation of production from the British government, to eventually accomplish the manufacture of porcelain successfully at Swansea in early 1815 using new kilns constructed by Samuel Walker. Later in 1815, Dillwyn commenced experiments on porcelain bodies, as described earlier (Dillwyn, *Notes of Recipes for Porcelain Bodies and Glazes 1815–1817*, reproduced in Eccles and Rackham, *Analysed Specimens of English Porcelain in the Victoria & Albert Museum Collection*, 1922) but he then sold the lease of the Cambrian Pottery to Timothy and John Bevington in 1817 to concentrate instead upon the management of his recently deceased father-in-law's estate, which later passed to John Dillwyn Llewelyn, at Penllergaer and Ynysygerwn. However, the Bevingtons were only interested in manufacturing earthenwares and when Dillwyn offered them the chance to purchase or lease the adjacent Swansea China Works later in 1820, they wanted the site only for their expansion of the Cambrian Pottery business and they decorated the surplus remnants of Swansea porcelain at the Swansea China Works for sale at the final auction in 1826 to recoup some finance. Whether the Bevington's did make porcelain at Swansea is rather conjectural as several pieces are known of rather poor translucency which bear a Bevington mark, such as the *Gibbins* service. Afterwards, Lewis Weston Dillwyn came back onto the site to run the Cambrian Pottery and in 1831 he appointed his younger son, Lewis Llewellyn Dillwyn, to help manage the earthenware business. Lewis Llewellyn Dillwyn took over the Cambrian Pottery in 1836 and ran it successfully until 1848; thereafter, from 1850 it transferred to the proprietorship of Evans & Glasson (D.J. Evans & Co.) until the Cambrian Pottery finally closed in 1870. A special earthenware known as Dillwyn's Etruscan Ware, which was decorated only in terracotta red and black and based upon ancient Greek ceramic designs comprising vases, amphorae and tazzas, was partially successful as a novel venture in the later years during 1847 and 1850, but this was not sufficient to save the Cambrian Pottery factory from its final sale to Evans & Glasson (Elis Jenkins, *Dillwyn's Etruscan Ware 1847–1850*, 1971).

3.2.2 *The Swansea China Works*

The co-ownership of the Cambrian Pottery and the Swansea China works by Lewis Weston Dillwyn in the second decade of the nineteenth century meant that he had access to the factory records, the clientele and commissions of both enterprises. Whilst it is recognised that Dillwyn needed Samuel Walker to construct his porcelain firing kilns in early 1815, he soon started experimenting with various porcelain paste compositions which were documented in his workbooks using a special code for the raw materials (Dillwyn, *Notes of Recipes for Porcelain Bodies and Glazes 185–1817*, reproduced in Eccles and Rackham, *Analysed Specimens of English Porcelain in the Victoria & Albert Museum Collection*, 1922). During these experiments he discovered his esteemed duck-egg porcelain, (Edwards, *Swansea Porcelain: The Duck-Egg Translucent Vision of Lewis Dillwyn*, 2017b) which soon became the enviable trademark of the finest Swansea porcelain and displaced the early, heavier glassy body paste. It is thought by some authors that the earliest Swansea porcelain was conceived possibly using the formulation of Billingsley and Walker's Nantgarw recipe, but only a few rare examples from this phase of operation have survived (Edwards, *Porcelain Analysis and Its Role in the Forensic Attribution of Ceramic Specimens*, 2021) and these have not been subjected to chemical analysis to determine their composition. However, Dillwyn's attempt to increase the robustness of his duck-egg porcelain body by substitution of the calcined phosphatic bone ash with steatite (soapstone, a magnesium silicate based on talc) decreased the translucency and changed the texture significantly and he soon lost his client base. Hence, he then terminated the production of porcelain at Swansea and the affairs of the Swansea China Works and the Cambrian Pottery were taken over by Timothy and John Bevington as mentioned above, who were not interested in the continuation of porcelain manufacture but they needed the site for expansion of their earthenware business. At Swansea the decoration of stocks of residual porcelain by the Bevingtons went on apace between about 1818 and 1825, using the local workshop decorators such as Henry Morris, David Evans, Thomas Baxter and William Pollard. The Swansea china output, like that of Nantgarw, was distributed between local commissions and London orders, but, unlike Nantgarw, they did not have a sole agency in the capital and their sales were distributed through five porcelain showrooms and decorating ateliers in London, such as those of John Mortlock of Oxford Street, Apsley Pellatt of St Paul's Churchyard and Bradley's of Pall Mall. The Bevingtons were keen to dispose of the remnant stocks of Swansea porcelain, preferably decorated which brought the best sale price, but Godden (*Chamberlain-Worcester Porcelain, 1788–1852*, 1982) has discovered evidence of the purchase of Swansea porcelain in the white for decoration at Chamberlain's manufactory in Worcester between 1819 and 1824.

It is hence appropriate here to evaluate the armorial earthenwares and creamwares that were produced by the Cambrian Pottery, Swansea, which existed some 50 years or so prior to the foundation of the Swansea China Works on an adjacent site. It is also relevant that the Dillwyns were proprietors both the Cambrian Pottery and the Swansea China Works during the first two decades of the nineteenth century, which certainly encompasses the time frame that we should anticipate for the production of the Swansea armorial porcelain services. Furthermore, it has been alleged, perhaps rather apocryphally, by several authors that the Cambrian Pottery produced armorial ceramics to a significantly greater extent than did its neighbour at the Swansea China Works and this statement also needs to be examined and verified as, if true, it might shed some light on some crests that could have potentially appeared on both the earthenwares and the porcelain. There could be several possible reasons for ceramic historians making such a statement: firstly, the Cambrian Pottery, which operated between 1767 and 1848 initially, had a lifespan of 81 years which was significantly longer than that of the Swansea China Works, which effectively made porcelain only during the 5 years between 1815 and 1820. If we include the Evans and Glasson period of tenure at the Cambrian Pottery this time frame for earthenwares is further extended to 104 years. Secondly, earthenwares and creamwares were considerably cheaper than fine porcelain to purchase and with the increase in size of the services required during the Georgian era to accommodate multiple courses and extended social dinners the lower tax imposed on their purchase, which was a function of their total weight, would be a significant advantage for economy-conscious purchasers and would materially assist in their commissioning of larger services. Thirdly, fine porcelain was delicate and consequently subject to greater damage during use, especially during the washing and stacking of plates, tureens and bowls in the kitchen, which encouraged porcelain manufactory proprietors to seek compositional changes which would render their products more robust without sacrificing the texture and translucency of the porcelain body. Lewis Weston Dillwyn had this point in mind at Swansea when he marketed his trident magnesian soaprock porcelain involving a replacement of the phosphatic bone ash component with steatite, but unfortunately his discerning clientele did not accept the inferior translucency and textural changes that resulted from the adoption of this new Swansea body. The author has theorised that titled local families who initially sponsored the Cambrian Pottery through their purchase of creamwares and armorial services might have also contemplated the purchase of the Swansea porcelain equivalents when their own economic and social standing increased in the community in vogue with the desirable Nantgarw and Swansea porcelain purchases which were then being acquired by their London-based contemporaries. Thus, a parallel research survey of the literature and listings of the coats-of-arms and crested ceramics that are known for the Cambrian Pottery might prove rewarding for comparison with those that have been derived already for the Swansea China Works and from which some similarities in family heraldic achievements may then perhaps become apparent.

Hence, we would expect that armorial porcelain services at both Swansea and Nantgarw may have been commissioned and decorated both in London and locally at each site. It is reasonable to assume that local families would perhaps order directly from the manufactory whereas other clients would most certainly have used the London agents. For Nantgarw, we should anticipate that because of John Mortlock's acquisition of most of the manufactory output directly from the Nantgarw China Works in the white (which has been stated to be >90% of the production) then a substantial amount of Nantgarw china surviving today would have necessarily been London decorated, whereas at Swansea we should still perhaps expect to see more evidence of some locally decorated services as a comparative arrangement was not in place with their London agents. This relevant information will input directly to our investigation of the sourcing of armorial services for Swansea and Nantgarw porcelains and perhaps feed into an assessment of the part played by locally entitled aristocratic and arms-bearing individuals and their families in their commissions.

3.3 Armorial Earthenwares from the Cambrian Pottery

The classic text on the Swansea earthenwares is that of Ernest Morton Nance (*The Pottery and Porcelain of Swansea and Nantgarw*, 1942) which is copiously illustrated with monochrome pictures and a survey of the Cambrian Pottery examples he has described therein is very revealing: of the 581 pieces of Cambrian Pottery earthenwares and creamwares illustrated in this book, some 52 belong in the category of inscribed commemorative pieces, 11 others are in the class of monogrammed, initialled or ciphered pieces, 12 more are strictly armorial according to our definition and 506 pieces do not conform to any of these three categories and can therefore be assigned as being "anonymous" artefacts. Hence, based upon these figures and the sampling base of Morton Nance's coverage of Cambrian Pottery earthenwares and creamwares, the genuine armorials would account for only 2% of the output of the Cambrian Pottery in its 81 (or more widely, 104) years lifespan. The earthenware armorials, therefore, comprise just one more in number than the armorial porcelains, which number 11 distinct exemplars, that we have identified thus far for the Swansea China Works over its 5-year lifespan. This analytical fact immediately refutes the suggestion that the Cambrian Pottery was significantly more prolific in its armorial output than its sister and neighbouring porcelain manufactory at the Swansea China Works! It is suspected that what may have occurred here is that the non-armorial, inscribed and commemorative pieces produced by the Cambrian Pottery have been termed "armorial" in a generic sense rather indiscriminately by earlier historians and chroniclers, in which case their collective number then dramatically increases to 75 pieces (according to Morton Nance's assemblage), an increase of 650% on the genuine armorials that were produced there. This latter figure then means that the inclusion of what we have earlier defined as commemoratives and pseudo-armorials in the Cambrian Pottery earthenwares and creamwares production then does far exceed the number of analogous armorials produced by the Swansea China Works in porcelain by 700%. So, then the historical allegation that the Cambrian Pottery was more prolific in the production of its armorials does bear some credibility but in reality this is only achieved by the inclusion of strictly non-armorial items as we have defined earlier in Chap. 2. The importance of our analytical definition of what constitutes an armorial piece and what does not in establishing the baseline for our studies now assumes an even greater significance: otherwise, many pseudo-armorials apparently emanating from the Cambrian Pottery significantly distorts the true picture in comparison with their porcelain congeners and analogues from the Swansea China Works. Of even greater interest and significance for our analytical research here are the truly armorial pieces (i.e. those which bear the partial coats-of-arms and crests or other heraldic devices) that would have been commissioned from the Cambrian Pottery when Lewis Weston Dillwyn was formally co-proprietor both of the Cambrian Pottery and of the Swansea China Works, i.e. between about 1810 and 1817, and there are several of these on record. The valid question then can be posed: did these same families who ordered armorial pieces from the Cambrian Pottery also do so from the Swansea China Works when the opportunity or change in their economic circumstances arose, in which case we may be able to access some new information relating to the as-yet unconfirmed or "unidentified" crests that appear currently in our Swansea porcelain armorial lists? In fact, it could be argued that Lewis Weston Dillwyn, with his prior knowledge of the identity of the titled and armigerous sponsors for his Cambrian Pottery earthenwares, would be in an especially advantageous position to encourage the same persons locally to invest in armorial porcelains from the Swansea China Works? Lewis Dillwyn would have been in a rather privileged position to encourage local entitled families and landed gentry to do this as he was himself arms-bearing, a local landowner and moved in the same level of Georgian society. In this context, it is of interest that William Duesbury, proprietor of the Derby China Works in the late eighteenth century, is on record as purposefully visiting the homes of his potential aristocratic and land-owning clients with his agent, Joseph Lygo, to actively foster his porcelain business and to encourage their purchase of his porcelain. Duesbury also opened a showroom in Bath for his porcelain, which was staffed by some of his relatives, during the summer season when London society decamped there to "take the waters" and socialise together, in an attempt to advertise his wares to potential purchasers.

3.3.1 List of Armorial Earthenwares from the Cambrian Pottery

From a survey of the literature, the 12 armorial earthenwares known to have been produced by the Cambrian Pottery, Swansea and identified during this research are listed in Table 3.1. The key pieces in this survey are three large punch bowls with full armorial bearings produced by the Cambrian Pottery and it is appropriate here to digress and review the popularity of drinking punch in the early nineteenth

Table 3.1 Cambrian Pottery, Swansea, Earthenware Family Armorials and Crests

Family	Coat-of-arms	Crest	Motto	Identified[a]	References
Richardson	Yes	Pelican	Praetio prudential praestat	Yes	Morton Nance (1942)
Duke of Beaufort	No	Portcullis and gold chain	Mutare vel timere sperno/Nid da lle gelli'r gwell	Yes	Morton Nance (1942)
Alcock	Yes	Cock, erminois, beaked/ Demi-swan, wings expanded	Vigilate	Yes	Gray (2005)
Turbervill	No	Spread eagle	Avi numerantur avorum	Yes	Morton Nance (1942 VIIIF)
Galton	No	Bull's head, horned	Gaudet luce videri	Yes	Morton Nance (1942 XXVIID)
Owen	No	Lion rampant	Honestas optimia politia	Yes	Morton Nance (1942)
Royal	No	Prince of Wales' feathers	Ich dien	Yes	Holdaway (2005) and Morton Nance (1942 XXIXCE)
?	Yes	Coronet	–	No	Morton Nance (1942)
"J.C."	No	Lion's head erased	–	No	Meager (1949) and Morton Nance 1942 XLIIF
"J.G."	No	No	Bydd gyfiawn ac nac ofna	No	Meager (1949)
French/ Italian?	Yes	No	–	No	Morton Nance (1942 XXX?)
	No	Pelican	–	No	Morton Nance (1942 VIIIB)

[a]The Column headed "Identified" summarises the literature conclusions and does not imply that these have been researched further and confirmed

century, which required the creation of large ceramic punchbowls; the Cambrian Pottery punchbowls that have survived are important in that they together represent some 25% numerically of the armorial output of the manufactory.

3.3.1.1 The Origin of Punch and Punchbowls

In the seventeenth century, punch was a very popular and convivial drink which was served using ladles from very large bowls: the satirist William Hogarth (1697–1764) depicted the drinking of punch very graphically as a socially debauched gathering of very inebriated people (Harvey 2008). An example of this portrayal is Hogarth's "*A Midnight Modern Conversation*" etching by T.E. Nicholson in 1732, and subtitled the "*Punch Party*", which is shown in Fig. 3.1; this scene shows a panelled

Fig. 3.1 "*A Midnight Modern Conversation*", 1732, print by William Hogarth, subtitled "*The Punch Party*", set in a panelled room in St John's Coffee House, Temple Bar, London. The large punchbowl has been re-filled and is of several gallons capacity (1 Imperial gallon is 8 pints ~5 litres). Bottles are strewn around the floor and the chamber pot is also overflowing, so the gathering has been in place for some time: 11 gentlemen are in an extreme state of drunkenness and stupor. (Metropolitan Museum of Art, New York, Open Access; Harris Brisbane Dick Fund, 1932, Accession Number 32.35 (26))

room, it is believed in St John's Coffee House, Temple Bar, London, at 4 am with 11 men in various states of advanced torpor and drunkenness and a very large ceramic punchbowl with a ladle on the table forming the centre of attention. The drink was first introduced to Britain from India by sailors of the HEIC in the early seventeenth century; an accredited first mention in this context is given to Robert Addams, who wrote to a friend on the 28th September, 1632:

> *I hop you will keep good house together and drinke punch by no allowance.*

The origin of the name "punch" is ascribed to the Hindi "*panc*" (and perhaps also the Dutch *paantsch*), meaning "five", as the five principal ingredients were alcohol (rum, brandy, arrack), sugar, citrus juice (lemon or lime), water and spices (nutmeg or cloves). An alternative derivation of the word is thought to come possibly from the "puncheon", a large cask of up to 500 litres capacity that was used for carrying alcohol (rum or brandy) on sailing ships. Usually served after dinner, punch was a potent drink termed a "wassail", whose main component was alcohol-based, usually rum, brandy or sometimes, wine. The first recorded recipe for a punch based on the above ingredients was noted in 1638 by Johan Albert de Mandelslo in Surat (Souratte), a city in Gujarat, India, which was the first recorded outpost of the Honourable East India Company in India during the early seventeenth century (Wondrich, *Punch: The Delights and Dangers of the Flowing Bowl – An Anecdotal History of the Original Monarch of Mixed Drinks with More than Forty Historic Recipes, Fully Annotated and a Complete Course in the Lost Art of Compounding Punch*, 2010).

Punch houses were certainly in existence in towns and cities by 1671 for the public consumption of punch and these soon became popular because of their tax exemption, unlike the tax imposed upon tea consumption; in early eighteenth century Britain the tax imposed on tea importation was 119% and this was only reduced to 12.5% after the introduction by William Pitt of the Tax Reform Commutation Act in 1784. When served before a meal, punch was alternatively known as a "Cup" – hence "*Pimm's No.1 Cup*", which is still a very popular summer cocktail aperitif today. Punch gatherings required the provision of very large bowls, firstly made of wood then usually in ceramics and these were quite expensive because of their very large size. In 1690, a three-quarts punch bowl held 6 pints, approximately 4 litres, and the punch held therein cost half-a-crown (2 s 6d, correspondingly equivalent to the modern 12.5 pence, and converting to about £20 value in present day money, some $25) which represented half a week's wage for a highly skilled manual worker at that time. The largest punch bowls were termed "Alderman" bowls and held up to 5 gallons of liquid (40 pints, 25 litres). A large English delftware Liverpool punchbowl, *ca.* 1750–1770, of diameter 11″ (28 cm), decorated in blue and white enamels depicts a transfer print of Hogarth's debauched "*Punch Party*" on the base of the inside, which is only seen when all the punch therein has been consumed! Punch bowls were also used at gatherings to promote nationalistic and political themes: a classic example of this is a Chinese porcelain punchbowl in the Winterthur Museum, Delaware, *ca.* 1750, bearing the coat-of-arms of the Grand President of the Anti-Gallican Society depicting St George on horseback with lion and eagle supporters, with the St George's cross banner of England and with the motto "*For Our Country*" (Fuchs, *Made in France*, 2005, p. 135). The Anti-Gallican Society was founded in the mid-eighteenth century "*to discourage the introduction of French modes and the importation of French commodities*". A Wedgwood armorial creamware plate from a dinner service with the heraldic device of the Anti-Gallican Society prominently displayed in black pigment monochrome is shown in Fig. 3.2.

3.3.1.2 The Sir Watkin Williams-Wynn Punchbowl, Jesus College, Oxford, 1732

One of the largest punchbowls was presented to Jesus College, Oxford, by an alumnus, Sir Watkin Williams-Wynn (1692–1749), in 1732 (Figs. 3.3 and 3.4). It is silver gilt and is an armorial bowl as it displays both the College arms and those of Watkin Williams-Wynn; it weighs 200 oz. (5.7 kg) and holds 10 Imperial gallons (80 pints, 45 litres). It was presented to the College by Watkin Williams-Wynn MP on the

Fig. 3.2 A Wedgwood six-lobed creamware armorial plate displaying the arms of the Anti-Gallic Society, with their motto "*For Our Country*", *ca.* 1780–90, decorated in monochrome black pigment. (*Reproduced courtesy of Kingschina Ltd, Mayfair, London*)

Fig. 3.3 The very large silver gilt punch bowl presented to Jesus College, Oxford, by an alumnus, Sir Watkin Williams – Wynn of Wynnstay, Denbighshire, in 1732, on the occasion of his receiving the Honorary Doctor of Civil Law, D.C.L., degree of the University of Oxford. The arms of Sir Watkin Williams-Wynn are shown engraved on the side of the bowl comprising an escutcheon but with no helm, crest, supporters or motto. (*Reproduced with the permission of Jesus College, Oxford*)

Fig. 3.4 The very large silver gilt punch bowl in Jesus College, Oxford, shown in Fig. 3.3 which portrays the arms of the College on the other side to those of Watkin Williams-Wynn. (*Reproduced with the permission of Jesus College, Oxford*)

occasion of his award of the Honorary Degree of Doctor of Civil Laws by the University of Oxford in 1732. The inscription running around the bowl is "D.D. WATKIN WILLIAMS WYNN DE WYNNSTAY IN COM. DENBIGH. LL.D. OLIM HUIJUS COLLEGII SOCIO COMMENSALIS 1732", which can be translated as "Watkin Williams Wynn of Wynnstay in Denbighshire, Doctor of Law, at one time a Scholar and Table Companion at this College gave this offering in 1732". Watkin Williams-Wynn succeeded to the 3rd Baronetcy at Wynnstay in October 1740, 8 years after the punchbowl was gifted to Jesus College. The inscription on the punchbowl includes the letters LL.D for his Honorary Doctorate at the University of Oxford; technically, the Honorary Degree of DCL (Doctor of Civil Law) would have been conferred at Oxford but this would have appeared synonymously as LLD in a Latin inscription. He died 9 years after succeeding to the Baronetcy in 1749 when thrown from his horse whilst out hunting. This punchbowl was later used at a famous dinner in the Radcliffe Camera, Oxford, held in 1814 to celebrate the defeat of Napoleon, rather prematurely it is evident in hindsight (!), which was attended by Frederick William II the King of Prussia, Alexander I the Tsar of Russia, the Prince Regent, the Duke of York, Field Marshal Blucher, Klemens von Metternich the Foreign Minister of the Austrian Empire and the Duke of Wellington . There is a tradition that whoever could encircle the bowl with their arms (it is 5′ 2″ or 1.6 metres in circumference) and then drain its contents of a strong punch fully would be presented with the gift of the bowl – it is understood that the first challenge has been achieved several times but the second has never been accomplished!

The two armorial bearings borne by the Jesus College punchbowl are themselves interesting for the information they convey through the heraldic emblazoning and it is worthwhile considering this specific topic as an illustration of what can be deduced historically from the analysis of heraldic arms. Firstly, the arms of Watkin Williams-Wynn comprise the bare escutcheon, which is devoid of supporters, helm, lambrequins, motto, crest and torse. Heraldically, the arms are described as first and third Quarters: three eagles in fesse or, wings spread and beaked with talons exposed (Wynn); second and fourth: two foxes, arg., rampant, sinister and dexter, counter salient in saltire, the dexter surmounted by the sinister (Williams). However, there is another escutcheon enclosed within this main shield, known heraldically as an "*inescutcheon of pretence*", which signifies that the arms bearer has married an heiress who is the sole daughter of an armigerous father, who herself has no male siblings. Hence, rather than impaling his wife's paternal arms as would be normally expected upon marriage, the paternal arms of his wife are now placed in in the centre of his own shield as a pretence (i.e. a claim) of headship of his wife's family, which will become or has become extinct heraldically on the male side. Any son of the marriage would then be expected to quarter his arms accordingly to absorb those maternal arms. So, who was the wife of Watkin Williams-Wynn, whose arms appear on the Jesus College punchbowl? Jane Thelwall (1665–1759) was the daughter and heiress of Edward Thelwall and Lady Sydney Wynn of Plas-y-Ward, Denbighshire, and she married Sir William Williams, 2nd Baronet, of Llanforda, Oswestry, in 1687 at the age of 21. Their son, Watkin, born in 1692, graduated from Jesus College, Oxford, and became MP for Denbighshire in 1716. Another interesting Jesus College link genealogically is through Jane Thelwall, who was related to Sir Eubule Thelwall of Bathafarn Park, Ruthin, who was himself the Principal of Jesus College, Oxford, in the mid-seventeenth century. In 1719, Watkin Williams became Williams-Wynn and acquired the large estate of Wynnstay, Ruabon, Denbighshire, through his mother's inheritance from her relative, Sir John Wynn of Gwydir, thereby assuming the surname of Wynn as well, in addition to his succession to his father's title and lands upon his death in 1740. He married Ann Vaughan, from whom he inherited further extensive estates in North Wales at Llwydiarth and Llangedwyn. On the death of Ann in 1748, he married his god-daughter Frances Shakerly, and their son, also Sir Watkin Williams-Wynn (1749–1789), then became the 4th Baronet. He died in a hunting accident in 1749 when his young son was only a few months old: an ardent Jacobite supporter of the Young Pretender, Charles Stuart, Watkin Williams-Wynn is portrayed in an oil painting hanging in the Jesus College SCR wearing a symbolic pale blue silk waistcoat signifying his support for the Jacobite cause. A descendant, Sir Watcyn Williams-Wynn (1772–1840), the 5th Baronet, will feature again in this book when we discuss Welsh porcelain armorials as he was a supporter of the Welsh porcelain manufactories and a crest on a Swansea armorial porcelain service can be potentially assigned to him as a result of this research. Watcyn and his two brothers Charles and Henry were known as "Pip, Squeak and Bubble". The Williams-Wynn crest of a spread eagle armed was that of the ancient line of Owen Gwynedd and confirmed their ancestry to the ancient house of Aberffraw through Sir John Wynn of Gwydir, as did the motto "*Eryr Eryrod Eryri*", which translates as "*The Eagle of the Eagles of Snowdonia*". The inescutcheon of Thelwall that is represented on the Jesus College punchbowl contains quartered arms as follows: first and third, a chevron with three boar's heads erect and erased (Thelwall); second and fourth, a stag trippant facing sinister (Powers). It is revealing that the official crest of Williams-Wynn, an eagle displayed with wings spread and beaked and talons drawn, does not appear in the arms on the punchbowl but it is represented in the first and third quartering of the arms. It is this same crest that has been assigned to a Swansea porcelain armorial service here that is hypothesised to be that of Sir Watcyn Williams-Wynn. It seems apparent, therefore, that the arms in the Jesus College punchbowl are those of the Williams-Wynn family and of his father, the second Baronet, as in 1732, Watkin Williams-Wynn had not yet succeeded to the baronetcy and claimed the title.

The second armorial shown on the Jesus College punchbowl displays the arms of the College, namely, an escutcheon vert, with three stags trippant argent attired or. This is represented in Fig. 3.5. The College was founded by Queen Elizabeth I in 1571 as "*Jhesus College within ye City and University of Oxforth of Quene Elizabethes foundation*" but the origin of its coat-of-arms is rather conjectural: a possible origin seems to be that of Bishop Rotherham, who had already established a theological Jesus College in his home town of Rotherham earlier and who was a supporter of Jesus College in Oxford. There are several stories circulating which attempt to derive the origin of these arms but none have yet been adopted as the most tractable solution to their origin. It appears that the original arms comprised the three stags trippant argent on an azure field, as demonstrated in Fig. 3.6, which shows a W.D. and H.O. Wills' cigarette card of the Arms of the Oxford and Cambridge Colleges of 1922, numbered 28 in the series of 42 cards, in which the original arms of Jesus College, Oxford, are depicted with the three stags upon an azure (blue) field. A green field was used in the early years for these arms from about 1605 until about 1619, when a reversion to blue was adopted until a reversion to a green field was made again in the late seventeenth century. The armorial College bearing on the Jesus College punchbowl seems to go against this trend chronologically somewhat as, although the polychromatic representation of arms cannot obviously be used in the engraving of the silver artefact, nevertheless, a horizontal cross-hatching is seen to be used in the field of the escutcheon, which is the official representation of a blue heraldic tincture for either a monochrome or colourless

Fig. 3.5 The arms of Jesus College, Oxford, comprising an escutcheon with three stags trippant argent, attired or, in a vert (green) field, as they appear today

Fig. 3.6 A cigarette card from the W.D. and H.O. Wills' 1922 series of the *Arms of The Oxford and Cambridge Colleges* (42 cards) of which this is number 28. Note that the arms of Jesus College, Oxford, are now rendered as three stags trippant argent in an azure (blue) field, which it has been proposed represents the earliest heraldic field tincture for the arms of the College

representation, as used in engravings and prints (this has been mentioned earlier in Chap. 2 and was known as "tricking"). Alternatively, in such engravings a green heraldic tincture is officially represented by a diagonal cross-hatching (adopting a 45° line declination from top left to bottom right), so implying that the Jesus College arms used on the bowl were out of context chronologically with their general adoption otherwise and elsewhere of a green field for these arms by the early 1700s. It is appropriate to stress this particular point regarding the monochrome representation or engraving of heraldic arms here as thereby one can lose an important item of information that is relevant to the attribution of the arms: this will surface again particularly when we come to consider the representation of heraldic crests that have been made wholly in gilt on armorial porcelain services, wherein the important information that can be derived from heraldic tinctures is thereby otherwise lost. Perhaps, the support for the Jacobite cause given by Watkin Williams-Wynn, as evidenced in his oil portrait hanging in the Senior Common Room at Jesus College where he is depicted wearing a sky-blue waistcoat, generated his reversion to the College arms being portrayed on an azure blue field. A light blue was the traditional colour for the supporters of Bonnie Prince Charlie (The Young Pretender, Prince Charles Stuart) on the battlefield: the Jacobite clansmen wore pale blue bonnets, as did his leading Jacobite regimental clan, The Stewarts of Appin, who carried his blue saltire standard at his final stand during the Battle of Culloden in 1746 against the English forces of the Duke of Cumberland.

3.3.1.3 The Consumption of Punch

Large though many punch bowls are, however, all pale into insignificance next to the volume of punch consumed at the "punch party" given by Admiral Edward Russell, Commander of the British Mediterranean Fleet based in Spain, on the 15th October, 1694, where a marble fountain was used as the "punchbowl" and contained a phenomenally large quantity of punch based on the standard recipe given above. In modern measures, converting hogsheads and pipes (the standard measures of volumes of alcoholic liquid in the seventeenth century) into gallons/litres, this fountain "punch bowl" contained 240 gallons of brandy, 90 gallons of Malaga wine, 20 gallons of lime juice, 400 gallons of water, 2500 lemons, 1400 lbs. of sugar, 5 lbs. grated nutmeg and 300 toasted biscuits, wherein 1 gallon is 4.5 litres and 1 lb. is approximately 0.5 kilo. This vast quantity of punch was served to 6000 guests at the punch party, testifying to the popularity of punch in the seventeenth century – a truly Hogarthian scene must have ensued!

The consumption of punch continued through into the early nineteenth century but waned in the mid-nineteenth century and Charles Dickens included the article "*A Bowl of Punch*" by Peter Cunningham in his weekly magazine *Household Words* (Cunningham 1853) which dealt with a variety of social issues of the time. Dickens himself commented upon the drinking of punch in his books – in *A Christmas Carol* of 1843 he described Scrooge's vision of a Christmas feast topped with "*seething bowls of punch that made the chamber dim with their delicious steam*" (Dickens, *A Christmas Carol*, 1843). In the 1850s it was noted that punch drinking had completely fallen out of favour: the large ceramic punch bowls were then no longer required and these could be seen stacked up high, forlorn and unused in taverns (Harvey 2008). It is remarkable then that three of the most noteworthy surviving and descriptive armorial porcelains from the Cambrian Pottery, Swansea, comprise large punch bowls, namely the Richardson-Francis bowl, the Swansea Corporation bowl and the Lliw (Llwchwr) Corporation bowl. An interesting link between the Cambrian Pottery and the Swansea China Works is a large earthenware punch bowl (diameter 14.5″, 37 cm) (Fig. 3.7) from the Cambrian Pottery, T & J Bevington period, *ca.* 1817–1820, which has been decorated with the transfer and polychromatic infilled chinoiserie *Mandarin* pattern which was popular in Dillwyn's Swansea China Works for porcelain at the same time and was used on some of his finest duck-egg porcelains (where it was labelled as Pattern 164). The letters IW are incised inside the footrim, the mark of Isaac Wood, a modeller at Swansea. A good example of this same chinoiserie pattern on porcelain is provided by the Thomas Lloyd of Bronwydd armorial service that will be discussed later in Chap. 5, where

Fig. 3.7 A large Cambrian Pottery, Swansea, earthenware punchbowl, T & J Bevington period, ca. 1817–1820, diameter 38 cm, decorated with the transfer chinoiserie polychrome infilled *Mandarin* Pattern 164 which was popular on duck-egg porcelain from the Swansea China Works. (*Reproduced courtesy of Bonhams Auctioneers and Valuers, London*)

the armorial crest occupies a vignette in the reserve of this *Mandarin* infilled chinoiserie pattern (Figs. 5.3 and 5.4) on the finest duck-egg porcelain from the Swansea China Works. It has been suggested by Fergus Gambon of Bonhams Auctioneers and Valuers, New Bond Street, London, that this Cambrian Pottery punchbowl was made in earthenware to supplement an order for a *Mandarin* pattern service in porcelain, which he maintains reasonably would have been difficult to fire in such a large example without incurring some distortion in the porcelain biscuit kiln.

3.3.2 *Armorials from the Cambrian Pottery*

3.3.2.1 The Richardson-Francis Punchbowl

This large Cambrian Pottery, Swansea, punch bowl manufactured by Lewis Llewellyn Dillwyn, of diameter 14.5″ and height 8″, displays the escutcheon of the arms of John Richardson, marking his position as Mayor of Swansea in 1845: his coat-of arms is sable on a chief argent, three lions heads erased ermines langued gules (Fig. 3.8). His crest in the armory is not shown on the bowl. Under the escutcheon is a statement "*Maiori vi(lle)ae Swansea*". An inscription on a continuous scroll inside the bowl (Fig. 3.9) reads "*This bowl is presented by Jno. Richardson Esqr. Mayor of the Antient Borough of Swansea to his grandson Jno. Crow Richardson-Francis, as a birthday gift this Vth day of February in the year of our Lord MDCCCXLV. Gaily still our moments roll whilst we quaff the flowing bowl*". The exterior is coloured with a deep cobalt blue ground underglaze with overglaze enamelling of the arms of the County of Glamorgan, Swansea, and of John Richardson. Underneath the bowl is written in script a description of the arms and "Made at Dillwyn's Pottery, Swansea" and "*Painted by Stephen Dingley at the charge of Jno. Richardson Esqr. Designed by Geo. Grant Francis FSA*". Figure 3.9

Fig. 3.8 The large armorial *Richardson-Francis* punchbowl in earthenware from the Cambrian Pottery, Swansea, manufactured by Lewis Llewellyn Dillwyn, son of Lewis Weston Dillwyn, in 1845, showing the escutcheon of the arms of John Richardson, comprising three lion's heads sable on a chief fess argent, with a mace and sceptre and the legend MAIOR.VILL.AE DE SWANSEA to celebrate his becoming Mayor of Swansea. The crest is absent. The bowl has a deep cobalt blue underglaze ground colour with overglazed enamels of the arms of Richardson, co. Glamorgan and Swansea. (*Copyright and reproduced with permission from the National Museum of Wales, Amgueddfa Cymru, Cardiff*)

Fig. 3.9 The bowl is dedicated to John Crow Richardson-Francis, his grandson, through the marriage of his daughter to Colonel George Grant Francis. Here the coat-of-arms of the town and borough of Swansea are shown: an embattlemented gateway of twin towers surmounted by an osprey devouring a fish. Underneath is enamelled a statement signifying it was made at Dillwyn's Pottery, Swansea, and "Painted by Stephen Dingley at the charge of Jno. Richardson Esqr. and rdesigned by Geo. Grant Francis FSA". (*Copyright and reproduced with permission from the National Museum of Wales, Amgueddfa Cymru, Cardiff*)

shows the punch bowl from the other side depicting the coat-of-arms of the town of Swansea comprising a fortified embattlemented gateway and portcullis surmounted by an osprey devouring a fish. John Richardson-Francis was the eldest son of (Colonel) George Grant Francis who was a respected early historian of the Swansea China Works and Pottery and whose personal interview of Henry Morris in the 1850s has provided anecdotal information about the Swansea China Works and its personnel, occasioning much debate and discussion from later ceramics historians (Edwards, *Porcelain Analysis and Its Role in the Forensic Attribution of Ceramic Specimens*, 2021). His wife, Sarah, was the eldest daughter of John Richardson. It is believed that Stephen Dingley was responsible for the enamelling of the actual armorial as the painting is visibly superior to that executed for the accompanying border of grapes and vine leaves (Morton Nance, *The Pottery and Porcelain of Swansea and Nantgarw*, 1942). Cross-checking the armorial bearings in Burke (*Burke's General Armory—The General Armory of England, Scotland, Ireland and Wales: Comprising a Registry of Armorial Bearings from the Earliest to the Present Time*, 1884) and in Fairbairn (Fairbairn, *Crests of the Families of Great Britain and Ireland and Their Kindred in Other Lands*, 1905) reveals that John Richardson JP, Pantygwydyr, co. Carmarthen, was descended from the Richardsons of Durham and had exactly the same escutcheon as that which is displayed on the bowl, with a crest (Fairbairn, *Crests of the Families of Great Britain and Ireland and Their Kindred in Other Lands*, 1905, Plate 48/12) of a mural crown or a lion's head erased and with a motto "*Praetio prudentia praestat*" which means "*Wisdom is beyond Price*". No fewer than 71 Richardsons appear in Fairbairn's lists with various crests and mottoes and 84 in Burke (*Burke's General Armory—The General Armory of England, Scotland, Ireland and Wales: Comprising a Registry of Armorial Bearings from the Earliest to the Present Time*, 1884) but the grandson, Richardson-Francis, does not appear entitled to bear arms and is missing from these lists. Burke is more descriptive of the genealogy of Richardson, which definitively identifies him: his descendants settled in Glanbrydan Park, Manordeilo, co. Carmarthen. It is most unusual to discover an armorial piece that is so descriptive of the entitled bearer of the arms and the reason for its commissioning. Fairbairn (Fairbairn, *Crests of the Families of Great Britain and Ireland and Their Kindred in Other Lands*, 1905) gives the Francis family crest as an eagle perched on a fructed tree stump – which appears very similar to another crest of a bird, similarly perched and appearing on Swansea porcelain.

3.3.2.2 The Llwchwr (Lliw) Corporation Punchbowl

Another large punchbowl from the Cambrian Pottery was presented to the Llwchwr (Lliw) Corporation on the 1st January 1825 and enamelled by William Pollard with sprays of flowers (Fig. 3.10) flanking the Llwchwr coat-of-arms (Fig. 3.11): William Pollard was an esteemed decorator at the Swansea China Works, which had now closed formally, and it is interesting that he had been assigned this role by Timothy and John Bevington, who were then running the Cambrian Pottery and also

Fig. 3.10 Another large punchbowl ostensibly from the Cambrian Pottery was presented to the Llwchwr (Lliw) Corporation on the 1st January 1825 and enamelled by William Pollard with sprays of flowers painted by William Pollard locally at Swansea flanking the Llwchwr coat-of-arms. This Llwchwr punchbowl was described as Swansea porcelain at auction in Sotheby's and if confirmed it would then provide the largest exemplar in size of Welsh armorial porcelain. It is certainly porcelain but of poor translucency and does not match the three Swansea bodies, glassy, duck-egg and trident, possibly therefore being a rare example of Bevington's porcelain or perhaps having been acquired elsewhere. (*Copyright and reproduced with permission from the National Museum of Wales, Amgueddfa Cymru, Cardiff*)

disposing of the remnant Swansea China Works porcelains after their decoration. The Llwchwr coat-of-arms (Fig. 3.11) is displayed as an escutcheon with three ravens below a chained portcullis and a crest of a black raven on a torse; elsewhere there is a demi-dragon gules holding in its forepaws a millwheel and cog, with leek supporters on each side. The interior of the bowl was painted with a border of red -winged dragons amongst purple and gilt foliate scrolls. The inscription beneath the bowl in red script enamel reads: "This bowl is respectfully presented to the Corporation of LLWCHWR by JOHN BEVINGTON of Swansea, January 1st 1825" (Fig. 3.12). Morton Nance (*The Pottery and Porcelain of Swansea and Nantgarw*, 1942) states that this bowl was still used for civic gatherings in Llwchwr in the 1940s and the bowl is illustrated in his book on page 136, Plate LIII, E, F and G. At the dissolution of the Llwchwr Corporation the bowl was acquired by Mr. F.W. Gibbins, a descendant of Joseph Gibbins, a relative of Timothy and John Bevington and a partner in the Cambrian Pottery's bankers Gronow, Gibbins and Eaton.

An interesting statement regarding this Llwchwr punchbowl has been made by Andrew Renton (Renton, personal communication, 2021), Head of Design Collections in The National Museum of Wales, Amgueddfa Cymru, Cardiff, which has very recently acquired this Llwchwr punchbowl (Fig. 3.9) from a prior auction at Sotheby's in 2003. In the sale catalogue it was described as Swansea *porcelain,* namely:

Fig. 3.11 The Llwchwr coat-of-arms on the bowl shown in Fig. 3.10 is displayed as an escutcheon with three ravens below a chained portcullis and a crest of a black raven on a torse. Elsewhere a demi-dragon gules is shown holding in its forepaws a millwheel and cog with leek supporters. The interior of the bowl was painted with a border of red-winged dragons amongst purple and gilt foliate scrolls. (*Copyright and reproduced with permission from the National Museum of Wales, Amgueddfa Cymru, Cardiff*)

Lot 171: A large and important Swansea porcelain bowl dated 1825.

Andrew Renton believes that it should now be credited as being porcelain – it is confirmed as being porcelain but of a rather poor translucency. It is not of a trident, glassy nor duck-egg Swansea body composition and could either be a rare example of the porcelain made by the Bevington's at Swansea or perhaps acquired from elsewhere by external purchase. It is possible that analytical data may assist in verifying its attribution. In any case, it cannot be a product of the Cambrian Pottery and, therefore, only possibly of the Swansea China Works, although it is clearly not of the remnant stock after the departure of Lewis Dillwyn, although it was decorated there locally and disposed of by the Bevingtons during their tenure. The date of 1825 borne by the punchbowl would be compatible with it being prepared along with remnant stock for the final auction sale of Swansea porcelain decorated by local artists at the Swansea China Works site by the Bevingtons which was held in 1826. Since it is believed doubtful by historians that the Bevingtons actually made porcelain during their tenure at the Swansea China Works site then this could be regarded as a rather unique and possibly rare example of a genuine surviving Swansea porcelain punchbowl and it would demand re-classification! This is being subjected to further research currently and if its classification as "Swansea porcelain" transpires to be correct then another example of a "Swansea porcelain" armorial, somewhat unique, has been discovered during the progress of this current research, and perhaps fittingly it also then would provide the largest exemplar in size of a Welsh armorial porcelain that is still extant. It will also then be most unusual in that it is the only example of a Welsh porcelain armorial artefact which gives the date of its presentation, the donor and the purpose of the gift – indeed a

Fig. 3.12 The inscription on the base of the bowl in Fig. 3.10 reads: "This bowl is respectfully presented to the Corporation of LLWCHWR by JOHN BEVINGTON of Swansea, January 1st 1825". (*Copyright and reproduced with permission from the National Museum of Wales, Amgueddfa Cymru, Cardiff*)

unique exemplar in all respects! However, because there is still the possibility that it is an interloper from another porcelain manufactory, the number of Welsh armorial porcelains in our database must still stand currently firmly at 18.

3.3.2.3 The Bevington Punchbowl

A third, large Cambrian Pottery punchbowl was presented by John Bevington to Swansea Corporation and was described by Colonel Grant Francis as a "large wassail bowl painted by Henry Morris": this is now apparently lost, but the armorial devices in previous pictures of this artefact show the escutcheon, crest and supporters of the Borough of Swansea, the Prince of Wales and the Duke of Beaufort, who was Lord of the Manor. The arms of the latter showed a chained portcullis surrounded by floral wreaths and five spearheads (pheons), with his Welsh motto: "*Nid da lle gelli'r gwell*", which translates as "*Not good where better can be*". The

Swansea arms surmount a motto: "*Floreat Swansea*" – "*Let Swansea Flourish*"! The supporters are sinister a red dragon with coronet and dexter a red lion. The crest is an osprey on a black and white torse, representing the older crest of the town of Swansea.

3.3.2.4 Other Armorials and Non-armorials

Several other Cambrian Pottery non-armorial punchbowls are illustrated in the literature, in addition to the one described earlier in the *Mandarin* pattern, for example, one decorated by Thomas Pardoe (1802–1810), and another called the *Helston Bowl*, marked "*The Star Inn, Helston, 1820*", a Cornish *Stannary* piece which bears an inscription and a toast in Cornish (Morton Nance, *The Pottery and Porcelain of Swansea and Nantgarw*, 1942).

Although the literature relating to the manufacture of earthenwares and creamwares at the Cambrian Pottery is quite extensive, little reference is made to truly armorial pieces, but many pseudo-armorials bearing inscriptions or slogans are found and a few also from its neighbouring pottery, the Glamorgan Pottery (1812–1834), which was under the proprietorship of Baker, Bevans and Irwin (Baker was the son-in-law of George Haynes, who managed the Cambrian Pottery for many years prior to Lewis Weston Dillwyn's tenure in 1810). An analytical survey of the existing literature on Welsh ceramics reveals the following statistics concerning the mention of armorials from the Cambrian Pottery:

W. Turner: *The Ceramics of Swansea and Nantgarw. A History of The Factories with Biographical Notices of the Artists and Others, Notes on the Merits of the Porcelain, the Marks Thereon Etc.*, Bemrose & Sons Ltd., The Old Bailey, London, 1897. No mention at all is made of armorial Swansea pottery or porcelain.

E. Morton Nance: *The Pottery and Porcelain of Swansea and Nantgarw*, Batsford, London, 1942. As mentioned above, Ernest Morton Nance describes a wide range of Cambrian Pottery and Glamorgan Pottery earthenwares, citing almost 600 examples, of which only 12 are genuine armorial pieces according to our definition. Several of these armorial pieces and the arms depicted thereon have already been discussed elsewhere here, including the three important Swansea punchbowls, namely, the Richardson-Francis punchbowl, the Bevington punchbowl and the Llwchwr (Lliw) Corporation punchbowl. Another unique exemplar in Morton Nance is the Cambrian Pottery creamware plate bearing an "armed" eagle with spread wings and talons drawn (*The Pottery and Porcelain of Swansea and Nantgarw*, Plate VIIIF) which was attributed to the Turbervills of Ewenny Priory by Ernest Morton Nance from information he received from a Major Edmondes. Other crested china included an unassigned "pelican in piety" crest (Plate VIIIB), a bull's head facing full frontal and horned assigned to Galton (Plate XXVIID), and the jug celebrating Sir John Owen's election and elevation to the peerage that was discussed earlier. An armorial plate shown in Plate XXX of Morton Nance was attributed to either French or Italian nobility: it certainly appears to be of a rather strange composition and poses a few questions which shall be considered later here. The unusual heraldic arrangement of the supporters – two lions were depicted but only one was actually supporting the escutcheon, the other adopting a lying down stance in front of the shield, heraldically adopting a *couchant* pose with its head raised – has led historians to believe that this is most likely an Italian heraldic depiction but some other interesting heraldic features have been noted which will be amplified later.

Other specimens of Cambrian Pottery armorials mentioned include an unidentified coronet with no other accompanying heraldic devices and a lion's head erased crest with the initials "J.C". A crest of a lion rampant was assigned to Sir John Owen but the basis of this was not stated and it is a rather common crest that is seen to be borne by several families. Finally, the Duke of Beaufort's crest of a portcullis and gold chain (fetterlock) is also mentioned and this is found illustrated on the Cambrian Pottery punchbowl presented to the Swansea Corporation by the Bevingtons.

W. Elis Jenkins: *Dillwyn's Etruscan Ware 1847–1850*, Crown Printers, Morriston, Swansea, 1971. No mention of armorial Swansea pottery is given but as stated in the title, the text concentrates on Lewis Llewellyn Dillwyn's later Etruscan wares based on Greek vases and dishes which have a terracotta red decoration of Greek scenes on black basalt, dating from between 1847 and 1850, and it appears that armorials were not manufactured in this medium.

Kildare Meager: *Swansea and Nantgarw Potteries. Catalogue of the Collection of Welsh Pottery and Porcelain on Exhibition at the Glynn Vivian Art Gallery, Swansea*, Swansea Corporation, Swansea/Sergeant Bros. Ltd., Abergavenny, 1949. Eighty pieces of Swansea Pottery in the Exhibition are described, comprising products from both the Cambrian and Glamorgan Potteries, of which two armorial and five inscribed and ciphered and initialled pieces are listed. One of the armorial pieces, a jug with the initials "*JG*" and a garter inscribed with the legend "*Bydd Gyfiawn ac Nac Ofna*", translating as "*It will be justified and not feared*", numbered 214 in Kildare Meager's list and the enamelling attributed to William Pollard, is very unusual in that if this is a motto then it is not accompanied by any other heraldic achievement such as an escutcheon or a crest and should be marked as especially unique for that reason. The motto is actually assigned in the armory to that of Sir John Williams Bt., of Plas Llanstephan, who with his wife Mary owned a famous named Nantgarw china service decorated in London and painted by the ex-Swansea artist "de Juinnie" or "de Junic" (see Edwards, *Porcelain Analysis and Its Role in the Forensic Attribution of Ceramic Specimens*, 2021) but Sir John Williams was born in 1840 and was ennobled only later in the nineteenth century, so chronologically he would seem to be excluded from being responsible for this particular "armorial piece". An earlier Sir John Williams, High Sheriff of Flintshire and later Baronet of Bodelwyddan, Denbighshire, would fit chronologically (d. 1830) but no motto is given for him. This motto is also shared by Vaughan and there are several possibilities here but without having the distinction of an accompanying crest it is impractical to assign. The second armorial cited, numbered 215 in Meager's list, has a crest of a couped lion's head erased in sepia along with the initials *JC* and underneath along with an enamelled signature for T. Pardoe is a SWANSEA impressed mark. No attribution has been forthcoming thus far for this armorial.

The inscribed and monogrammed pieces include a jug celebrating the military victories of the Duke of Wellington in Spain against the French armies of Napoleon: this jug is dated 1813 with a monogrammed *WR*, slightly later than the one referred to earlier as it includes the military victories in the battles at Vittoria and Madrid. The Madrid battle was a fierce one which centred on the Buen Retiro porcelain factory, which had been heavily fortified by the defending Napoleonic French occupying forces and which was destroyed in the ensuing battle (Edwards, *Porcelain Analysis and Its Role in the Forensic Attribution of Ceramic Specimens*, 2021). The jug also has an inscription "*Success to the Navy*" and there is another jug referring to "*The Golden Fleece, Bideford*".

J. Gray: "The Cambrian Pottery Before 1802", in *Welsh Ceramics in Context, Part I*, Royal Institution of South Wales, Swansea, pp. 19–38, 2003. A seminal chapter which discusses early wares from the Cambrian Pottery, particularly those made before 1790. The first recorded flask inscribed to Morgan John is shown and is dated 28th March 1768. Others are inscribed tea canisters and inn jugs, and ship plates such as "*Success to the Neptune of Clovelly*", and "*Milford 1788*". In all, some 22 early earthenwares are described for the Cambrian Pottery but no genuine armorials as have been defined here are mentioned therein.

M. Holdaway: "Early Swansea Printed Earthenwares, Part 1", in *Welsh Ceramics in Context, Part I*, Royal Institution of South Wales, Swansea, pp. 52–74, 2003. Some 45 pieces of early earthenwares are described for the Cambrian Pottery but only 15 are monogrammed or inscribed and there are no genuine armorial pieces cited.

H. Hallesy: "The Glamorgan Pottery, Swansea, and some Comparisons with the Cambrian", in *Welsh Ceramics in Context, Part I*, Royal Institution of South Wales, Swansea, pp. 99–119, 2003. An interesting comparison of the products of the Glamorgan and Cambrian Potteries is undertaken in this chapter. The Glamorgan Pottery was founded by George Haynes in 1812 on an adjacent site to that of the Cambrian and he placed his son-in-law William Baker in charge of the operations, trading as Baker, Bevans & Irwin with an impressed Prince of Wales' feathers mark on the underside of the ceramic item. The Glamorgan Pottery fell into financial troubles and was sold to Lewis Weston Dillwyn, then the owner of the rival Cambrian Pottery, in 1838 which would then be managed by his son Lewis Llewellyn Dillwyn thereafter. The production of ciphered, initialled or inscribed wares by the Glamorgan Pottery concentrated upon the "Reform" jugs with political messages concentrated around the Reform Act of 1832 and several relating to the monarchy, such as the death of Queen Caroline on August 7th 1821. Of 64 items referred to in the article only two can be assigned to ciphered or inscribed pieces and there are no armorials cited. In contrast with the Cambrian Pottery, the Glamorgan Pottery ship plates, for example, do not refer to either the ships' names or their captains and are therefore more anonymous. An interesting snippet of information contained in this article is the fact that in 1819, no fewer than 140,000 pieces of "pottery" (which could have included porcelain decorated by the Bevingtons?) were shipped out of Swansea harbour and destined for Bristol and Cornish ports such as Falmouth; some of these pieces survive with Cornish legends and attachments and others were exported to the USA, where Dillwyn already had established family contacts in the old American colonies and with whom he had traded previously.

J. Gray: "The Cambrian Pottery under Coles and Haynes, 1789–1802", in *Welsh Ceramics in Context, Part II*, Royal Institution of South Wales, Swansea, pp. 45–70, 2005. This article covers an important period of the factory prior to William Dillwyn taking it over in 1802. Several pieces are described of inscribed wares such as "Elizabeth Quick, Latant (St Ives, Cornwall)", "Jane Davies 1792", "Miss Isabella Rowlands, Llantrisent, 1795" and "Ann Bowen 1800". For the purposes of our research here, however, the most interesting illustration is an earthenware plate with an escutcheon showing three cock's heads in argent embattlemented with an inner chevron azure bearing three scythes and a motto "*Vigilate*" meaning "*Watchful*" or "*Watch ye*". Extremely useful for the identification of the owner of the escutcheon, the underside of the plate has the legend "*Mr A. Alcock, painted Swansea, September 1798*". Fairbairn (*Crests of the Families of Great Britain and Ireland and Their Kindred in Other Lands,* 1905) assigns these arms as being borne by Alcock of Ridge, co. Chester: argent a fesse azure between three scythes and motto "*Vigilate*" with a crest of a cock erminois, beaked, or alternatively a demi-swan with wings expanded. Again, we have the possibility of four potential families of individuals using this motto alone: namely, Alcock, Edwardes, Gael and Leeds, but the name inscribed on the base of the artefact and the arms on the escutcheon suffice to identify the entitled individual unequivocally. This is the only armorial cited in this article and other commemorative pieces refer to the Swansea Volunteers, a militia unit raised between 1796 and 1802, and a "*Mr Husbands, Swansea 1800*", painted by Thomas Pardoe and bearing the legend "*Success to NP*".

M. Holdaway: "Early Swansea Printed Earthenwares Part 2", in *Welsh Ceramics in Context, Part II*, Royal Institution of South Wales, Swansea, pp. 93–113, 2005. Several interesting examples are illustrated of inscribed Cambrian Pottery artefacts, but one artefact of note is a jug inscribed with a "*View of La Guillotine or the modern beheading Machine in Paris by which Louis XVI the King of France suffered on the Scaffold, January 21st, 1793*", and a true armorial piece, a cabbage leaf jug, showing a Garter Star with motto "*Honi Soit Qui Mal y Pense*" inside the eight-pointed star with the Prince of Wales' feathers at the side, painted by Thomas Pardoe. This is the same type of garter star as exhibited on the Nantgarw porcelain plate decorated by Pardoe but which is missing its motto (Figs. 4.10 and 4.11), so it appears that the allegation that it was adopted as a motif for porcelain decoration by Pardoe could well be correct (see amplified discussion below)!

A. Renton: "The Swansea Diaspora: The Later Careers of David Evans, Henry Morris and William Pollard", in *Welsh Ceramics in Context, Part II*, Royal Institution of South Wales, Swansea, pp. 209–234, 2005. Although covering the decorating of porcelains sourced outside Swansea by these three Swansea artists, two inscribed earthenware jugs are noted – one from the Cambrian Pottery with an inscription "*D&C Evans Sept 17 1826*" is painted by Henry Morris, and the other inscribed

"*JMJ 1829*" is painted by William Pollard. No armorials are mentioned in this article.

3.3.2.5 The Garter Star Motif

Thomas Pardoe worked at the Cambrian Pottery between 1802 and about 1810, when Lewis Weston Dillwyn employed him and William Weston Young to decorate earthenwares and creamwares: after 1812 he was based at Bristol independently decorating china for several factories until he joined William Weston Young at Nantgarw in 1821. At Swansea there is mounting evidence that he adopted a gilt signature motif based on the eight-pointed Star of the Garter enclosing the cross of St George but minus the Garter band with its motto "*Honi Soit Qui Mal y Pense*". Ernest Morton Nance (*The Pottery and Porcelain of Swansea and Nantgarw*, 1942) affords some material evidence for this motif belonging to Pardoe as several items of Cambrian Pottery earthenware decorated during this period bear his script signature, *T. Pardoe*, as well as this motif and are accompanied by a geometric border gilding that is also characteristic of his work. Pardoe maintained this signature motif whilst at Bristol, where he marked some items with the script mark "*Pardoe Fecit, Bristol*". Morton Nance has identified the following artefacts as unequivocally belonging to Pardoe's hand and displaying the same modified eight-pointed garter star device whilst he was working at Swansea, Bristol and Nantgarw:

- A tea service, possibly Staffordshire porcelain, with the following script underneath, "*Painted for the Honble. Mrs T. King by T. Pardoe, Bristol*", as shown in Morton Nance (*The Pottery and Porcelain of Swansea and Nantgarw*, 1942, Plate CLXXXVIIIA) in the form of a plate with the central gold eight-pointed star motif and a similar gilding design found on his earlier Cambrian Pottery work.
- A large earthenware barrel-shaped cider jug painted with a tiger prowling through rocks and trees and bearing the gold star motif on the handle (Morton Nance, *The Pottery and Porcelain of Swansea and Nantgarw*, 1942, Plate XXIIA and B).
- A similar cider jug painted with a girl carrying a bunch of grapes bearing a gold star on its handle (Morton Nance, *The Pottery and Porcelain of Swansea and Nantgarw*, 1942, Plate XXIE).
- A gold star centrally placed on an earthenware service marked SWANSEA impressed from the Haynes, Dillwyn & Co. period, 1802–1810, with Pardoe's characteristic intersecting elliptical gilding pattern at the edge (Morton Nance, *The Pottery and Porcelain of Swansea and Nantgarw*, 1942, Plate XXVIIC).
- A Nantgarw 12-lobed indented porcelain plate with groups of flowers in Pardoe's hand, impressed NANT-GARW C.W. mark, and a gold garter star centrally placed (Morton Nance, *The Pottery and Porcelain of Swansea and Nantgarw*, 1942, Plate CLXXXVIIC). This plate is similar to that shown and discussed in Figs. 4.10 and 4.11 later.

It may safely be concluded, therefore, that the eight-pointed "Garter Star" motif, which is termed a "gold cross with rays" by some historians, is definitively the work of Thomas Pardoe and that this can truly be described as his signature motif on his decorative work. We shall encounter the Garter Star again when we discuss Nantgarw armorials and we shall also consider it as potentially an armorial device that could belong to a Garter Knight, before finally concluding that it is in fact a Pardoe motif rather than a genuine heraldic symbol.

3.3.3 *A Summary of the Cambrian Pottery Armorials*

Reference to Table 3.1 will show the 12 Cambrian Pottery armorials and the types of heraldic device which they display, excluding the two punchbowls that are Corporation arms inspired. Of these 12 items, 7 have been attributed to the family bearers of the arms, namely, the Duke of Beaufort, Richardson, Alcock, Turbervill, Galton, Owen, and the Prince of Wales – a success rate of 60%. This is greater than that which has been achieved for Swansea porcelain heraldic identification and can be ascribed in no small measure to the associated script or legend which is normally found to be accompanying the armorial devices on the earthenwares. Three partial coats-of-arms are quite definitive, namely those of Richardson, Owen and Alcock, in that they comprise the escutcheon and the crest and in the case of Owen and Alcock include the motto also. Four crests have been identified, namely those of the Duke of Beaufort, Turbervill, Galton and the Royal insignia of the Prince of Wales' feathers. Still remaining unidentified and awaiting further research are five earthenware Cambrian Pottery armorials, comprising a crest of a lion's head erased and the initials *J.C.*, unusually a motto and initials *J.G.* but no crest depicted, a crest of a pelican in piety , an anonymous coronet , and finally an armorial which displays an escutcheon, a crest, helm, mantlings and two supporters, which theoretically should more easily facilitate its identification but it is believed that it is representative of foreign nobility, designated possibly French or Italian by Ernest Morton Nance (*The Pottery and Porcelain of Swansea and Nantgarw*, 1942), and has remained unidentified thus far (but see the analysis below).

3.3.3.1 Coat-of-Arms on Cambrian Pottery Earthenware: Foreign Nobility?

In the coat-of-arms mentioned above, illustrated in the armorial plate under investigation in Fig. 3.13, which has been assumed to belong to foreign nobility (Morton Nance, *The Pottery and Porcelain of Swansea and Nantgarw*, 1942, Plate XXX), the two supporters are lions but one is *rampant regardant* whilst the other is lying down and *couchant* with its head raised in front of the escutcheon – a rather strange combination of heraldic "supporters" when it is apparent that one of them is clearly not actually supporting the escutcheon and arms! The original function of supporters heraldically was to guard and protect the armour and arms of the noble involved in

Fig. 3.13 Cambrian Pottery earthenware armorial plate, *ca.* 1810, with a rather unusual heraldic depiction of an escutcheon, coronet, helm with visor and supporters which has been the cause of some controversy in the Welsh armorial literature relating to the Cambrian Pottery. It was formerly assigned to foreign nobility, probably Italian, but here a hypothesis is advanced for it belonging to Mrs. Hester Thrale Piozzi Salusbury and that it contains some anecdotal commentary on her family lineage. Now in the Victoria and Albert Museum, South Kensington, London, the gift of J. H. Fitzhenry, Accession number 604–1906. (*Reproduced by courtesy of the Victoria and Albert Museum, South Kensington, London*)

battle or in jousting tournaments, hence they were usually depicted in rampant, segreant or aggressive poses, and it would not be possible to do this effectively if the supporter was lying down (couchant) beneath the arms. A search of other national armories reveals that the design of the coronet shown on this armorial Cambrian Pottery, Swansea earthenware plate immediately rules out Danish, Swedish and Spanish nobility, but the pearls visible could match the designation of a Vicomte (France) or a Visconti/Nobile (Italy) – equivalent to a Baron in English nobility. Further than that, the only clues to a potential assignment would be the closed visor helm and lambrequins, but minus a crest and torse, and the winged figure of a rampant Pegasus below it which is surmounted by three estoiles (*etoiles rayonnant*): it is interesting that the emblem of Pegasus is extremely rarely found in an escutcheon, being usually featured as a crest, and the heraldic symbolism is also intriguing– the estoiles are emblems of God's goodness and the eminence of the bearer, with the rampant Pegasus representing fame through brilliance of the mind, deep friendship and faithfulness. Ernest Morton Nance's (*The Pottery and Porcelain of Swansea and Nantgarw*, 1942) attribution of the flower painting to Thomas Pardoe's work on the basis of the rather limited associated painting comprising scattered single floral sprays would apparently date it to before the end of the first decade of the nineteenth Century as Pardoe had then left Swansea in about 1809/10 and set up a ceramics decorating business in Bristol. This plate is now in the Victoria and Albert Museum (No. 604, acquired in 1906; donated by J.H. Fitzhenry), where it was examined by Bernard Rackham, at one time the Curator of Ceramics there, who believed it to have been decorated in London for an Italian noble family at the Cambrian Pottery Warehouse prior to its closure in 1807 – in which case it would debar Thomas Pardoe from its enamelling since he had never worked in London. Also, this would definitively have dated the commissioning of the armorial service to before 1807. The plate has a spade device impressed on its reverse, a characteristic mark of the Cambrian Pottery. Morton Nance (*The Pottery and Porcelain of Swansea and Nantgarw*, 1942, pp. 64, 83) also quotes an extract from the *Cambrian* newspaper, dated December 18th, 1807, which advertises the closure of the Cambrian Pottery warehouse of Haynes, Dillwyn & Co. in 64, Fleet Street, London, and of the sale of its stock of "*an elegant assortment of every kind of service and article in earthenware ... to be sold off at a 10 % discount for ready money*".

Ernest Morton Nance has found a record that John Mortlock of 47, Oxford Street, London, then took over the agency for the Cambrian Pottery, Swansea, earthenwares and creamwares in London and were fully operating as its agents on this basis by 1808 – which itself must be a tribute to the quality and desirability of the earthenwares and creamwares then being offered for sale by the Cambrian Pottery in the capital. Morton Nance also noted that there were several titled people interested in acquiring some of these ceramics from Mortlock's, including a Mrs. Hester Lynch Salusbury Piozzi, a noted connoisseur of ceramics at that time. It is intriguing to speculate upon the foreign nobility source of this armorial as this period was at the height of the Napoleonic Wars and the commissioning of this service by foreign nobility at either Swansea or London would suggest strongly that the noble family concerned were possibly émigrés who had escaped the French "Terror" in 1792 or equally the invasion of Italy by Napoleon, as at this time the French, Italian and Spanish ports were being blockaded by the Royal Navy, which would have interfered with the delivery of such a service when completed if the noble family were still based in France or Italy. Napoleon, between 1796 and 1799, had acquired and occupied in the name of the French Revolution most of the Italian states, such as Lombardy, Sardinia, Emilia Romagna, Tuscany, Venice, Piedmont, Genoa, Rome and Milan, as satellites of France, whilst appointing himself Emperor of France and King of Italy by 1805. Perhaps realistically it would be appreciated that such an émigré family would have been based in London and this would then reinforce Bernard Rackham's conclusion, but still keep open the potentially French sourcing of the coat-of-arms. Alternatively, the mention of the name Piozzi, clearly being of an Italian origin, caused the author to examine Burke's armory in this context, which gives the following citation for Mrs. Hester Lynch Piozzi:

Piozzi Salusbury: Bachergraig, co. Denbigh, descended from Sir John Salusbury of Lleweny Hall, Denbighshire. Hester Lynch, daughter and heiress of Thomas Salusbury, of Brynbella, co. Flint, married first Henry Thrale, then Signor Grambattista Piozzi of Brescia in Lombardy, who assumed by royal licence in 1813 the surname of Salusbury and became high sheriff of Flint in 1816, receiving a knighthood from the Prince Regent in 1816. Arms: gules a lion rampant or ducally crowned between three crescents of the last a canton erm. Crest, a demilion rampant couped ar. collared gules and ducally crowned holding in its dexter paw a crescent gold. Motto: "Sat est prostrasse leoni", which means "It is enough to have laid the lion prostrate".

This last statement surely provides a graphic description of the strange configuration of the two lion supporters of the coat-of-arms depicted here, where one rampant lion is visibly triumphant over another prostrate lion, heraldically *couchant*, at the front—this could be an example perhaps of a "pictorial motto" .....? Further research into Hester Lynch Salusbury Piozzi is very revealing and demonstrates how fascinating and intriguing the background can be to armorial ceramic artefacts. Hester Lynch Salusbury Piozzi (1741–1821) was born at Bodfel Hall in Caernarfonshire, the daughter of Hester Lynch and Sir John Salusbury, one of the most influential and wealthiest Welsh landowners of his time, whose seat was at Lleweni Hall, Llysmarchweithian, Denbighshire. Sir John Salusbury founded Halifax in Nova Scotia, Canada between 1749 and 1753, and named it in honour of his patron, Lord Halifax, President of the Board of Trade, but he then lost most of his fortune in several failed Canadian venture investments. After the death of her father in 1761, Hester and her mother then moved to live with her uncle, Sir Thomas Salusbury, at Offley Park in Herefordshire. In 1763, aged 22, she married a wealthy London brewer, John Thrale, who in 1765 became MP for Southwark. The same year she was befriended by Dr Samuel Johnson, who often enjoyed the company of his well-educated hostess and her family and friends at their London home of Streatham Park. In this context, she came into contact with Susana Dawson Dobson, whose husband was Mrs. Thrale's physician, and whom we have cited in Chap. 2 as an historian and writer on the origins of heraldry and chivalry (Dawson Dobson, *Historical Anecdotes of Heraldry and Chivalry: Tending to Shew the Origin of Many English and Foreign Coats of Arms, Circumstances and Customs*, 1795). Mrs Dobson had ambitions to join Mrs. Thrale's eminent literary group, which included James Boswell and Oliver Goldsmith as well as Dr. Johnson. However, Hester was not keen at all on her joining the group and said, "*Mrs Dobson persecutes me strongly as if with a violent and undesired friendship .. she is jealous*". John Thrale died in 1781, upon which Hester sold the Anchor Brewery, Southwark, and increased her wealth; it was widely speculated that Hester would then marry Dr. Samuel Johnson but instead she married Gabriel Mario Piozzi, an impecunious musician and Italian music teacher to her 12 children (of whom only four daughters survived infancy), in July 1784. Her marriage to Piozzi was not favoured at all by London society and she complained that the newspapers gave her some bad press:

The English newspapers are full of gross insolence towards me by raising an obscure and penniless Fiddler into sudden Wealth.

She became estranged from Dr. Samuel Johnson and other *literati* upon her marriage and with Gabriel Piozzi she travelled widely on the Continent, after which they both then retired to her country home, Brynbella, on her Bach-y-graig estate at Tremeirchion in North Wales. Her marriage to Piozzi was strained but she did strike up a renewed correspondence with Samuel Johnson just before he died in December 1784. In 1786 she wrote her "*Anecdotes of the Late Samuel Johnson LLD During the Last Twenty Years of His Life*" (Hester Lynch (Thrale) Piozzi, 1786), which was well received and judged to be the best personal account written of Samuel Johnson in an new anecdotal style termed *"a new fashioned biography"* which excited much literary comment as it had never been experienced before in literary circles (Clifford, *Hester Lynch Piozzi(Mrs Thrale)*, 1987; Birkbeck-Hill, *Hester Thrale*, 1897; D'Ezio, *Hester Lynch Thrale Piozzi: A Taste for Eccentricity*, 2010). The anecdotal nature of her writing (anecdote definition: a short amusing story about a real incident or person, possibly unreliable hearsay!) must have been a major reason for Susana Dobson wanting to join her literary circle as Mrs. Dobson's survey of heraldry and chivalry was itself written in a similar anecdotal form!

The entry for Hester Lynch Thrale Piozzi Salusbury in Burke's armory is incorrect in several aspects, in the spelling of her country estate (Bach-y-Graig), in the spelling of her husband's name, which is given as Grambattista Piozzi, instead of Gabriel Mario Piozzi, and in his appointment as High Sheriff of Flintshire dated 7 years after he had died (!), but an important message is that she was the one entitled to bear arms under the name Hester Lynch Piozzi Salusbury. Grambattista (*sic.* Giambattista, Giovanni Battista) was actually Gabriel Piozzi's brother and in 1798, Hester and Gabriel Piozzi adopted the 5-year old son of Giambattista and renamed him John Salusbury Piozzi. Gabriel died of gout in 1809 and in 1814 Hester gifted her estate to her adopted son John upon his marriage to Harriet Maria Pemberton of Ryton Grove; John then acquired the additional family surname, with an accompanying knighthood, and became Sir John Salusbury Piozzi Salusbury, becoming the High Sheriff of Flintshire in 1816 (which now agrees with Burke's armory)! Hester then went to live in Bristol in 1814 and died there in Sion Row, Clifton, in 1821 (Franklin, *Piozzi Hester Lynch (1741–1821)*, 2004). This is a most interesting discovery, as Thomas Pardoe was then in business in Bristol from 1812 up to 1821 decorating ceramics, when thereafter he left to work with William Weston Young at Nantgarw. It would be a reasonable supposition that Hester Lynch Piozzi Salusbury would have been in an ideal situation to engage Thomas Pardoe to locally decorate Cambrian Pottery ceramics in Bristol (perhaps shipped over the Bristol Channel from Swansea), where they were then both resident. This scenario would have created an armorial Cambrian Pottery service with her arms and even using an "anecdotal" pun, for which she was famous in literary circles (!), based upon her motto in having the most unusual situation of one of her lion supporters lying prostrate before the other. This situation also means of course that both Bernard Rackham and Ernest Morton Nance were both partially correct in their supposition that the coat-of-arms were possibly Italian in origin and that it would be perfectly possible for Thomas

Pardoe to have decorated the service, as alleged by Morton Nance, but this would have occurred in Bristol and not in Swansea nor in London. This reasoning is supported by Pountney (Pountney, *Old Bristol Potteries*, pp. 115–116, 1920), who stated that Pardoe definitely decorated Swansea "china" (earthenwares and creamwares) in Bristol (see Appendix 2) up to 1821 – this having probably been shipped directly in small coasters from Swansea into Bristol docks. The historical research evidence then is clearly supportive of Mrs. Hester Lynch (Thrale) Piozzi Salusbury as being the commissioner and originator of the Cambrian Pottery armorial plate under investigation, probably around the middle of the second decade of the nineteenth Century, for which a celebratory date would surely be the award of a knighthood to her adopted son John in 1814/15 upon his accession to her estate? The rather unusual form of the armorial depicted would surely have appealed to someone who has been cast as the doyenne and awoved mistress of the literary anecdote in Georgian England. Does this then perhaps the constitute a "pseudo-armorial" according to our definition in that it contains elements of a true armorial but also some unusual attributes, although of course the lady had every right to bear arms? Other small heraldic clues place this particular armorial in a distinctly different category to others: the escutcheon is oval (as befits a European continental influence), whereas an armigerous lady's escutcheon in English heraldry is lozenge shaped, the helm is in the style of Italian armour with its gilt visor, and the coronet is accompanied by a wreath or torse in the helm, which is unusual in English heraldry. Crests are rarely used in Italian heraldry as are supporters and these are therefore not regulated (Von Volborth, *Heraldry of the World*, 1973). The coronet depicts five strawberry leaves which in Italian heraldry signifies a Ducal status of non-regal birth, which Hester Piozzi certainly was not. The presence of a two-handled ewer above the arms in the verge is also strange as this could be assumed to be a crest, which exists only rarely in Italian heraldry. The consensus must be therefore that this is an enigmatic armorial ceramic piece which displays elements of English and Italian heraldic symbols but is clearly anecdotal in its construction.

In contrast to this artefact, with its rich association of historical information, the coronet presented on its own as depicted on a Cambrian Pottery piece without a motto or a monogram, cipher or initials or any other associated heraldic devices cannot be identified as it is too anonymous, unlike the *Viscount Cremorne* Derby porcelain service discussed earlier and shown in Fig. 1.17. Unfortunately, this particular piece has also not been illustrated or described further in Morton Nance (*The Pottery and Porcelain of Swansea and Nantgarw*, 1942) so there is no possibility of even identifying the rank of its bearer in the peerage, e.g. Baron, Viscount, Earl, Marquess or Duke, from an analysis of the composition of its strawberry leaves and spiked pearls.

It will be appropriate in our later discussion to evaluate the identification of these Cambrian Pottery crests in comparison with those that still remain unidentified on the Swansea China Works porcelain armorials.

3.4 Decorators of Nantgarw and Swansea Porcelains

It will be appreciated that the decorators, enamellers and gilders of Swansea and Nantgarw porcelains had a vital role in the marketing and sale of the porcelain: the subtle distinction between the two manufactories in this respect arises from the fact that most of the Nantgarw China Works output was acquired by John Mortlock who received it in the white, it is believed glazed, in London for decoration in his chosen ateliers of Robins & Randall in Islington and John Sims in Pimlico. This decoration was of a very high quality, accompanied by fine and often profuse gilding, but often the name of the artist is unknown, except for notable exceptions such as Moses Webster and James Plant with the added name of the mysterious de Junic (Edwards, *Porcelain Analysis and Its Role in the Forensic Attribution of Ceramic Specimens*, 2021). Little Nantgarw porcelain in comparison would have been decorated locally by William Billingsley during the Billingsley/Walker period and some most certainly would have been carried out by junior hands in the workshop. After 1820, most of the Nantgarw locally decorated porcelain would have been accomplished by Thomas Pardoe.

In contrast, the Swansea porcelain output was extensively locally decorated in the manufactory by well-known and accomplished artists such as Thomas Baxter, David Evans, Henry Morris, George Beddow, William Pollard and occasionally William Billingsley himself. At the closure of the manufactories, the stocks of remnant porcelain left in the white were decorated locally by the above artists at Swansea and by William Weston Young and Thomas Pardoe (and it has been suggested perhaps also by William Henry Pardoe) at Nantgarw. This aspect of local and London decoration of Welsh porcelains has been amplified in Appendix 2 and in Appendix 3, respectively. The distinction between London and local decoration is also important for armorial pieces and especially where the heraldic symbols of the crest or escutcheon are accompanied by floral painting which can serve to place the artefact in its proper chronological context. An example of this is provided by the two *Ramsay of Straloch* services in Nantgarw porcelain, both containing the appropriate and correct family crest and motto, the first of which has a fine gilded verge associated with a London decoration from the Billingsley/Walker period, *ca.* 1817–1819, whilst the second service has sprays of garden flowers randomly applied in the distinct hand of Thomas Pardoe, which firmly dates it to the period *ca.* 1820–1822. Hence, the identification of the decorator of the armorial accompaniment establishes unequivocally the relative chronology of the commissioning of these two services and has suggested that the first was London-decorated through John Mortlock's agency and the second chronologically was locally decorated a few years later by Thomas Pardoe on site at Nantgarw. It would be extremely useful here to attempt the non-destructive molecular spectroscopic analysis of the glazes on each of these two *Ramsay of Straloch* service plates as the different glazes used in the Billingsley/Walker and the Young/Pardoe periods at Nantgarw have definitive spectral signatures which would confirm their source and attribution (Edwards, *Porcelain Analysis and Its Role in the Forensic Attribution of Ceramic Specimens*,

2021) and would thus provide confirmatory analytical scientific evidence of this assertion. A final comment relates to the relatively small number of established artists employed at the Nantgarw and Swansea China Works; from a statement by Richard Millward, who was employed at Nantgarw there were perhaps a dozen people employed in the decorating workshop and most of these were young "hands" who would be utilised for general and fairly mundane tasks and only William Billingsley and perhaps William Pegg the Younger were recognised first class enamellers there, with lesser duties being carried out by reasonably proficient artists such as Lavinia Billingsley and someone called "Kitty" . The situation at the Swansea China Works according to Henry Morris' statement to Colonel Francis was marginally better with six premier artists identified including William Billingsley and Thomas Baxter, along with Henry Morris, William Pollard, David Evans and George Beddow, again with younger "hands" to carry out the more menial tasks such as transfer printing. In contrast, the author has found that the Derby China Works employed 85 named artists, including enamellers of the calibre of William Billingsley, Zacariah Boreman, Moses Webster, John Brewer, George Complin, William "Quaker" Pegg, George Robertson and Edward Withers and 30 gilders; some of the latter have escaped listing as the main gilders at Derby were accorded numbers to signify their work, of whom Thomas Soar was 1, William Billingsley was 7, and John Whitaker was 37 (Twitchett, *Derby Porcelain: An Illustrated Guide, 1748–1848*, 2002).

References

Sir J. Bernard Burke, *Burke's General Armory—The General Armory of England, Scotland, Ireland and Wales: Comprising a Registry of Armorial Bearings from the Earliest to the Present Time* (Harrison & Sons, Pall Mall, London, 1884)

G. Birkbeck-Hill, Hester Thrale: *Anecdotes of the Late Samuel Johnson LLD During the Last Twenty Years of His Life*, in *Johnson Miscellanies, Volume 1*, (Clarendon Press, Oxford, 1897), pp. 141–351

J.L. Clifford, *Hester Lynch Piozzi (Mrs Thrale)* (Columbia University Press, New York, 1987)

P. Colomban, H.G.M. Edwards, C. Fountain, Raman spectroscopic and SEM/EDAXS analyses of highly translucent *Nantgarw* soft-paste porcelain. J. Eur. Ceram. Soc. **40**, 4664–4675 (2020)

P. Cunningham, A bowl of punch, in *Household Words*, ed. by C. Dickens, Volume VIII, Magazine No. 168, 11th June 1853 (Publisher Bradbury & Evans, London, 1853), pp. 346–349

M. D'Ezio, *Hester Lynch Thrale Piozzi: A Taste for Eccentricity* (Cambridge Scholars Publishing, Newcastle-upon-Tyne, 2010)

S. Dawson Dobson, *Historical Anecdotes of Heraldry and Chivalry: Tending to Shew the Origin of Many English and Foreign Coats of Arms, Circumstances and Customs* (Holl and Brandish, Worcester, 1795)

C. Dickens, A. Christmas, *Carol. In Prose. Being a Ghost Story of Christmas* (Chapman & Hall, London, 1843)

L.W. Dillwyn, *British Conferva or Coloured Figures and Description of British Plants Referred by Botanists to the Genus Conferva* (W. Phillips, George Yard, London, 1809)

L.W. Dillwyn, *Notes of Recipes for Porcelain Bodies and Glazes 1815–1817* (Library of the Victoria & Albert Museum, South Kensington, London) Deposited by John Campbell Esq. in

1920. Reproduced in Eccles and Rackham, *Analysed Specimens of English Porcelain in the Victoria and Albert Museum Collection,* see reference below

H. Eccles, B. Rackham, *Analysed Specimens of English Porcelain in the Victoria & Albert Museum Collection* (South Kensington, Victoria & Albert Museum Publications, London, 1922)

H.G.M. Edwards, *Nantgarw Porcelain: The Pursuit of Perfection*, ed. by M. Denyer, (Penrose Antiques, Thornton, 2017a)

H.G.M. Edwards, *Swansea Porcelain: The Duck-Egg Translucent Vision of Lewis Dillwyn* (Penrose Antiques Ltd. Short Guides, Thornton, 2017b) ISBN:9780244325787

H.G.M. Edwards, *Porcelain to Silica Bricks: The Extreme Ceramics of William Weston Young, 1776–1847* (Springer-Nature, Dordrecht, 2019)

H.G.M. Edwards, *18th and 19th Century Porcelain Analysis: A Forensic Provenancing Assessment* (Springer-Nature, Dordrecht, 2020)

H.G.M. Edwards, *Porcelain Analysis and its Role in the Forensic Attribution of Ceramic Specimens* (Springer-Nature, Dordrecht, 2021)

W. Elis Jenkins, *Dillwyn's Etruscan Ware 1847–1850* (Crown Printers, Morriston, Swansea, 1971)

J. Fairbairn, *Crests of the Families of Great Britain and Ireland and Their Kindred in Other Lands,* vols. I and II (T.C. and E.C. Jack, Edinburgh and London, 1905 and New Orchard Publishing, London, 1986)

M.J. Franklin, *Piozzi, Hester Lynch (1741–1821), Oxford Dictionary of National Biography* (Oxford University Press, Oxford, 2004)

R. Fuchs, *Made in France* (Delaware, The Winterthur Museum, 2005), p. 135

Gartre'n Ol: Coming Home – An Exhibition of the Finest Nantgarw Porcelain, Nantgarw China Works Museum, Tyla Gwyn, Nantgarw, 2019, Nantgarw China Works Trust, Rhondda Cynon Taff, Glamorgan, 2019

G.A. Godden, *Chamberlain-Worcester Porcelain, 1788–1852* (Barrie & Jenkins, London, 1982)

J. Gray, The Cambrian pottery before 1802, in *Welsh Ceramics in Context, Part I*, ed. by J. Gray, (Swansea, Royal Institution of South Wales, 2003), pp. 19–38

J. Gray, The Cambrian pottery under Coles and Haynes, 1789–1802, in *Welsh Ceramics in Context, Part II*, ed. by J. Gray, (Royal Institution of South Wales, Swansea, 2005), pp. 45–70

H. Hallesy, The Glamorgan pottery, Swansea, and some comparisons with the Cambrian, in *Welsh Ceramics in Context, Part I*, ed. by J. Gray, (Royal Institution of South Wales, Swansea, 2003), pp. 99–119

K. Harvey, Barbarity in a teacup? Punch, domesticity and gender in the 18th century. J. Des. Hist. **21**, 205–221 (2008)

M. Holdaway, Early Swansea printed Earthenwares, part 1, in *Welsh Ceramics in Context, Part I*, ed. by J. Gray, (Royal Institution of South Wales, Swansea, 2003), pp. 52–74

M. Holdaway, Early Swansea printed Earthenwares, part 2, in *Welsh Ceramics in Context, Part II*, ed. by J. Gray, (Swansea, Royal Institution of South Wales, 2005), pp. 93–113

M. Keating, *The Hunt for White Gold* (Hodder & Stoughton, London, 2011)

K.S. Meager, *Swansea and Nantgarw Potteries. Catalogue of the Collection of Welsh Pottery and Porcelain on Exhibition at the Glynn Vivian Art Gallery, Swansea* (Swansea Corporation/Sergeant Bros. Ltd, Swansea/Abergavenny, 1949)

E. Morton Nance, *The Pottery and Porcelain of Swansea and Nantgarw* (Batsford, London, 1942)

W.J. Pountney, *Old Bristol Potteries* (J.W. Arrowsmith, Bristol, 1920), pp. 115–116

A. Renton, The Swansea diaspora: The later careers of David Evans, Henry Morris and William Pollard, in *Welsh Ceramics in Context, Part II*, ed. by J. Gray, (Swansea, Royal Institution of South Wales, 2005), pp. 209–234

A. Renton, Personal Communication to the Author (2021)

H.L. Thrale Piozzi, *Anecdotes of the Late Samuel Johnson LLD during the Last Twenty Years of his Life* (T. Cadell, The Strand, London, 1786)

W. Turner, *The Ceramics of Swansea and Nantgarw. A History of the Factories with Biographical Notices of the Artists and Others, Notes on the Merits of the Porcelain, the Marks Thereon Etc* (Bemrose & Sons Ltd, The Old Bailey, London, 1897)

J. Twitchett, *Derby Porcelain: An Illustrated Guide, 1748–1848* (Antique Collectors Club, Woodbridge, Suffolk, 2002)

C.A. von Volborth, *Heraldry of the World, Italy* (Batsford Books, London, 1973), pp. 126–133 and 213–216

E. Wilson, *Against Happiness: In Praise of Melancholy* (Farrar, Straus & Giroux//Macmillan / Holtzbrinck Publishing Group, New York/London/Stuttgart, 2008), p. 110

D. Wondrich, *Punch: The Delights and Dangers of the Flowing Bowl – An Anecdotal History of the Original Monarch of Mixed Drinks with More than Forty Historic Recipes, Fully Annotated and a Complete Course in the Lost Art of Compounding Punch* (Tarcher Perigree/Penguin Putnam Inc, New York/London, 2010)

W.W. Young, *The Diaries of William Weston Young, 1776–1847 (1802–1843)*, 30 volumes (West Glamorgan Archive Service, Swansea, SA1 3SN). https://arcgiveshub.jisc.sc.uk/data/gb216-d/dxch/ddxch/i/hub

Chapter 4
Nantgarw Armorial Porcelain

A little knowledge is a dangerous thing.
Alexander Pope (1688–1744), in An Essay on Criticism, 1709.
A little knowledge is a dangerous thing. So is a lot
Albert Einstein (1879–1955)

Nobel Laureate in Physics, 1921.

Abstract Detailed in-depth discussion of the assignment and attribution of all known Nantgarw armorial porcelain exemplars in the literature, including the "Garter Star" motif, which it is concluded is not a true armorial but rather a signature device adopted by Thomas Pardoe to designate his decoration on earthenwares and porcelain firstly at the Cambrian Pottery then at the Nantgarw China Works and also in between at his decorating studio in Bristol. All 7 Nantgarw armorial bearings are deemed to have been identified.

Keywords Nantgarw armorials · Full coat-of-arms · Escutcheon · Crests · "Garter star" motif · Thomas Pardoe

As far as can be ascertained from a detailed survey of the literature and historical documentation relating to the Nantgarw China Works there are only seven distinct armorials in existence that are portrayed for Nantgarw porcelain and one other in addition which is perhaps not strictly acceptable for this category, the Garter Star motif, as it is probably representative of a stylised motif based upon an order of heraldry rather than a genuine example of an heraldic achievement and coat-of-arms, hence, it can be classified as a pseudo-armorial. These Nantgarw China Works armorial exemplars will now be considered individually.

H. G. M. Edwards, *Welsh Armorial Porcelain*,
https://doi.org/10.1007/978-3-030-97439-8_4

4.1 The *Viscount Weymouth* Service

Dr. W.D. John, G.J. Coombes and Katherine Coombes in their book (*Nantgarw Porcelain Album*, 1975, Plate 56) illustrate a very fine armorial Nantgarw plate (diameter 9″), impressed NANT-GARW C.W., *ca.* 1817–1820, depicting the full coat-of-arms (illustrated earlier in Fig. 2.1) which they assigned to Viscount Weymouth, Lord Henry Frederick Thynne (1797–1837), who became the third Marquess of Bath in 1837. His father, Thomas Thynne, was created the second Marquess of Bath in 1796. Lord Henry was the second son but his elder brother, also Thomas Thynne, and Viscount Weymouth predeceased him in January 1837, so Henry inherited the title, firstly of Viscount Weymouth in January 1837 and then became the third Marquess of Bath in March 1837, just 4 months before he died. The escutcheon is described in the armories as quarterly, first and fourth, a barry of ten or and sable; second and third, or a lion rampant tail nowed gules and erected. Crest: on a wreath of gules and or, a reindeer statant or collared sable. Supporters: dexter, a reindeer or gorged with a plain collar sable; sinister, a lion tail nowed and erect, gules. Motto: "*J'ay bonne cause*" which translates as "*I have good reason*". Family seat, Longleat House, Wiltshire.

There is extensive gilding applied to this armorial plate with its 12-lobed moulded border, which with its very white glaze clearly belongs to the Billingsley /Walker period of production at the Nantgarw China Works and would probably have been commissioned in London through John Mortlock's agency in 47, Oxford Street, and decorated at either Robins & Randall's or John Sims' ateliers. This is apparently an unequivocal attribution of a coat-of-arms, yet all is not straightforward on a closer examination and there are several very curious heraldic anomalies being portrayed here:

- Firstly, the escutcheon is not quartered as expected for the Bath/Weymouth arms but comprises the ten bars component indicated solely on the shield and these are coloured white and green (argent and vert) rather than black and gold (sable and or) as were specified in the heraldic armory literature. The description given in the armories is confirmed in the monochrome bookplates of Viscount Thynne shown in Figs. 4.1a and 4.1b, where the "tricking" is clearly indicative of the barry being yellow (dotted background) and sable (hatched background) rather than argent (clear) and vert (diagonal lines, top left to bottom right) as portrayed on the Nantgarw plate in Fig. 2.1. Burke's Armory (Burke, *Burke's General Armory—The General Armory of England, Scotland, Ireland and Wales: Comprising a Registry of Armorial Bearings from the Earliest to the Present Time*, 1884) does not differentiate between the arms borne between Viscount Weymouth and the Marquess of Bath, who appears under the family name of Thynne, and this is seen in the two versions of the Viscount Weymouth bookplate shown in Figs. 4.1a and 4.1b. This raises the question as to whether there is some special significance or message in this amendment of the barry heraldic tinctures – green signifies hope and loyalty in love, white signifies sincerity and peace, whereas yellow signifies generosity and black, knowledge and constancy?

Fig. 4.1a Bookplate of Thomas Thynne, Viscount Weymouth, showing his arms comprising a ten-fold barry of or and sable, yellow and black, which are represented by dots and cross-hatching, respectively, in what is termed heraldic "tricking"

Fig. 4.1b Bookplate of Thomas Thynne, Viscount Weymouth, showing his fuller escutcheon comprising the quartered arms of the Marquess of Bath

- Secondly, the shape of the escutcheon itself is not the standard heater shield as expected but is a distorted and asymmetric ellipse which is contained within a gadrooned baroque shield type frame that is referred to as "*baroquc cardidoid*". The escutcheons shown in the two bookplates in Figs. 4.1a and 4.1b are known as "*square-eared, indented top and lobed base*" in type.

- The construction of this coat-of-arms is also rather strange in several respects; firstly, the helm is shown facing dexter and not frontal as depicted on the Viscount Thynne arms from his bookplate and it also exhibits a full armoured breastplate and shoulder guards. The coronet upon which it sits has a chapeau and the dexter reindeer supporter is facing away from the arms and not placed in a supporting role. Likewise, the lion in sinister support is looking frontally and not dexter as it should be. The reindeer also does not have the correct coloured collar (sable) as specified in the armory.
- The number of inconsistencies in this portrayal of these coat-of-arms require an explanation which is not immediately forthcoming and would surely give cause for concern to the noble client as displaying his incorrect arms unless a hidden message is contained therein.
- A more serious genealogical anomaly arises in the assertion of Dr. John et al. that this plate is attributed to Lord Henry Frederick Thynne, because as the second son of the Marquess of Bath he was not entitled to call himself Viscount Weymouth until after his elder brother died childless in 1836; this was, of course, the title assumed by his elder brother Thomas Thynne in 1796 when his father succeeded to the Marquessate of Bath. Hence, the Nantgarw *Viscount Weymouth* service must surely be more correctly attributed further to Thomas Thynne for the period of its production between 1817–1820 and not to Henry Frederick Thynne. It is to be noted that Lord Henry Frederick Thynne would have attained the age of 21 in 1818 and this would quite possibly have merited the purpose of the original commissioning of this service in Nantgarw porcelain had he been truly Viscount Weymouth at that time, which he was not! This service does not seem to be referred to elsewhere in the literature and it is not clear if other examples are known to exist at this time; it is quite possible therefore that this plate could perhaps have been an experimental specimen or perhaps a cabinet sample for a service that was never fully commissioned. It is revealing to explore further the genealogy of the Thynne family at Longleat House as this uncovers some interesting information which should be considered in the correct attribution of this armorial plate and especially regarding the use of the incorrect heraldic tinctures in the barry of the escutcheon .

4.1.1 Marquess of Bath /Viscount Weymouth Genealogy

The Marquessate of Bath was created on the 18th August 1789 and was held first by Thomas Thynne, the third Viscount Weymouth. The first Viscount Weymouth was Thomas Thynne (1640–1714) created in 1682, followed by his son, also Thomas Thynne (1710–1751) as the second Viscount, and then by his son, Thomas Thynne (1734–1796), Baron Thynne of Warminster, as the third Viscount, who then became the first Marquess of Bath (Mearain, 1968). He was succeeded by his son, Thomas Thynne (1765–1837), as the second Marquess of Bath in 1796, having previously succeeded his father as the fourth Viscount Weymouth until this time. Thomas

Thynne was hence the second Marquess of Bath during the Nantgarw China Works production period from 1817–1820. The second Marquess of Bath had 11 children by his wife, Isabella, daughter of Viscount Byng, who died in 1830. The eldest son, also Thomas Thynne (1796–1837), held the title of fifth Viscount Weymouth during the second decade of the nineteenth Century and would therefore be the name that would naturally be associated with the armorial *Viscount Weymouth* Nantgarw service. He attended Eton College and St Johns College, Cambridge between 1814 and 1816, graduating with an Honorary MA, and then became a diplomat under the Earl of Clancarty, who was "*not impressed with his ability*". Then in 1818, he represented Weobley in Parliament as Tory MP, a rotten borough, which his father had previously represented before him (Lord Henry Cavendish-Bentinck was the other MP for Weobley along with Thomas Thynne) – but in 1820 he did not stand in the General Election that arose because of the death of King George III. Thomas, Viscount Weymouth, clashed with his father over his friends and heavy debts incurred in his lifestyle. On the 11th May 1820, Thomas, the fifth Viscount Weymouth, eloped with Harriet Matilda Robbins, a tollgate keeper's daughter, and they married in Marylebone; afterwards they resided in Paris and Florence, having been cut off by his father, the second Marquess of Bath, who disapproved of his son's lifestyle and socially unacceptable marriage and who attempted legally to disinherit him from succession to the title, but failing to do so in the courts under the law of primogeniture. Thomas returned to England in the late 1820s and reappeared on the social scene once more, as noted in columns such as *Fashionable Movements* and *Fashionable Intelligence* in society newspapers, which is indicative of his acceptance on the social circuit, although the Viscountess is rarely mentioned except that she travels from Paris occasionally. Thomas Thynne is recorded as selling objects of art and china that he had collected to raise money. His father meanwhile maintained that his only recourse was to outlive his son to prevent his inheritance. There was no reconciliation with his father in the years that followed but his mother did visit Thomas and Harriet in Paris before she died in 1830. Thomas died as tenant of Shanks House, Dorset, on the 16th January 1837, aged 40. Some 5 weeks later his father Thomas, the Marquess of Bath, died then the younger brother Henry Frederick, now Viscount Weymouth, became the Marquess of Bath until he too died on the 24th June 1837. Henry Frederick served in the Royal Navy, reaching the rank of Captain, and was also MP for Weobley between 1824–1826 and 1828–1832; Henry married a Harriet Baring on the 10th April 1830. John, Viscount Lonsdale of Lowther, a good friend of Thomas Thynne, Viscount Weymouth, died in 1818 and Viscount Weymouth was an executor, who was bequeathed the sum of £100 "*....to be bestowed in a piece of plate in remembrance of me*". A piece of plate could relate to silver but maybe this gift enabled Viscount Weymouth to commission the Nantgarw armorial porcelain service— his relationship with his father being so strained at this time it is difficult to see otherwise how an impecunious Thomas would have had the wherewithal to commission a Nantgarw service? Henry Frederick's son, John Alexander Thynne (1831–1896), then became the 4th Marquis of Bath and the 7th Viscount Weymouth in 1837 at the tender age of six. Thomas Thynne's widow, Harriet (1800–1873), remarried an Italian Count Inghirami but never had any children.

The coat-of-arms of the Marquess of Bath, which are shared with the Viscount Weymouth, are: Quarterly – 1st and 4th, a barry of ten or and sable (Botteville), 2nd and 3rd argent a lion rampant with tail nowed and erect gules (Thynne). Crest: a reindeer statant or. Supporters: Dexter – a reindeer or gorged with a plain collar sable; Sinister- a lion with tail nowed and erect gules. Motto: *"Jay bonne cause"*, meaning *"I have good reason"*. A portrayal of these arms is given in the bookplates of the Viscount Weymouth in Figs. 4.1a and 4.1b, from which it can be seen there is a match with the armorial records but certainly not with the armorial plate as depicted in Fig. 2.1!

It is apparent that the attribution of the Nantgarw armorial plate described here is incorrect as Lord Henry Frederick Thynne did not hold the title of Viscount Weymouth in the second decade of the nineteenth Century; this belonged to his elder brother, Thomas, who was living overseas and in disgrace for some years from 1820. The Viscount Weymouth service must, therefore, relate to Thomas Thynne. It is also realistic to postulate that it could have been ordered by the second Marquess of Bath, Thomas Thynne, possibly to celebrate the coming of age of his second son, Lord Henry Frederick, in 1818 and perhaps the anomalies in the coat-of-arms may reflect this. An alternative explanation could be that it was ordered by the elder brother Thomas Thynne, Viscount Weymouth, before he eloped with Harriet in 1820. Thereafter, the family differences and enforced separation were maintained but Isabella, Marchioness of Bath, did try to keep in touch with Thomas and Harriet by visiting them in Paris in the 1820s. What is still unexplained is the clear misplacement of the heraldic colours or tinctures in the escutcheon --- instead of a barry of ten or and sable (gold and black) we actually have a barry of ten white and green (argent and vert).

What is clear is that this apparently indisputable assignment historically still retains some unanswered questions – but at least it can be amended to exclude Lord Henry Frederick Thynne, who was not the Viscount Weymouth that had been previously assigned to this service; even if we accept the premiss that it could have been manufactured to celebrate his coming of age in 1818, it should not be referred to as the Viscount Weymouth service but preferably as the Lord Henry Frederick Thynne service! Alternatively, the assignment to Thomas Thynne, who was Viscount Weymouth at that time before his disgrace does then carry with it the complexity of the incorrect heraldic tinctures! In this case it must now be called the Thomas Thynne, Viscount Weymouth, service. As a final speculative thought, the service could have been commissioned by him in celebration of his impending marriage to his socially unacceptable Harriet and declaring his sincerity, hope and loyalty in love with the unusual heraldic tinctures, whilst symbolically adopting the odd stance of the reindeer supporter in his arms as his family turning against him?

In a link with the Nantgarw armorial service of Constantine Phipps, Viscount Normanby, which will be described later, it appears that Viscount Normanby and Viscount Weymouth are recorded in the Cambridge Reporter on October 22nd1814 as matriculating together into Cambridge University, so they must have known each other: Phipps into Trinity College and Weymouth into St John's College. After his graduation from Cambridge, Phipps was a frequent visitor to Florence in the 1820s and it is very likely that the two gentlemen preserved their acquaintance during this period. It is remarkable that both had Nantgarw porcelain armorial acquisitions, so rare are these.

4.2 The Wyndham Service

In Fig. 4.2 is shown a Nantgarw porcelain plate, *ca.* 1817–1822, with a coat-of-arms centrally located and a plain gilded rim which is thought to be possibly indicative of a locally decorated service; it has been alleged historically that much London-decorated Nantgarw porcelain was accompanied by dentil-edged gilding at the rim – but see Appendix III for a quantitative analytical evaluation of this statement. We have just seen above that the *Viscount Weymouth* plate (Fig. 2.1) also does not possess dentil-edge gilding and this has always been accredited as a London decorated piece, so dentil-edge gilding cannot be considered as necessarily a *de rigueur* requirement for London decoration! The plate shown in Fig. 4.2 is illustrated in Fergus Gambon's book (*Porslen Abertawe a Nantgarw*, 63/N25, 2016) and is currently in the Andrews Collection at Plas Glyn-y-Weddw, Llanbedrog, Abersoch, Gwynedd. The escutcheon shows three lions, one mullet and a chevron or on an azure ground and is a rococo shape shield . The crest is a lion couped enclosed in a chain circle or, the torse is blue and gold and the motto is "*Au Bon Droit*" which translates as "*With Good Right*". The coat-of-arms matches that exactly given in *Burke's Armory* and in Fox-Davies (*Armorial Families: A Complete Peerage, Baronetage and Knightage, and a Directory of Some Gentlemen Coat -Armour and being the First Attempt to Show Which Arms are in Use at the Moment are Borne by Some Legal Authority*, 1895) as: azure a chevron d'or, three lion's heads erased with gilt fetterlock chain. Motto "*Au bon droit*". Supporters: dexter a lion azure, ungorged or, sinister a griffin argent. Crest: a lion's head erased within a fetterlock and chain d'or for Wyndham (Fox-Davies, *Armorial Families: A Complete Peerage, Baronetage and Knightage, and a Directory of Some Gentlemen Coat -Armour and being the First Attempt to Show Which Arms are in Use at the Moment are Borne by Some Legal Authority*, 1895: 30–8). For Quin, a crest: a wolf's head erased (Fox-Davies, Plate 19.4). Motto: "*Quae sursum volo videre*" translates as "*I wish to see all the things which are above*". Henry Quin, later Earl of Dunraven and Mount Earl, was granted by royal licence the additional surname of Wyndham in 1815. With this grant he assumed the arms of Wyndham of Orchard Wyndham. He lived in Dunraven Castle, Southerndown, Glamorgan. Thomas Wyndham, MP for Glamorgan 1789–1814, died in 1814 and then Windham Henry Quin (1782–1850) became Henry Windham Wyndham- Quin in April 1815 and Viscount Mount Earl of Adare in Co. Limerick, in February 1816. He was created second Earl of Dunraven in 1822. A locally decorated service such as this would indicate a recognition of this advancement in the peerage at Nantgarw perhaps during the tenure of William Weston Young and Thomas Pardoe or possibly this could have been commissioned earlier either locally during the Billingsley/Walker period for local decoration or it

Fig. 4.2 Dessert plate from the Nantgarw China Works, *ca.* 1817–1820, impressed NANT-GARW C.W., bearing the escutcheon, crest, torse and motto of Wyndham; the supporters, helm and mantling are absent. The motto "*Au Bon Droit*" translates as "*With Good Right*". It is believed that this was a locally decorated Nantgarw armorial service. Henry Windham Wyndham-Quin (1782–1850) became Viscount Mount Earl of Adare in Co. Limerick, in February 1816 and became the Earl of Dunraven in 1822. A locally decorated service would indicate a recognition of this advancement at Nantgarw during the tenure of William Weston Young and Thomas Pardoe. Family seat, Dunraven Castle, Glamorgan, and Uffords Manor, Norfolk. *Reproduced with permission from Gwyn Jones, Director of the John Andrews Charitable Trust Collection, Oriel Plas Glyn-y-Weddw, Llanbedrog, Abersoch, Gwynedd*

may have been decorated in London. Family seats, Dunraven Castle, Glamorgan, and Uffords Manor, Norfolk. It would be possible analytically and non-destructively to determine if the glaze is of the type *Nantgarw No.1* (Billingsley/Walker period) or *Nantgarw No.2* (Young/Pardoe period) as the spectral characteristics of the two are quite distinctive as shown in earlier studies carried out by the author (Edwards, *Porcelain Analysis and Its Role in the Forensic Attribution of Ceramic Specimens,* 2021). This could confirm analytically the sourcing of the service locally or in London.

4.3 The Ramsay Services (2)

There are two distinct versions of the *Ramsay* armorial service known in Nantgarw porcelain, both bearing the same crest and motto and both with artefact(s) carrying the impressed NANT-GARW C.W. mark but involving a different supporting decoration to the armorial crest and motto which actually places them in different Nantgarw China Works periods of decoration. The commissioning family is that of Ramsay of Straloch House, Barra, near Aberdeen; motto "*Migro et Respicio*", meaning "*I come forth and look back*". Crest: an eagle rising regardant sable, armed and membered or, langued gules (Fairbairn, *Crests of the Families of Great Britain and Ireland and Their Kindred in Other Lands,* Volumes I and II, 1986, Plate 77/4; Fox-Davies, *Armorial Families:A Complete Peerage, Baronetage and Knightage, and a Directory of Some Gentlemen Coat -Armour and being the First Attempt to Show Which Arms are in Use at the Moment are Borne by Some Legal Authority*, 1895) . John Ramsay of Barra and Straloch, Inverurie, Aberdeenshire, b.1786, married Susan Innes of Pitmedden in 1831 and died in 1832, so he would have been aged in his early to mid-thirties when the Nantgarw services were ordered and the chronology is not correct for these being commissioned on the occasion of his marriage. Susan Ramsay later married Captain Nares RN in 1844 and died in 1887. The first *Ramsay* service (Fig. 4.3) is illustrated in John (*Nantgarw Porcelain*, 1948, page 115, Plate 34B) and also, in the *Gartre'n Ol Exhibition Catalogue* (2019, page 34) and shows the crest and motto centrally located on the plate and the verge gilded quite prolifically with a pattern of stylised acanthus leaves. It is dated to the Billingsley/Walker period of porcelain production and probably, therefore, it is London decorated. The second version comprises the same crest and motto but with an attractive polychrome floral decoration in the reserve and accompanied by minimal gilding, which is readily attributed to Thomas Pardoe, therefore confirming that it must have been locally decorated at Nantgarw between 1820 and 1822 and is clearly a later service commission (Fig. 4.4). Earl Dalhousie is the current chief of Clan Ramsay with his seat at Brechin Castle; it is interesting that the Ramsay of Straloch and Barra crest is not the same as that of Clan Ramsay, whose motto is "*Ora et Labora*", meaning "*Work and Prayer*". There are no fewer than 41 Ramsays and Ramseys cited in Fairbairn (*Crests of the Families of Great Britain and Ireland and Their Kindred in Other Lands,* Volumes I and II, 1986), but Ramsay of Straloch and Barra is unique in having this eagle regardant crest and motto.

4.4 The Baron Phipps of Normanby Service

A Nantgarw armorial dinner plate with C-scroll moulding, completely undecorated and carrying the impressed NANT-GARW C.W. mark, (Edwards, *Swansea and Nantgarw Porcelain; A Scientific Reappraisal*, 2018; Edwards, *Porcelain Analysis and Its Role in the Forensic Attribution of Ceramic Specimens,* 2021) is illustrated

Fig. 4.3 Nantgarw China Works porcelain plate, impressed NANT-GARW C.W. mark, from the *Ramsay* service, *ca.* 1817–1820, with the crest of Ramsay of Straloch House, Barra, near Aberdeen; motto "*Migro et Respicio*", means "*I come forth and look back*". The crest is of an eagle with wings extended, armed, facing sinister with beak open. This is one of two variants of a Ramsay Nantgarw service with a gilt border. From the *Gartre'n Ol Exhibition Catalogue* (2019, page 34) of the finest Nantgarw porcelains celebrating the 200th Anniversary of porcelain manufacture at the Nantgarw China Works. *Reproduced courtesy of Charles Fountain, Director of the Nantgarw China Works Museum Trust, Tyla Gwyn, Nantgarw, Glamorgan*

in Fig. 4.5. The crest is located at the verge in gilt and comprises a demi-lion rampant facing dexter holding in both paws a palm branch proper (Fairbairn, *Crests of the Families of Great Britain and Ireland and Their Kindred in Other Lands*, Volumes I and II, 1986; Fox-Davies, *Armorial Families:A Complete Peerage, Baronetage and Knightage, and a Directory of Some Gentlemen Coat -Armour and being the First Attempt to Show Which Arms are in Use at the Moment are Borne by Some Legal Authority*, 1895, p. 445: crest illustrated 12–7). William Phipps, heir to the barony of Phipps, which was created on 22 July 1656, was born in 1698 in London and married Lady Catherine Annesley. Their son Constantine, born in Lythe, North Yorkshire in 1722, married Lepell Hervey and had children called Constantine, Charles, Henry, Henrietta, Augustus and Edmund. General Henry Phipps (1755–1831) became the first Earl of Mulgrave and Viscount Normanby.

Fig. 4.4 Second variant of the Ramsay of Straloch House, Barra, Nantgarw service shown in Fig. 4.3. The crest is still placed centrally with the motto, but the reserve now contains an attractive floral decoration enamelled by Thomas Pardoe locally at the Nantgarw China Works between 1821 and 1823. *Reproduced by courtesy of Bonhams Auctioneers and Valuers, London*

Constantine Phipps (1797–1863) became the first Marquess of Normanby and Earl of Mulgrave (Fox-Davies, *Armorial Families:A Complete Peerage, Baronetage and Knightage, and a Directory of Some Gentlemen Coat -Armour and being the First Attempt to Show Which Arms are in Use at the Moment are Borne by Some Legal Authority*, 1895, pp.403 and 414) and he would have attained the age of 21 in 1818 — possibly an occasion to celebrate with the commissioning of this Nantgarw service? Coat of arms: quartered first and fourth, a trefoil slipped between eight mullets argent, second a bordure company argent and azure, third grant from King James II to his natural daughter Lady Catharine Darnley (who was the Duchess of Buckingham) being the royal arms of England (King James II) with a paling of 6 argent and azure a bend gules for Annesley. Supporters: dexter, a unicorn ermine; sinister, a goat ermine armed unguled or gorged with a chaplet of roses. Motto "*Virtute quies*" which means "*Rest through Valour*". Family seat: Mulgrave Castle, Lythe, Whitby, North Yorkshire. The crest illustrated in Fairbairn (Fairbairn, *Crests of the Families of Great Britain and Ireland and Their Kindred in Other Lands,*

Fig. 4.5 Nantgarw dinner plate, ***Baron Phipps***, Viscount Normanby, service, *ca.* 1817–1819, impressed mark NANT-GARW C.W., demonstrating the beauty of Nantgarw porcelain translucency at its very best with a simple gilt armorial crest decoration at the verge between moulded scrolls of foliage, flowers and stars and otherwise undecorated. Only two Nantgarw services are known of this type displaying just a crest and being otherwise completely undecorated with no enamelling, the other exemplar being the plain, unmoulded *Homfray* tea and coffee service of Penllyne Castle (Crest: an otter, pierced by an arrow in gold), as shown here in Fig. 4.7. Henry Phipps was Lord Chief Justice of Ireland, created Baron Mulgrave of York in 1794 and Viscount Normanby, Earl of Mulgrave in 1812. Marked impressed NANT-GARW C.W. His crest comprises a demi-lion rampant or holding in both paws a palm branch vert. *Private Collection*

Volume II, 1986, Plate 224/4) is reproduced here in Fig. 4.6: Fairbairn notes that the coronet worn by the rampant demi-lion shown in his depiction of this crest is absent for Phipps.

4.5 The Homfray Service

John Homfray (1793–1877) of Penllyne Castle, Cowbridge, Vale of Glamorgan, was head of the Penydarren Ironworks Dynasty. Motto, "*Vulneratur non vincitur*", translating as "*He is wounded but not defeated*"; an alternative motto is also given

Fig. 4.6 Crest of Phipps illustrated in Fairbairn, Volume II, Plate 224/4, a demi-lion rampant holding in both paws a palm branch (with coronet missing). Reproduced from Fairbairn, *Crests of the Families of Great Britain and Ireland and Their Kindred in Other Lands,* Volumes I and II, T.C. and E.C. Jack, Edinburgh and London, 1905 and New Orchard Publishing, London, 1986

in the armories, "*L'homme vrai aime son pays*" meaning "*The true man loves his country*". The crest is a speared otter, wounded in the shoulder (heraldically termed, imbrued). Dr. John (*Nantgarw Porcelain*, 1948) ascribes a Nantgarw tea and coffee service with this crest as being decorated locally by Billingsley—in gilt – and some items from this service are now in the National Museum of Wales Amgueddfa Cymru, in Cardiff. A tea bowl from this service is at the Nantgarw China Works Museum at Tyla Gwyn, Nantgarw and is shown in Fig. 4.7, with an enlargement of the crest shown in Fig. 4.8. The only other definitive and locally crested items, apart from the second *Ramsay* service painted by Pardoe and described above, are six Masonic emblemed beakers for which an invoice exists in the National Museum of Wales archive in William Billingsley's handwriting, specifying the delivery of six tapered side beakers with Masonic motifs to a Mr. Thomas Johnes of Cardiff in December 1819, which have been mentioned earlier and for which one is illustrated in Figs. 2.6 and 2.7. It is thought that this may well have been William Billingsley's last authenticated commission at Nantgarw before he and Samuel Walker left for the Coalport China Works on or before April 1820 (Morton Nance, *The Pottery and Porcelain of Swansea and Nantgarw,* 1942).

4.6 A Thomas Pardoe Decorated Armorial

A 12-lobed locally decorated Nantgarw plate dating from the period 1820–1822, during the tenure of William Weston Young as proprietor of the Nantgarw China Works, was offered for sale at Bonham's auction house in London recently,

Fig. 4.7 Nantgarw porcelain tea bowl from a tea and coffee service commissioned by John Homfray (1793–1877) of Penllyne Castle, Cowbridge, Vale of Glamorgan, head of the Penydarren Ironworks Dynasty. Motto, "*Vulneratur non vincitur*", translating as "*He is wounded but not defeated*". The crest is a speared otter. Dr. John (*Nantgarw Porcelain*, 1948) ascribes a Nantgarw porcelain tea and coffee service with this crest as being decorated in gilt locally at Nantgarw by William Billingsley. *Reproduced courtesy of Charles Fountain, Director of the Nantgarw China Works Museum Trust, Tyla Gwyn, Nantgarw, Glamorgan*

Fig. 4.8 Enlargement of Homfray crest on the Nantgarw China Works porcelain tea bowl shown in Fig. 4.7. *Reproduced courtesy of Charles Fountain, Director of the Nantgarw China Works Museum Trust, Tyla Gwyn, Nantgarw, Glamorgan*

containing a rather simple coat-of-arms in gilt at the verge comprising an escutcheon with a gold chevron and three gilt inverted vees (^) surmounted by a bow and ribbons and standing upon four fern fronds, two ending in gilt tapered leaves. Neither a motto nor a crest is depicted with this escutcheon. The solitary roses and

Fig. 4.9 Locally decorated Nantgarw porcelain plate, *ca.* 1821–1823, with floral groups and a gilt escutcheon in the reserve, a chevron or with three inverted vees, which as yet has not been unequivocally unidentified but there is strong evidence for its assignment to Nicholas Jeffreys of Brecon Priory. Enamelled by Thomas Pardoe during the tenure of William Weston Young at the Nantgarw China Works, *ca.* 1821–1822. *Reproduced by courtesy of Bonhams Auctioneers and Valuers, London*

birds painted at the verge are unmistakably the work of Thomas Pardoe as evidenced by similar enamelling illustrated in John et al. (*Nantgarw Porcelain*, 1948, Plate 54B) and John et al., (*Nantgarw Porcelain Album*, 1975, Plate 67), where the latter is signed in gold on the reverse, "*Pardoe, Cardiff, 1823*", and which therefore must have been accomplished literally just a few months before he died in July of that year and seemingly after William Weston Young had already left the Nantgarw China Works late in 1822. Bonham's described this armorial as "*....somewhat of an innovation for Pardoe*" and class it as "*very rare*", which it most certainly is. This escutcheon so far has remained unidentified. Illustrated here in Fig. 4.9, *Burke's Armory* (Burke, *Burke's General Armory—The General Armory of England, Scotland, Ireland and Wales: Comprising a Registry of Armorial Bearings from the Earliest to the Present Time*, 1884) states that a gold chevron between three spearheads (pheons) appear on the coat-of-arms of both John Bruce-Pryce of Duffryn, Aberdare, father of Lord Aberdare, and also on that of Nicholas Jeffreys of The Priory, Brecon. The latter is especially interesting as it is also incorporated through marriage into the arms of the first Viscount Camden, granted in 1812, see later.

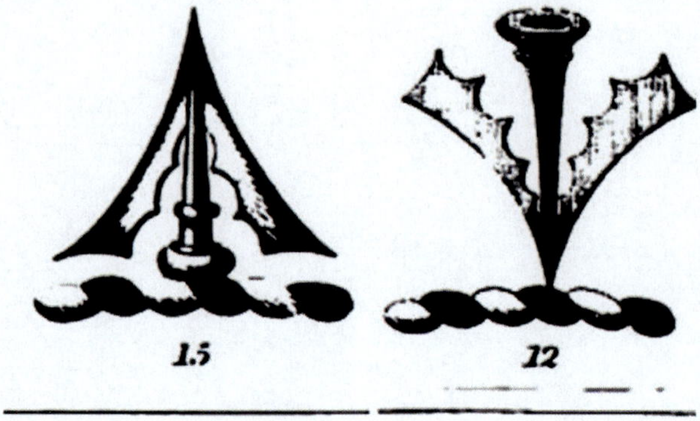

Fig. 4.10 Representation of pheons, spearheads, in heraldic crests: left, Plate 234/15, vertical pheon with serrated inner edge on torse; right, Plate 294/12, vertical, inverted pheon with serrated inner edge on torse. Reproduced from Fairbairn, *Crests of the Families of Great Britain and Ireland and Their Kindred in Other Lands,* Volumes I and II, T.C. and E.C. Jack, Edinburgh and London, 1905 and New Orchard Publishing, London, 1986

The inverted spearheads, simplistically drawn here in the escutcheon shown in Fig. 4.9 and enlarged later in Fig. 7.5 actually represent the pheon, an heraldic device which may be used vertically or inverted as shown in Fig. 4.10: here the broad spearheads have a serrated edge on the inner blades, a detail which is absent from the basically drawn inverted vees shown in Fig. 4.9.

4.7 Armorial Order of the Garter Star—Or Is It?

Several examples are known of armorial Welsh porcelain and pottery items which display a motif that is seemingly based on the Star of the *Order of the Garter*, technically *The Most Noble Order of the Garter*, which has the well-known motto "*Honi Soit qui Mal y pense*" and translates as "*Evil be to he who evil thinks*" or alternatively "*Shame on him who thinks this evil*". It is the world's oldest order of chivalry and is the highest British Order of Chivalry, being instituted by King Edward III in 1348. Today, the Order of the Garter comprises 26 Knights (K.G.) of whom the Queen as reigning monarch has place number 1 and the Prince of Wales as heir to the throne, place number 2. The Garter Knights have stalls in St George's Chapel, Windsor. The badge of the Order of the Garter is the Cross of St George, red on a white background surrounded by a blue garter circlet bearing the motto "*Honi Soit qui Mal y pense*" and superimposed on a silver eight-pointed star. The sash worn with the Order carries a medallion or badge bearing an image of St George on horseback slaying a dragon. The John Andrews Collection at Oriel Plas Glyn-y-Weddw, Llanbedrog, Gwynedd, features a locally decorated 12-lobed Nantgarw porcelain

Fig. 4.11 Nantgarw dessert plate, *ca.* 1821–1823, locally decorated with floral groups by Thomas Pardoe and containing a centrally gilded eight-pointed "Star of the Order of the Garter with the Cross of St George" but minus the Garter emblem and motto *"Honi soit qui mal y pense"*, which translates as *"Evil be to he who evil thinks"*, or alternatively *"Shame on him who thinks this evil"*. It is conjectural whether or not this really represents an armorial bearing for a Knight of the Garter when missing its important Garter emblem or is simply a stylistic motif identifying Pardoe's work, but current thinking is that it is best described as the latter. *Reproduced with permission from Gwyn Jones, Director of the John Andrews Charitable Trust Collection, Oriel Plas Glyn-y-Weddw, Llanbedrog, Abersoch, Gwynedd*

dessert plate with six floral groups characteristic of the painting of Thomas Pardoe and bearing the motif of a gilt eight-pointed Garter Star and Cross of St George centrally (Fig. 4.11) but this is minus the garter and motto encircling the cross as is seen clearly in the enlargement shown in Fig. 4.12. Fergus Gambon (*Porslen Abertawe a Nantgarw*, 2016, number N43 on page 72) attributes this decoration to Pardoe's "signature motif" and as a feature of his work which also appears on earlier examples of Pardoe's decoration on earthenware at Swansea (see earlier in this text, Chap. 3). It is surely not acceptable that an enameller would be permitted to use the badge of an ancient order of chivalry to which he was not entitled as a characteristic personal signature feature and a possible alternative explanation is that a newly invested Order of the Garter incumbent may have commissioned such an heraldic

Fig. 4.12 Enlargement of the eight-pointed "Garter Star" motif attributed to Thomas Pardoe on the Nantgarw dessert plate shown in Fig. 4.10. *Reproduced with permission from Gwyn Jones, Director of the John Andrews Charitable Trust Collection, Oriel Plas Glyn-y-Weddw, Llanbedrog, Abersoch, Gwynedd*

device in recognition of his elevation into this order of chivalry? However, it must be appreciated that this is not an exact portrayal of the Garter Star in that it is missing the garter and motto legend so perhaps it is really better described as a signature motif after all. We can define chronologically the narrow period in which Thomas Pardoe actually worked at Nantgarw with William Weston Young decorating porcelain which was left in stock following the departure of William Billingsley and Samuel Walker for Coalport to work with John Rose in mid-1820, probably around April of that year or perhaps just before. William Weston Young took over the tenure of the Nantgarw China Works later in 1820 and initiated the first and second sales of remnant porcelain in 1821 and 1822. In the meantime, he persuaded Thomas Pardoe to transfer his decorating business from Bristol to Nantgarw to enamel the remaining porcelain there. Pardoe appeared in Nantgarw in January 1821 and remained there until he died in July 1823, but since Young departed after the second sale later in 1822, then not much porcelain would have been decorated in the early part of 1823 before Pardoe died. Research into the archives of the Knights of the Order of the Garter who were alive between 1815 and 1822 reveals that approximately 55 nobles would have been eligible to use this badge of chivalry during that period. If we accept that this badge does rather loosely represent the portrayal of the Garter Star and the commissioning of a porcelain service would perhaps be in recognition or celebration of an investiture into the Order then we can chronologically narrow down the possible candidates who may have initiated this commission to just a few names, specifically with the dates of their investiture in our time frame:

namely, the Duke of Northumberland, 1819; the Marquess of Buckingham, 1820; the Marquess of Cholmondeley, 1822; the Marquess of Hertford, 1822 and the Marquess of Bath, 1822 . Other names from the list of KGs who were alive and entitled to bear the badge of the Order in this period and who were recorded as having commissioned porcelain services sometime earlier from the two Welsh factories were: George Augustus Frederick, Prince of Wales; William Frederick, Duke of Gloucester and Henry Paget, Marquess of Anglesey and Earl of Uxbridge. Noteworthy here is Henry Paget, who was second-in command to the Duke of Wellington at Waterloo in 1815 and, as the Marquess of Anglesey, he commissioned a combined Swansea and Nantgarw tea service around 1817 which was decorated in London. Invested with the Order of the Garter in 1818 he would perhaps have been ideally placed to commission another order of locally decorated Nantgarw porcelain – but we must remember that this is indeed a rather strange depiction of an heraldic Garter Star without its garter and motto.

Realistically, therefore, we must conclude that this is a motif, rather than being a true representation of the Garter Star and is much more likely to be a simulated "signature motif" adopted by Thomas Pardoe as has been proposed earlier by Fergus Gambon (*Porslen Abertawe a Nantgarw*, 2016, number N43 on page 72) and Ernest Morton Nance (*The Pottery and Porcelain of Swansea and Nantgarw*, 1942). An enlargement of this "Garter Star" motif is given in Fig. 4.12. Dr. John (John et al., *The Nantgarw Porcelain Album*, 1975, Illustration 65) shows a Pardoe-decorated plate which displays a "Garter Star" badge in the centre, which is identical to the motifs he has used similarly hitherto on porcelains from the Worcester and Caughley manufactories and earthenwares from the Cambrian Pottery that he had bought in and decorated in Bristol. Dr. John has also compared this star badge with that of the Coldstream Guards, the oldest unit in the British Army infantry, which was raised in 1650 by Colonel Monk as part of Oliver Cromwell's New Model Army: Colonel Monk was awarded the KG and Garter Star in 1660 and this was then adopted by the regiment as its badge. Dr. John also notes a similarity with the badge of the Salvation Army but this must surely be *a non sequitur* as the Salvation Army was not even extant in Pardoe's day, only being founded by William Ebdon in 1865 and, therefore, it really cannot be considered as an appropriate contender chronologically for this exercise.

In contrast, Dr. Holdaway (*Welsh Ceramics in Context, Part II*, pp. 93–113, 2005) has shown a large jug from the Cambrian Pottery which displays a Garter Star with the cross of St George which does include the motto in the garter, "*Honi Soit Qui Mal y Pense*", and the Prince of Wales' feathers and motto "*Ich Dien*", which therefore does constitute a genuine armorial bearing. An interesting story centres around the badge of the three ostrich feathers and the motto, which has been traced back to Edward the Black Prince (1330–1376), heir apparent of King Edward III, who bore a shield sable on which were emblazoned three ostrich feathers argent and bearing the motto "*Ich Dien*", meaning "*I serve*". Said to have been acquired by the Black Prince at the Battle of Crecy in 1346 from the defeated King John of Bohemia, the motto is a homophonic phrase for "*Eich dyn*", which is Welsh for "*Your Man*" and could well be a rendition of the Welsh phrase in recognition of the critical role played by the Welsh longbow archers who fought so decisively for Prince Edward in the battle.

It appears, therefore, that we have just the 7 distinct exemplars noted above of truly heraldic armorial services on Nantgarw porcelain and that we must exclude the modified gilt "Garter Star" motif adopted by Thomas Pardoe as his signature emblem from consideration as a genuine armorial decoration on Nantgarw porcelain. These seven armorials have all been confirmed and checked out properly with the armorial literature.

References

Sir J. Bernard Burke, *Burke's General Armory—The General Armory of England, Scotland, Ireland and Wales: Comprising a Registry of Armorial Bearings from the Earliest to the Present Time* (Harrison & Sons, Pall Mall, London, 1884)

H.G.M. Edwards, *Nantgarw and Swansea Porcelains: An Analytical Perspective* (Springer-Nature, Dordrecht, 2018)

H.G.M. Edwards, *Porcelain Analysis and Its Role in the Forensic Attribution of Ceramic Specimens* (Springer-Nature, Dordrecht, 2021)

J. Fairbairn, *Crests of the Families of Great Britain and Ireland and Their Kindred in Other Lands, Volumes I and II, T.C. and E.C* (Jack, Edinburgh and London, 1905 and New Orchard Publishing, London, 1986)

A.C. Fox-Davies, *Armorial Families: A Complete Peerage, Baronetage and Knightage, and a Directory of some Gentlemen Coat -Armour and Being the First Attempt to Show Which Arms Are in Use at the Moment Are Borne by some Legal Authority, Volumes I and II, T.C. and E.C* (Jack, Grange Publishing Works, Edinburgh, 1895)

F. Gambon, *Porslen Abertawe a Nantgarw (Swansea and Nantgarw Porcelain), Oriel Plas Glyn-y-Weddw* (Llanbedrog, Abersoch, Gwynedd, 2016)

Gartre'n Ol, *Coming Home - an Exhibition of the Finest Nantgarw Porcelain* (Nantgarw China Works Museum, Tyla Gwyn, Nantgarw, 2019)

M. Holdaway, Early Swansea Printed Earthenwares, Part 2. In *Welsh Ceramics in Context, Part II* (Royal Institution of South Wales, Swansea, pp. 93–113, 2005)

W.D. John, *Nantgarw Porcelain* (Ceramic Book Company, Newport, Gwent, 1948)

W.D. John, K. Coombes, G.J. Coombes, *Nantgarw Porcelain Album* (Ceramic Book Company, Newport, Gwent, 1975)

L.O. Mearain, The Bath Estate, 1777–1800. *Clogher Record/Clogher Historical Society* , **6/3**, pp. 555–573, (1968)

E. Morton Nance, *The Pottery and Porcelain of Swansea and Nantgarw* (B.T. Batsford & Son, London, 1942)

Chapter 5
Swansea Armorial Porcelain

Start with what is right rather than with what is acceptable.

Franz Kafka (1883–1924).

It is not what you look at that matters it is what you see.

Henry David Thoreau (1817–1862).

Abstract Detailed in-depth discussion of the assignment and attribution of all known Swansea armorial porcelain exemplars in the literature, including several which had been incorrectly attributed hitherto and showing the importance of cross-checking all available information (nothing is redundant!) as exemplified by the Clarke of co. Hereford impaling Parkinson escutcheon in the partial coat-of-arms where the escallop quartered crest is critically the key to the proper assignment of the arms. Several crests are hereby identified and one escutcheon and 4 crests remain unidentified thus far.

Keywords Clarke of co. Hereford · Parkinson of Kinnersley · Escallop quartered crest · Swansea armorials

From Table 2.1 it appears that we need to consider 11 distinct exemplars of different armorial representations on Swansea porcelain services; although this is almost 60% larger than the armorials we have already identified in total for Nantgarw there is another potential advantage for the attribution of Swansea porcelain armorials in that we have the possibility of using three basic body pastes in our analysis which will potentially input to the chronological time frame of factory production for the execution of the armorial service. This contrasts with Nantgarw porcelain where we only have the one body paste used during the Billingsley/Walker and Young/Pardoe

H. G. M. Edwards, *Welsh Armorial Porcelain*,
https://doi.org/10.1007/978-3-030-97439-8_5

periods, although we do have the possibility of adopting a recently discovered analytical chemical spectroscopic differentiation between the signatures of the *Nantgarw No.1* and *Nantgarw No.2 glazes* used during these two periods. For example, it is well known that Lewis Weston Dillwyn at Swansea first produced rather heavily potted glassy porcelain in 1815, which swiftly gave way to the production of his esteemed duck-egg porcelain between 1816 and 1817, and finally to his trident steatitic magnesian ware from 1817–1820 (Edwards, *18th and 19th Century Porcelain Analysis: A Forensic Provenancing Assessment,* 2020; Edwards, *Porcelain Analysis and Its Role in the Forensic Attribution of Ceramic Specimens,* 2021) . A second potential line of enquiry will involve the adjacent Cambrian Pottery site in Swansea, which as we have seen also manufactured several armorials between 1800 and 1848, prior to, during and after the commencement of porcelain manufacture at the Swansea China Works between 1815 and 1820, and carrying on thereafter from the closure of the porcelain manufactory site in the mid-1820s.

However, the situation is complicated further by the fact that Timothy and John Bevington, who it is believed probably did not manufacture porcelain themselves when they took over at Swansea, decorated the residual stock left by Dillwyn and marked it with a red enamelled SWANSEA stencil. Hence, we could still be looking at duck-egg porcelain decorated in the Bevington workshop at Swansea after the departure of Lewis Dillwyn in 1817, i.e. between 1817 and 1823, after which the sales auctions commenced, although this porcelain body would not have ostensibly been manufactured there after 1817. It must also be remembered that some Swansea porcelain during Dillwyn's tenure (and probably that of the Bevingtons) was sent to the London outlets, although not to the same extent as that practised at Nantgarw, and thence to the enamelling ateliers for which five retailers have been identified in the literature. These comprise the following retailing outlets:

John Mortlock, 47 and 250, Oxford Street.
Bailey, Neale & Bailey, 8, St Paul's Churchyard.
Apsley Pellatt & Green, 16, St Paul's Churchyard.
Boucher & Guy 271, Strand, or 128, Leadenhall Street.
Bradley & Co., 47 and 54, Pall Mall.

One occasionally finds some beautiful exemplars of London-decorated Swansea porcelains of which the retailers must have been very proud which have enamelled underneath the retailer's name in addition to the Swansea mark (Edwards, *Porcelain Analysis and Its Role in the Forensic Attribution of Ceramic Specimens,* 2021). One of these exemplars so marked and several others have been illustrated in Appendix III Fig. A3.1 from Bradley and Co., 47 Pall Mall, comprising a superb set of four dessert plates and a dessert dish which currently reside in the Andrews Collection at Plas Glyn-y-Weddw, Llanbedrog, Abersoch, Gwynedd (see Appendix III Figs. A3.1, A3.2, A3.3, A3.4 and A3.5).

5.1 The Clarke of Hereford Impaling Parkinson Service

Perhaps the best recorded example currently extant of a Swansea armorial porcelain artefact, this service has been considered by many authors to be unequivocally identified. However, all is not straightforward upon cross-checking with the armories in the literature as will soon be apparent. This service is cited in four books or articles that mention Swansea armorials: namely, Gambon (*Porslen Abertawe a Nantgarw*, 2016, 33/S40), John (*Swansea Porcelain*, 1958, page 135 and Plate 78C), Morton Nance (*The Pottery and Porcelain of Swansea and Nantgarw*, 1942, page 335) and Phillips (*Heraldic Swansea Emblems*, 2006, 18/50). Morton Nance ascribes the attribution of the coat-of-arms on this service to Abel Gower of London but the other three authors favour Clarke of Hereford impaling Parkinson. It is reasonable to propose that Morton Nance, the earliest of these authors, made his assessment primarily on the basis of the motto "*Frangas non Flectes*" meaning "*Thou may'st bend but shall not break me*": the Gower family is one of the thirteen families identified in Fox-Davies (*Armorial Families: A Complete Peerage, Baronetage and Knightage, and a Directory of Some Gentlemen Coat -Armour and being the First Attempt to Show Which Arms are in Use at the Moment are Borne by Some Legal Authority*,1895) as bearing this motto. The dinner plate of steatitic magnesian porcelain, with the impressed trident and Swansea mark (Fig. 5.1), and therefore chronologically belonging to the later Dillwyn manufacturing period dating from 1817 and afterwards, has applied small porcelain beads at the rim and is attractively decorated with a border of flowers, chrome green urns and chrome green enamel and gold scrolls and was painted locally by Henry Morris. Morton Nance (*The Pottery and Porcelain of Swansea and Nantgarw*, 1942) has ascribed the service to Abel Gower as mentioned above – whose family motto matches the one shown on the porcelain. The Gower family of wealthy London merchants moved to Corsygedor, Merioneth, from Boughton St John in Worcestershire in the 1700s to inherit the Vaughan mansion there. Abel Gower (1694–1788) inherited the estate and had two sons, Erasmus and Abel Anthony. Admiral Sir Erasmus Gower (1742–1814) became the Commander of the British expedition to the Chinese Imperial Court and later the Governor of Newfoundland. Abel Anthony (1768–1837) lived in the house in Clunderwen, also owning Castle Maelgwyn, and he became High Sheriff of Pembrokeshire. He would have been 50 in 1818, perhaps, therefore, it was thought a memorable occasion for commissioning a Swansea porcelain armorial service!! The family motto is precisely that which appears on the armorial porcelain plates, but unfortunately for Ernest Morton Nance's attribution the escutcheon displayed on the armorial plate is not displaying the Gower coat-of-arms. The floral decoration is assigned to Henry Morris, so implying that the service was definitely locally decorated at the Swansea China Works. Examination of the heraldic evidence immediately also raises a problem for the attribution of the arms to Clarke of Hereford, the alternative and currently favoured assignment for the other three literature citations; in particular, the crest of Clarke of Hereford is depicted in the

Fig. 5.1 This dinner plate of "trident" steatitic magnesian porcelain, with the impressed trident and SWANSEA mark, has applied small porcelain beads at the rim and is attractively decorated with a border of flowers, chrome green urns and chrome green enamel and gold scrolls painted locally at the factory by Henry Morris. The escutcheon and its quartering is definitively that of Clarke of co. Hereford impaling Parkinson of Kinnersley Castle, Hereford, as defined in *Burke's Armory*, and the escallop shell quartered crest in red and gold is also that of Clarke of co. Hereford. However, the motto *Frangas non Flectes*", meaning *"Thou may'st bend but shall not break me"*, is not that recorded for Clark of Hereford, which prompted Ernest Morton Nance (*The Pottery and Porcelain of Swansea and Nantgarw*, 1942) to assign the arms incorrectly to Abel Gower of Corsygedor, Merioneth. *Reproduced with permission from Gwyn Jones, Director of the John Andrews Charitable Trust Collection, Oriel Plas Glyn-y-Weddw, Llanbedrog, Abersoch, Gwynedd*

armories as a liver bird (cormorant) holding a branch of laver seaweed in its mouth whereas that of Parkinson of Kinnersley Castle, Hereford, is a cubit arm erect with vested erminous cuff, holding in the hand an erect ostrich feather (Burke, *Burke's General Armory—The General Armory of England, Scotland, Ireland and Wales: Comprising a Registry of Armorial Bearings from the Earliest to the Present Time*, 1884). Neither of these crests appear on the Swansea armorial service under investigation here; the crest depicted on the porcelain artefact is clearly an escallop shell quartered in red and gold on a red and gold torse. It is interesting that Clarke of Hereford also possesses the motto: "*Soyez ferme*", meaning "*Be Firm*", and not "*Frangas non flectes*". The motto of Parkinson is not given in the armories. The

Fig. 5.2 Enlargement of the escutcheon crest and motto on the Swansea armorial plate shown earlier in Fig. 5.1. *(Reproduced with permission from Gwyn Jones, Director of the John Andrews Charitable Trust Collection, Oriel Plas Glyn-y-Weddw, Llanbedrog, Abersoch, Gwynedd)*

arms on the plate, seen more clearly in the enlargement in Fig. 5.2, comprise the following: sinister, three ostrich feathers and chevron with three blue mullets; dexter, 1st and 3rd, five bars gules with three scallops or, 2nd and 4th two escallops azure and eagle with wings raised. These quite clearly are compatible with the arms of Clarke of Hereford shown on the plate impaling the correct arms of Parkinson (Burke, *Burke's General Armory—The General Armory of England, Scotland, Ireland and Wales: Comprising a Registry of Armorial Bearings from the Earliest to the Present Time*, 1884). The crest on the plate is an escallop shell quartered in or and gules. However, another Clarke of co. Hereford (or Herefordshire, rather than the city of Hereford) is listed as having these same arms and with the appropriate crest of an escallop shell quartered in gules and or, exactly as that portrayed here on the plate!

Clearly, here, the heraldic symbols must be precisely interpreted for what is a partial coat of arms: the escutcheon and its quartering is compatible with that of Clarke of co. Hereford (Herefordshire) impaling Parkinson of Kinnersley Castle as defined in *Burke's Armory*, and the scallop shell quartered crest is also that of Clarke of co. Hereford (Herefordshire) but this is not the same as Clarke of Hereford, who has a different crest entirely. Whereas the motto is not itself specific to, but nevertheless recognisable as that of Gower, the arms are definitively not his and 13 families share this same motto, none of whom bear any relation to the arms or the crest displayed on the porcelain. The specific forensic clue to a resolution of this situation, therefore, resides in the crest, which is a quartered escallop shell, gules and or, which we shall also mention in connection with another crest found on Swansea porcelain later. So, whereas the armorial escutcheon and the crest are matching those of Clarke of co. Hereford (Herefordshire) impaling Parkinson of Kinnersley Castle (Herefordshire) and support each other in that assignment, the motto has been unofficially acquired. Heraldic texts do advise that a motto, unlike the other components of the arms, is not registered formally to a particular arms-bearer in England, Wales and Ireland, but this is not so in Scotland. Hence, we can expect perhaps to see evidence of an unofficial motto usage and adoption by several unconnected families, and this seems to have happened here also.

The attribution status of this service is drawn further into question when the citation which accompanies the National Museum of Wales specimen artefact from this service attributes it alternatively to John Altham Graham-Clarke of Newcastle-upon-Tyne (1782–1862) and his wife Mary Elizabeth, daughter of Leonard Edmund Parkinson of Kinnersley Castle, Herefordshire, who were married on January 1st 1812. Hence, this plate according to this new attribution then strictly depicts the arms of Graham-Clarke of Newcastle impaling Parkinson and not Clarke of co. Hereford impaling Parkinson. The family of Graham-Clarke has an interesting literary connection in that John Graham-Clarke was the uncle of Elizabeth Barrett Browning, maiden name Elizabeth Moulton Barrett, and the Brownings were frequent visitors to Newcastle (*The Brownings Correspondence,* 1984). In 1803 Leonard Parkinson bought Frocester Manor in Gloucestershire, which passed to his daughter Mary on his death in 1818. John Altham Graham-Clarke died at Frocester in 1862, aged 80. Also, the crest of Graham is an escallop shell or, cited as being identical to that of Clarke of co. Hereford as described above (Fairbairn 117/4). It appears therefore that the impaling of the arms of Graham-Clarke of Newcastle with Parkinson of Kinnersley genuinely reflects their marriage in 1812, unlike that of Clarke of co. Hereford, but the reason for its commission would not be in celebration of their marriage as at that time Dillwyn had not even started to manufacture porcelain at Swansea – and the trident body which constitutes the porcelain of this service was the final body paste composition made at Swansea which commenced from the earliest date of 1817 until the final closure of the Swansea China Works. It is more reasonable to propose that the commissioning of this service would concur with John and Mary Graham-Clarke of Newcastle-upon-Tyne assuming the inheritance of Frocester Manor in Gloucestershire in 1818 from Leonard Parkinson who had left the estate to his daughter, Mary; chronologically this would fit perfectly with its steatitic body composition in trident porcelain. The similarity of the escallop shell crest and escutcheon for the two Clarke families in co. Hereford and Newcastle is therefore a red herring but a genealogical investigation as to which one married Mary Parkinson of Kinnersley Castle to give rise to the impaling of the arms would illuminate the problem. It is indeed fortunate that the Browning papers alluded to above give details of Mary Elizabeth Parkinson of Kinnersley Castle co. Herefordshire and her marriage to John Graham- Clarke in 1812 and of the death of Mary Elizabeth Graham Clarke in 1851. It seems, therefore, that Clarke of co Hereford is to be replaced by Graham-Clarke of Newcastle in the attribution of this service and the assumption of the title of Frocester Manor in 1818 would indeed have the correct chronology for its commission. This is a classic example of how all aspects of the heraldic devices displayed must be considered and how the consultation of family documentation and genealogy can be used to properly define the correct attribution of an armorial service. Kinnersley Castle was a very ancient foundation originally built by Hugh de Kynardsley in the 13th Century as a Marcher Lord's castle near Hay-on-Wye, Herefordshire.

In a rather anecdotal conclusion to the research undertaken on this particular and perhaps best- known Swansea porcelain armorial exemplar, although it has been incorrectly ascribed in the literature, an example from this same service recently was offered at auction in Devon from a well-known national collection where it was described as a product of Dyllwyn's (*sic.*!) Cambrian Pottery, Swansea, made in soft-paste porcelain between 1814–1826. Not only subscribing to a mixed metaphor, there are several incorrect statements and assumptions made here including the very broad dateline given chronologically for its manufacture.

5.2 The Lloyd of Bronwydd Service

Thomas Lloyd (1788–1845) of Bronwydd, Ceredigion, ordered a service of Swansea duck-egg porcelain in the *Mandarin* pattern (set pattern No. 164, Jones & Joseph, *Swansea Porcelain: Shapes and Decoration*, 1988) to celebrate his marriage to Anne Davies of Llwydcoed and Llettymawr, Carmarthenshire, in 1819 and this crested service has been mentioned above. This comprises a black transfer-printed service enamelled with a polychrome infill in the *famille rose* style typical of the Westernised *chinoiserie* decoration that was much in vogue at that time Jones and Joseph (*Swansea Porcelain: Shapes and Decoration,* 1988, page 172) allege that several different services of crested porcelain exist in this pattern but unfortunately do not give any further examples or details which can be accessed or researched further armorially in this context and there has been no other mention of additional crests in this pattern cited elsewhere in the literature. A Thomas Lloyd crested dessert comport in duck-egg porcelain with the characteristic attractive Swansea twigged handles is illustrated in Fairclough (*Welsh Ceramics in Context,* II, 2005, page 200), in Phillips (*Heraldic Emblems on Swansea*, 2006, 18/51), and also here in Fig. 5.3. This artefact is now in the collection of the National Museum of Wales Amgueddfa Cymru, Cardiff. What makes this armorial crested service rather special is the placement of the Lloyd crest in polychrome enamels, namely a boar chained to a holly tree, at the verge and accompanied by the family motto in Welsh "*I Dduw Bo'r Diolch*" – translating as "*Thanks be to God*" (as seen in the enlargement of the crest in Fig. 5.4) – one of only two such services now known on Welsh porcelain which bear a Welsh inscription, although several presentation pieces with Welsh inscriptions and mottoes are known on Swansea earthenwares originating from the Cambrian and Glamorgan Potteries as we have seen in Chap. 3. It has been maintained hitherto by ceramic historians that this Lloyd of Bronwydd service is the only example of an armorial service to bear a Welsh inscription but a second example has now surfaced and has been assigned during this research. Thomas Lloyd was a descendant of the ancient Norman Lords of Dyfed, the 22nd in line of the Barony of

Fig. 5.3 Thomas Lloyd (1788–1845) of Bronwydd, Ceredigion, ordered a service of Swansea duck-egg porcelain in the *Mandarin* pattern (set pattern No. 164) to celebrate his marriage to Anne Davies of Lwydcoed and Llettymawr, Carmarthenshire, in 1819. This crested dessert comport with the characteristic attractive Swansea twigged handles shows the Lloyd of Bronwydd crest, a boar chained to a holly tree, at the verge and accompanied by the family motto in Welsh "*Y Dduw Bo'r Diolch*" – "*Thanks be to God*" – one of only two such services known on Welsh porcelain which bear a Welsh inscription. (*Reproduced Courtesy of Bonhams Auctioneers and Valuers, London, and the National Museum of Wales, Amgueddfa Cymru, Cardiff*)

Kemes, Pembroke, from Martin de Tours, a companion of William the Conqueror. He was High Sheriff of Cardiganshire (Ceredigion) in 1814. His son, also Thomas Lloyd (1820–1877), was created a baronet in 1845. Although *Burke's Armory* (Burke, *Burke's General Armory—The General Armory of England, Scotland, Ireland and Wales: Comprising a Registry of Armorial Bearings from the Earliest to the Present Time*, 1884) lists 94 Lloyds entitled to bear arms with a variety of crests, this Thomas Lloyd is unique for his crest and associated motto in Welsh, given in the armory as: Arms: azure a wolf rampant ar. a bordure erminois; crest: in front of a holly tree ppr. a boar passant ar. semee of estoiles az. collared and chained to the tree or. Motto "*I Dduw Bo'r Diolch*", which translates as "*Thanks be to God*". All previous literature records assign this service to be special because of its unique motto in Welsh: in fact, as will be seen below, there is another service with a crest which also bears a Welsh motto, which has been assigned here to Wyndham Lewis of Greenmeadow, Llantrisant.

Fig. 5.4 Enlargement of the crest of Thomas Lloyd of Bronwydd showing a boar chained to a holly tree. (*Reproduced Courtesy of Bonhams Auctioneers and Valuers, London, and the National Museum of Wales, Amgueddfa Cymru, Cardiff*)

5.3 Boar's Head Crest

The Glynn Vivian Exhibition Catalogue of Swansea and Nantgarw Porcelains comprised the collections of Richard Glynn Vivian, Herbert Eccles, Lord Swansea and F.E Andrews and also items from the Leonard Rodolphe Warrant Bequest. Here, 356 items of porcelain were displayed and have been described in Kildare Meager's booklet (*Swansea and Nantgarw Potteries*, 1949) but only one armorial boar's head crested plate in Swansea duck-egg porcelain, with Paris fluting, shell decoration, and the crest centrally located (Meager, *Swansea and Nantgarw Potteries*, 1949, page 38 no. 111) was exhibited and was there ascribed by Kildare Meager to Sir Watcyn Williams-Wynn of Wynnstay. No reason was proffered for this assignment but it has been accepted apparently without question by later historians. This crest is actually incorrect for Williams-Wynn according to the armories — the Williams-Wynn crest is an armed eagle with wings spread, dexter and sinister (Fairbairn, *Crests of the Families of Great Britain and Ireland and Their Kindred in Other Lands,* 1905, page 519, plates 35/8 and 48/11). The crest on the fluted plate seen here is of a boar's head, erased or couped at the neck; unfortunately, this is rather a common heraldic symbol which can be ascribed to several potential arms bearers (see later)! The Swansea example is illustrated in Fig. 5.5, accompanied by a shell decoration in gilt at the verge with the attractive and popular Paris fluting body moulding, and is shown in enlargement in Fig. 5.6. It appears in this form in several publications, generally without attribution: Gambon (*Porslen Abertawe a Nantgarw*, 2016, 27/S26); John (*Swansea Porcelain*, 1958, Plate 84C); Phillips (*Heraldic Symbols on Swansea*, 2006, 18/49) and Jones & Joseph (*Swansea Porcelain,* 1988, 213/1).

Burke's Armory (Burke, *Burke's General Armory—The General Armory of England, Scotland, Ireland and Wales: Comprising a Registry of Armorial Bearings from the Earliest to the Present Time*, 1884) lists 63 potential family candidates for

Fig. 5.5 Swansea deep dish in duck-egg porcelain with boar's head erased / couped crest in gilt centrally located inside a circlet of gilt leaves and circle of escallop shells at the verge. The dish is moulded in the popular Paris fluting design. (*Reproduced with permission from Gwyn Jones, Director of the John Andrews Charitable Trust Collection, Oriel Plas Glyn-y-Weddw, Llanbedrog, Abersoch, Gwynedd*)

this crest, including the Clans Campbell and Innes in Scotland: with the boar's mouth pointing slightly upwards and teeth bared as indicated here (Fig. 5.6), the heraldic description of this stance being "erect" (Fairbairn, *Crests of the Families of Great Britain and Ireland and Their Kindred in Other Lands,* 1905, plate 42/2). Three families could be the potential owners of this particular version of the crest, namely, Ingilby of Ripley, North Yorkshire, Vennor and Tyrrell and others more local to Swansea are explored later. Scottish versions tend to have rather more of the boar's neck on view and it is cut off with a jagged edge, whereas here the head is cut higher up on the neck and fused with the torse. It is seen here that an associated motto would have assisted in the narrowing down and elimination of several possibilities. One thing is clear, that this is definitely not the crest of Sir Watcyn Williams-Wynn as assigned by Kildare Meager and further adopted in the literature; the original reason for Kildare Meager's assignment of this crest to Sir Watcyn Williams-Wynn of Wynnstay is not revealed in his exhibition survey nor by later historians.

Fig. 5.6 Enlargement of the boar's head crest from Fig. 5.5. (*Reproduced with permission from Gwyn Jones, Director of the John Andrews Charitable Trust Collection, Oriel Plas Glyn-y-Weddw, Llanbedrog, Abersoch, Gwynedd*)

5.4 Dolphin Crest

A complete armorial dessert service is illustrated in John (*Swansea Porcelain*, 1958, Plate 33) with a central dolphin *naiant* on a torse, which also has the unique addition of dolphin-moulded handles to the tureens and lid knops. An eight-lobed plate in duck-egg porcelain showing this dolphin crest is described in Gambon's book (*Porslen Abertawe a Nantgarw*, 2016, 25/S19) and there it is labelled as "unidentified": this specimen is illustrated here in Fig. 5.7 and is in the Andrews Collection at Plas Glyn-y-Weddw, Llanbedrog, Abersoch, Gwynedd. The crest is mounted on a torse in gold and red and is accompanied by sprigs of blue cornflowers with foliage and a simple gilding at the edge and is shown enlarged in Fig. 5.8. This crested service was ascribed by Dr. John (*Swansea Porcelain*, 1958) to Garnett-Orme of Abbeytown, Co. Mayo, Ireland; motto "*Fortis et Fideles*", which is translated as "*Strong and Faithful*". It is interesting that an identical dolphin crest appears as one of Admiral Lord Horatio Nelson's accoutrements – and was paraded as 11th in line at his State Funeral in January,1806. However, on close inspection of the armories, Garnett-Orme of Abbeytown, Crossmolina, Tirawley, Co Mayo, Ireland, has a crest showing a dolphin *naiant* which is surmounted by a poleaxe—and this is not the crest shown on this Swansea service. The motto of Garnett-Orme is "*Diligentia et Honore*", meaning "*With Diligence and Honour*". However, according to *Burke's Armory* (Burke, *Burke's General Armory—The General Armory of England, Scotland, Ireland and Wales: Comprising a Registry of Armorial Bearings from the*

Fig. 5.7 An eight-lobed, cruciform plate in duck-egg Swansea porcelain showing a dolphin naiant crest is described in Fergus Gambon's book (*Porslen Abertawe a Nantgarw*, 2016, 25/S19) and labelled there as unidentified. The crest is mounted on a torse in gold and red and is accompanied by sprigs of blue cornflowers with foliage and with a simple gilding at the edge. The crest is unequivocally assignable here to Orme of Northampton. (*Reproduced with permission from Gwyn Jones, Director of the John Andrews Charitable Trust Collection, Oriel Plas Glyn-y-Weddw, Llanbedrog, Abersoch, Gwynedd*)

Earliest to the Present Time, 1884), Orme of Northampton has a crest of a dolphin *naiant* azure, fins, tail and tusk or which matches that shown in Fig. 5.9. The motto "*Fortis et Fideles*", translates as "*Strength and Fidelity*" or "*Strong and Faithful*". Arms 1st and 3rd: 2nd and 4th quartered eagle or and poleaxe on azure, chevron argent and three escallops gules; mantling sky blue with gold. This armorial plate crest matches exactly that shown in in Fairbairn's book (Fairbairn, *Crests of the Families of Great Britain and Ireland and Their Kindred in Other Lands,* 1905), so we can safely ascribe definitively for the first time the attribution of this Swansea armorial service to Orme of Northampton and not to Garnett-Orme of Abbeytown, Co, Mayo, Ireland.

Fig. 5.8 Enlargement of dolphin naiant crest on the plate shown in Fig. 5.4. *(Reproduced with permission from Gwyn Jones, Director of the John Andrews Charitable Trust Collection, Oriel Plas Glyn-y-Weddw, Llanbedrog, Abersoch, Gwynedd)*

Fig. 5.9 Depiction of the dolphin naiant heraldic crest (from Fairbairn, *Crests of the Families of Great Britain and Ireland and Their Kindred in Other Lands,* Volumes I and II, T.C. and E.C. Jack, Edinburgh and London, 1905 and New Orchard Publishing, London, 1986)

5.5 Spread Eagle Crest

A C-scroll border moulded Swansea duck-egg porcelain service with a gilt crest of an eagle with wings outstretched at the centre has been assigned historically to the *Venn* service: the origin of this attribution is unclear to the author and a search of the written records has failed to find the basis for this attribution, but it has appeared several times in the literature, for example, most recently in Phillips (*Heraldic Symbols on Swansea*, 2006, 18/46). This armorial representation seems to match better the crest of Sir Watcyn Williams-Wynn of Wynnstay, Merioneth, *ca.* 1815–17

(Fairbairn, *Crests of the Families of Great Britain and Ireland and Their Kindred in Other Lands,* 1905, page 519, Plate 48/11). The description of the Venn crest as cited in Fairbairn employs either a lion passant on a mount erminous dexter resting on an escutcheon azure charged with a fleur-de-lys or an eagle's head proper on a ducal coronet (Fairbairn, *Crests of the Families of Great Britain and Ireland and Their Kindred in Other Lands,* 1905, Plate 4/12–4/14 and Plate 20 /12). Neither of these would satisfy its current attribution to an eagle with outspread or extended wings as shown here in Fig. 5.10. The Venn motto is "*Fide et Integritate*" translating as "*With Fidelity and Integrity*" (which is solely ascribed to Venn) whereas the Williams-Wynn of Wynnstay motto is "*Y Cadarn y Cyprwyna*" meaning "*The Mighty and Cunning*" (which again is a unique motto for Williams-Wynn). This again provides an exemplar where the use of an appropriate motto in association with the crest would facilitate the potential cross-checking of the attribution of a crest if one was deemed to have been available.

A more detailed search of the writings of Morton Nance reveals an illustration of a Cambrian Pottery Swansea creamware plate (Morton Nance, *The Pottery and Porcelain of Swansea and Nantgarw*, 1942, Plate VIIIF facing page 39), dating from 1800–1810, marked SWANSEA impressed and with decoration attributed by Ernest Morton Nance to Thomas Pardoe, which contains a similar crest of an eagle displayed armed, wings extended, and centrally located. Major Edmondes has identified the crest to Ernest Morton Nance as belonging to the Turbervill family of Ewenny Priory. This seriously puts the assignment of such a crest to Venn in doubt without an inspection of the evidence upon which that assertion was based. Ewenny Priory was fortified by William de Londres in 1200 as part of the Triangle of Fortresses comprising Ogmore, Newcastle and Coity in what is now mid-Glamorgan.

Fig. 5.10 The spread-eagled gilt crest assigned historically to the *Venn* service of Swansea duck-egg porcelain. Here shown on a serving dish of Nantgarw-type C-scroll moulding. *(Reproduced with permission from Gwyn Jones, Director of the John Andrews Charitable Trust Collection, Oriel Plas Glyn-y-Weddw, Llanbedrog, Abersoch, Gwynedd)*

Fig. 5.11 A Swansea Cambrian Pottery creamware plate from a service, *ca.* 1800–1810, (illustrated in Morton Nance, Plate VIIIF facing page 39, *The Pottery and Porcelain of Swansea and Nantgarw*, 1942), marked SWANSEA impressed and attributed by Ernest Morton Nance to Thomas Pardoe, which contains an identical crest of an eagle displayed armed and centrally located. This crest has been identified by Major Edmondes as that of the Turbervill family of Ewenny Priory. Ewenny Priory was an ancient Norman manor foundation and passed from the Carnes in the 16th Century to the Turbervills and thence to the Picton-Turbervills in 1867

After the Dissolution of the Monasteries in 1545, Ewenny was purchased by Sir Edward Carne and it passed through inheritance to the Turbervills in 1741. The Cambrian Pottery, Swansea creamware plate with this armorial crest is shown in Fig. 5.11.

It appears, therefore, that the crest allocated to Venn, however it came to be assigned, does not match the heraldic records and should probably now on the basis of the current research and the entry in Morton Nance be better re-assigned to either the Turbervills of Ewenny or to Williams-Wynn of Wynnstay.

5.6 Paschal Lamb Crest

The crest (Philips, *Heraldic Symbols on Swansea*, 2006, 18/47) of a Paschal lamb in gilt facing dexter and holding in its right foreleg a pole with the cross of Saint George on a pennant and with the motto "*Duw ar fy rhan*", translating as *"God for my Protection*" (Fig. 5.12), is recorded on a C-scroll moulded Swansea dish in duck-egg porcelain, *ca.* 1817–1820. The emblem of St John the Baptist, the Paschal lamb, appears on the coats-of-arms of several English towns and cities as he has been adopted as the patron saint of wool-workers. As yet unidentified hitherto, this current research has now established that this is the crest of the Davie family of

Fig. 5.12 The crest (Philips, *Heraldic Symbols on Swansea*, 2006, 18/47) of a Paschal lamb in gilt facing dexter and holding in its right foreleg a pole with the cross of Saint George on a pennant and the motto in Welsh *"Duw ar fy rhan"*, translating as *"God for my protection"*, is recorded on this C-scroll moulded Swansea dish in duck-egg porcelain, *ca.* 1817–1820. This is one of two crests recorded for Wyndham Lewis of Greenmeadow, Llantrisant, who was a descendant of Price (Pryce) and is one of only two crests and coats-of-arms on Welsh porcelain to bear an inscription in Welsh, the other being that of Thomas Lloyd of Bronwydd (Figs. 5.3 and 5.4). (*Reproduced with permission from Gwyn Jones, Director of the John Andrews Charitable Trust Collection, Oriel Plas Glyn-y-Weddw, Llanbedrog, Abersoch, Gwynedd*)

Devon, of Price/Pryce, of an Evan Lewis of Llandaff and one of two crests of Wyndham Lewis of Greenmeadow, Llantrisant, who was a descendant of Price / Pryce. The *Greenmeadow* service is a famous, although not an armorial, service in Nantgarw porcelain commissioned by Wyndham Lewis and it is reasonable to propose that this crested Swansea service, if indeed this be his, would bear one of his crests as a supporter of the Welsh porcelain manufacturing industry. The Lewis crest is a lion sejant ar. or lion rampant or; motto: *"Patriae fidus"* meaning *"Faithful to my Country"*. This service, depicting a Paschal lamb crest, is important in Welsh armorials as it is one of only two crests and coats-of-arms on Welsh porcelain to bear an inscription in Welsh, the other being that of Thomas Lloyd (Figs. 5.3 and 5.4) which has been discussed above. Wyndham Lewis (1780–1838), a supporter of the Nantgarw China Works identified by William Weston Young, was a descendant of the Price/Pryce family and this crest is logged in the armories as belonging to Price, Pryce, Davie, Basset Jones and Evan Lewis of Llandaff.

It is likely, therefore, that a committed and prominent supporter of Welsh porcelain manufacture with an estate in South Wales would have commissioned such an armorial service in the finest Swansea porcelain and therefore its potential attribution here to Wyndham Lewis would not be amiss.

5.7 Escallop Shell Crest

A Swansea saucer dish in duck-egg porcelain in the popular Paris fluting moulding simply gilded with a band of gold at the edge and an escallop shell crest centrally placed is shown in Fig. 5.13, with an enlargement in Fig. 5.14. This crest is cited by Gambon (*Porslen Abertawe a Nantgarw*, 2016, 25/S21) and by Phillips (*Heraldic Symbols on Swansea*, 2006, 18/48) and both have recorded it as unidentified. However, following on from our previous discussion of Clarke of Hereford, Clarke of co. Hereford and Graham-Clarke of Newcastle-upon-Tyne (see Sect. 2.3.1. above), the escallop shell crest is recorded in *Burke's Armory* (Burke, *Burke's General Armory—The General Armory of England, Scotland, Ireland and Wales: Comprising a Registry of Armorial Bearings from the Earliest to the Present Time*, 1884) as being that of Clarke of co. Hereford (Herefordshire) or Graham-Clarke of Newcastle-upon-Tyne and appears above the arms in the escutcheon there, although quartered in gules and or, as was the torse. Here the gilt escallop shell and torse would suffice to record the family crest, especially on a simply decorated service – it should be pointed out that the Paris fluting moulding was very popular at Swansea and was used especially for breakfast and tea services, which were generally supplied with a simple gilt decoration and this crest would obviously fit in with that category in contrast with the more highly decorative polychrome dinner and dessert services. An example of a tea cup and saucer from a tea/breakfast service in this same Paris fluting moulding and with an identical gilding employed at the verge is shown in Fig. 5.15, with the attractive type 5 Swansea tea cup (as tabulated in Jones & Joseph, *Swansea Porcelain*, 1988).

5.8 Bird on a Tree Stump Crest

This crest, which symbolises an unidentified bird, possibly a dove or a martlet, sitting on a sprouting (heraldically, a fructed) tree stump in gilt, is centrally located on a 12-lobed deep saucer dish of Swansea duck-egg porcelain with the Nantgarw-type C-scroll border and is otherwise simply gilded at the edge of the verge and the reserve (Fig. 5.16) with an enlargement shown in Fig. 5.17. Several possibilities emerge for an assignment of the match of a family to this crest can be found after

Fig. 5.13 A Swansea saucer dish in duck-egg porcelain in the popular Paris fluting moulding simply gilded with a band of gold at the edge and a gilt escallop shell crest centrally placed is assigned to Clarke of co. Hereford (see Figs. 5.1 and 5.2) where the same escallop shell is seen as the crest sitting upon the escutcheon, quartered in gules and or, as is the torse. The Paris fluting moulding was very popular at Swansea and was used especially for breakfast and tea services, which were generally supplied with simple gilt decoration and this crest would obviously fit in with that category in contrast with the more highly decorative polychrome dinner and dessert services. (*Reproduced with permission from Gwyn Jones, Director of the John Andrews Charitable Trust Collection, Oriel Plas Glyn-y-Weddw, Llanbedrog, Abersoch, Gwynedd*)

diligent searching in either *Burke's Armory* or in *Fairbairn's Crests* (Burke, *Burke's General Armory—The General Armory of England, Scotland, Ireland and Wales: Comprising a Registry of Armorial Bearings from the Earliest to the Present Time*, 1884; Fairbairn, *Crests of the Families of Great Britain and Ireland and Their Kindred in Other Lands,* 1905). Several representations of different birds sitting on similar tree stumps are found illustrated in the armories, some of which are sprouting (signifying a re-growth or re-birth heraldically), but in most cases the birds are identified as Cornish choughs, ravens, falcons or eagles and in several instances these are also carrying prey; it will be seen later that the Clarke crest of Graham-Clarke of Newcastle-upon-Tyne is that of a lark on a fructed tree stump, wings expanded ppr., with two ears of wheat in its beak. The porcelain exemplar illustrated here is from the John Andrews Collection at Oriel Plas Glyn-y-Weddw, Llanbedrog, Abersoch, which has also been illustrated in Gambon (*Porslen Abertawe a*

Fig. 5.14 Enlargement of the escallop shell crest shown on the plate in Fig. 5.13. *Reproduced with permission from Gwyn Jones, Director of the John Andrews Charitable Trust Collection, Oriel Plas Glyn-y-Weddw, Llanbedrog, Abersoch, Gwynedd*

Fig. 5.15 An example of a tea cup and saucer from a tea/breakfast service in the same Paris fluting moulding and with an identical gilding at the verge as shown in Fig. 5.13, with the attractive type 5 Swansea cup with a curved loop handle (Jones & Joseph, *Swansea Porcelain: Shapes and Decoration*, 1988). *Private Collection*

Fig. 5.16 This crest, which symbolises a dove or a martlet sitting on a sprouting (fructed) tree stump in gilt is centrally located on a 12-lobed Swansea deep saucer dish of duck-egg porcelain with the Nantgarw -type C-scroll border and is otherwise simply gilded at the edge of the verge and the reserve. (*Reproduced with permission from Gwyn Jones, Director of the John Andrews Charitable Trust Collection, Oriel Plas Glyn-y-Weddw, Llanbedrog, Abersoch, Gwynedd*)

Nantgarw, 2016, 25/S22); another example is shown in Phillips (*Heraldic Symbols on Swansea*, 2006, 18/48) but here the image of the crest is not as clearly shown. An enlargement of the crest does provide a better image for research purposes (as shown in Fig. 5.17), from which it can be deduced that the bird depicted on the fructed tree stump is possibly a dove or maybe a martlet. It certainly does not have the exposed talons of a bird of prey exhibited heraldically by an eagle or falcon, such as those described earlier for an eagle or for the falcon depicted on Goss china (Appendix I Fig. A1.2).

5.8.1 *Hill* Versus *Lockwood*

A further enlargement of the crest shown in Fig. 5.17 reveals that the bird has a closed beak and is shown with very small feet: the martlet in heraldry represents a swallow or swift, birds that were observed to be constantly in flight and for a time it

Fig. 5.17 Enlargement of the bird, a dove or a martlet, on a tree stump crest shown in Fig. 46 from which it is seen that the bird depicted is probably a dove. (*Reproduced with permission from Gwyn Jones, Director of the John Andrews Charitable Trust Collection, Oriel Plas Glyn-y-Weddw, Llanbedrog, Abersoch, Gwynedd*)

was believed that they could not perch as did other birds because they did not possess feet! The bird shown in Fig. 5.17 could be a dove but the suggested presence of extremely small feet is indicative of it being a martlet: the martlet is invariably depicted heraldically in a black tincture, but of course this would not be used in a gilded crest such as that we have shown here in Fig. 5.16 and Fig. 5.17. A rather unique assignment for a martlet sitting on top of a fructed oak tree stump is provided by the armories for Thomas Lockwood of Dews Hall, Essex. The detail in the crest provided here is insufficient to determine if the tree stump is an oak, but it certainly is a possibility. An alternative suggestion for an assignment of this crest is to that of Richard Hill of the Plymouth Ironworks in Merthyr Tydfil, but in this case his crest clearly depicts a falcon (Fairbairn, *Crests of the Families of Great Britain and Ireland and Their Kindred in Other Lands,* Volumes I and II, 1905, Plate 78/8). The entry in Fairbairn clearly states that the crest of Hill is "*On the trunk of a tree or, a falcon ppr. beaked and belled of the first*": this means that the falcon's feet are exhibited and have bells attached. In addition, the description of this crest indicates that the bird is "beaked" a heraldic term meaning that the raptor has an open beak exposing the tongue. In the same vein, the crest of Lockwood states "*On the stump of a tree erased ppr. a martlet sable*": in contrast with Hill, for whom there are 77 entries in Burke's Peerage, only 3 entries are provided for Lockwood. The Lockwood, of Dews Hall, Lambourne, Essex, had an extinct ancestry linked with Wood, of Gayton Park, where the crest in deference to the name is "*an oak tree fructed vert*", with a preponderance of martlets and oak trees in evidence on the arms. Clearly the type of bird exhibited in Hill's crest is not manifest here and the consensus judgement must therefore be that this crest is much better attributed to Lockwood, of Dews Hall, Essex, with a martlet sitting on a fructed tree stump (whether this is specifically an oak tree is rather difficult to discern from the crest on the armorial plate as shown): the motto of Lockwood is "*Esperance*", meaning "*Hope*".

5.9 An Unidentified Escutcheon

A rather diffuse coat-of-arms depicted on a Swansea plate illustrated in Jones & Joseph (*Swansea Porcelain*, 1988, page 121/3) was unfortunately not able to provide enough detail for the identification process to be accomplished successfully – the central escutcheon comprises a quartered coat-of-arms: 1st two martlets, 2nd indecipherable, 3rd three bars and 4th one martlet. The embossed Swansea duck-egg porcelain armorial plate bearing these arms is shown in Fig. 5.18. The black and white monochrome picture shown in Jones & Joseph (*Swansea Porcelain: Shapes and Decoration*, 1988, page 121/3) does not enable a colour match to be undertaken for detailed verification and as a result no attribution could be provided at this stage. The porcelain substrate is beautifully moulded with floral emblems and the gilding is a simple thin band at the edge of the plate. An enlargement of the escutcheon (Fig. 5.19) from the monochrome picture is still not sufficient to aid the attribution of the coat-of-arms and access to the original artefact was clearly required before any progress could be made in the attribution.

After some months of fruitless research attempting to investigate the escutcheon depicted on this artefact, the author is indebted to Andrew Renton (National Museum

Fig. 5.18 The embossed Swansea duck-egg porcelain plate is shown here in a black and white monochrome picture of the coat-of-arms depicted in Jones & Joseph (*Swansea Porcelain*, 1988, page 121/3) unfortunately does not provide enough detail for the identification process to be accomplished – the central escutcheon comprises: 1st two martlets, 2nd indecipherable, 3rd three bars and 4th one martlet. The porcelain substrate is beautifully moulded with floral emblems, a rather rare moulding for Swansea, and the gilding is a simple thin band at the edge

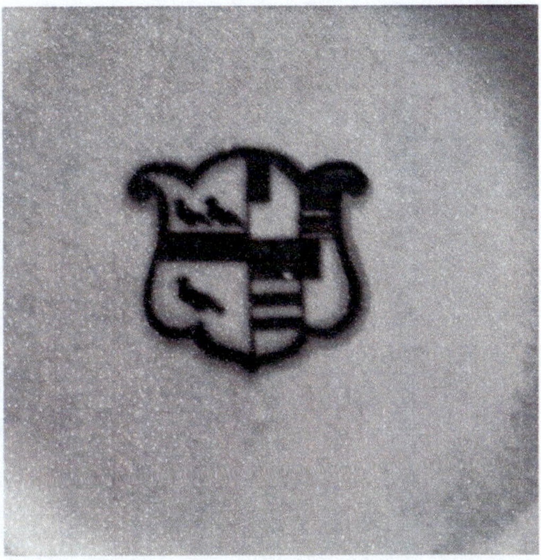

Fig. 5.19 Enlargement of the escutcheon depicted in Fig. 5.18

of Wales, Amgueddfa Cymru, Cardiff) who sent a polychrome picture of a plate from this same service which now resides in the Welsh Ceramics Collection at the National Museum of Wales in Cardiff, and this is reproduced in Fig. 5.20. Accompanying this figure is its heraldic assignment to the arms of Lockwood impaling Manners-Sutton, hence this definitive assignment can now be offered for this Swansea armorial as described further in the author's research below.

5.9.1 Arms of Lockwood Impaling Manners-Sutton

The notification of an unidentified partial coat-of-arms on Swansea porcelain in Jones & Joseph (*Swansea Porcelain: Shapes and Decoration*, 1988) initially caused some concern and difficulty in progressing the research since it was not mentioned elsewhere in the literature, it was depicted only in a black and white monochrome illustration with rather low resolution and it was not referenced regarding its location so that any further detailed examination and research could not be effected. Then, a polychrome image was located and some further information was provided (courtesy of Andrew Renton, National Museum of Wales, Amgueddfa Cymru, Cardiff) which facilitated its ongoing investigation. The colour image of the partial coat-of-arms is shown in Fig. 5.20 and the arms have been identified as those of Lockwood impaling Manners-Sutton. The family of Lockwood of Dews Hall, Essex, has already featured earlier (Sect. 5.8.1) in the consideration of their crest of a martlet sitting on top of a fructed tree stump .

John Manners, the 3rd Duke of Rutland, also entitled the Marquess of Granby, acquired the estate of Roseby, Lord Lexington of Averham, who died at Averham Park in 1723, through his daughter Bridget, Duchess of Rutland, which was then

Fig. 5.20 A polychrome picture of a plate from the same service as that shown in monochrome in Fig. 5.18 and its enlargement in Fig. 5.19, wherein the clear depiction of the coat of arms has assisted in its decipherment with an unequivocal attribution to the arms of Lockwood impaling Manners-Sutton. (*Reproduced with Permission from the National Museum of Wales, Amgueddfa Cymru, Cardiff*)

passed on to Lord Robert Manners with his assumption of the name of Sutton, then becoming Manners-Sutton with a seat at Kelham Hall, near Newark in Nottinghamshire. Lord Thomas Manners-Sutton, the first Baron Manners (1756–1842) became Baron of the Exchequer and Lord High Chancellor of England with estates at Haddon Hall in Derbyshire and at Oakham Castle in Rutland. The Manners-Suttons were an influential family: the Most Reverend Charles Manners-Sutton (1755–1828), brother of Lord Thomas Manners-Sutton, was Archbishop of Canterbury between 1805 until 1828, when he died aged 73. His son, also Charles, became the Speaker of the House of Commons and the first Viscount Canterbury (1780–1845); the father, Charles, eloped with and married his cousin Mary Thoroton in 1778, aged 23. Lord George Manners-Sutton, the 3rd son of the Duke of Rutland, had three daughters, Louisa Bridget, who married Edward Lockwood Perceval of Bishop's Hall, Essex, in 1790 and who died in 1800, Mary, who married the Rev., Richard Lockwood of Fifield, Essex in 1799 and died in 1829, and Charlotte Manners-Sutton, (born 1764) who married Thomas Lockwood of Dews Hall,

Lambourne, in co. Essex and Gayton, Northampton, in June 1789 and died in 1827. It is interesting that all three Manners-Sutton daughters married Lockwoods from Essex! In 1812, Thomas Lockwood (born 1768, died 1836) leased land he owned in Swansea to Baker, Bevans and Irwin, who founded the Glamorgan Pottery on the site in 1814. The armorial bearings on the Swansea plate of the finest duck-egg porcelain match those of Thomas Lockwood impaling Manners-Sutton. What is striking is that the three daughters of Lord George Manners-Sutton all married into branches of the Lockwood family within the space of ten years. A potential reason for commissioning the service between 1815 and 1825 is not obvious as it cannot be to celebrate a wedding, but perhaps a special birthday or birth of a child? Research into the genealogy of Thomas Lockwood and Lady Charlotte Manners-Sutton reveals that Thomas was born in 1768 so he would have attained the age of 50 in 1818, but their only daughter, Lady Horatia Charlotte Lockwood (born 1797, died 1838) married Viscount Hood, and their daughter Mary Isabella was born in 1818! So, 1818 would indeed have been a memorable year for celebration in the Lockwood household to occasion the commissioning of a fine porcelain service!

The Swansea armorial plate illustrated in Fig. 5.20 shows a centrally located escutcheon depicting the arms of Lockwood impaling Manners-Sutton within an attractive moulded floral and foliate border and bearing the SWANSEA stencil mark on the reverse. A representative plate from this service apparently was shown in an exhibition at the Glynn Vivian Art Gallery in Swansea in 1969: the Glynn Vivian Art Gallery was founded at the bequest of Richard Glynn Vivian (1835–1910), the fourth son of John Henry Vivian of Singleton Park and Sketty Hall, Swansea, the owner of the largest copper works in Swansea at the Hafod. John Vivian was an enthusiastic collector of Swansea porcelain and an ardent supporter of the Swansea China Works: he acquired the *Marino Ballroom* service of fine Swansea china, comprising more than 260 pieces, for evening entertaining at his *Marino* residence in Singleton Park. The arms of Lockwood (*Burke's General Armory—The General Armory of England, Scotland, Ireland and Wales: Comprising a Registry of Armorial Bearings from the Earliest to the Present Time*, 1884) depict quarterly, 1st and 4th, a fesse argent between three martlets sable for Lockwood and 2nd and 3rd, a bend engr. Sable with three plates for Cutts. His crest (Lockwood) is, on the stump of an oak tree erased proper a martlet sable, as shown in Fairbairn (*Crests of the Families of Great Britain and Ireland and Their Kindred in Other Lands,* 1905. Plate 30/5 and Plate 111/5 for the martlet). Richard Lockwood, a wealthy merchant dealing in Turkish goods, bought Dews Hall in Lambourne in 1735 from Pate Thorogood and he married Susannah Cutts, thereby acquiring other estates in Lambourne. A bookplate showing the arms of Richard Lockwood is shown in Fig. 5.21, which shows this crest clearly. Richard died in 1756, and the estate passed to his son, also Richard, who died 1797, aged 84, and having no children the estate passed to his brother the Rev. Edward Lockwood, who died in 1802, the estate then passing to William Lockwood, a grandson, in 1802. In 1842, the estates of Dews Hall and Lambourne were reunited in time for the inheritance of Amelius Richard Lockwood (1847–1928), who became the first Baron Lambourne. For Manners-Sutton, the arms depict for Sutton quarterly, 1st and 4th, argent a canton sable, 2nd and 3rd, or two bars azure

Fig. 5.21 A book plate of Richard Lockwood Esq. (1676–1756), Dews Hall, Essex, showing clearly his crest of a martlet sitting atop a fructed tree stump (possibly showing acorns?), and his helm facing dexter with closed visor and lambrequins surmounting his escutcheon. Neither supporters nor a motto is depicted. The escutcheon gives the arms of Lockwood impaling Cutts. He married firstly, Susannah Cutts, daughter of Edward Cutts of Maldon, Essex, then Matilda Vernon, daughter of George Vernon of Sudbury, Derbyshire, in 1713 – so this bookplate predates his second marriage

a chief quarterly, azure and gules, and for Manners, 1st and 4th, quarterly, two fleur-de-lys and in 2nd and 3rd, a lion passant gardant or. Not shown are the supporters which are two unicorns argent, armed, maned, tufted and unguled or. The crests for Manners -Sutton are: Manners – a wolf's head erased gules (Fairbairn *Crests of the Families of Great Britain and Ireland and Their Kindred in Other Lands,* 1905; 14/6), or on a chapeau gules turned up erminois a peacock in pride ppr. Charged with a crescent (Fairbairn, *Crests of the Families of Great Britain and Ireland and Their Kindred in Other Lands,* 1905; 50/15). For Sutton, the crest is a greyhound's head couped, erminois, collared and ringed or, on a collar three annulets, gold (Fairbairn, *Crests of the Families of Great Britain and Ireland and Their Kindred in Other Lands,* 1905; 43/11). Motto: *"Pour y Parvenir*", which translates as "*To Attain*".

It is very interesting that the crest for Lockwood of a martlet sitting on a fructed tree stump affords a very good description and match for our previously unidentified crest shown in Figs. 5.16 and 5.17, where the bird was tentatively identified possibly as a dove or a martlet. It therefore seems reasonable to propose that this particular crest is in fact attributable to Lockwood, whose crest of a martlet sitting on a fructed tree stump is seen quite clearly in Fig. 5.21. The heraldic symbolism of a martlet is in its relationship to a swallow, which for a time was believed to have no feet as it never rested on the ground but was constantly in flight and therefore assumed to be relevant to heavenly matters rather than earthly ones. The migration and return of swallows to and from warmer climates was thereby assigned to resurrection and rebirth, hence the location of a martlet on a fructing tree stump, reinforcing this concept. Also in heraldry, the adoption of a sable martlet on a coat-of-arms escutcheon signified the arms belonged to the fourth son, who was unlikely to inherit anything and therefore had to make his own way in life, aptly related to the phrase applied to the swallow or martlet "flying constantly in the world".

5.10 Demi-Leopard Crest

The plate shown in Fig. 5.22 is again of a moulded Paris fluting body in duck-egg porcelain, but the gilding at the edge is now much more detailed and prolific: the crest shows a gilt demi-leopard rampant with both forepaws raised. It is sometimes rather difficult to distinguish between a leopard and a lion heraldically when the crest is simply gilded as the spots on a leopard can occasionally be only very faintly perceived but the absence of a lion's mane can then be significant. Here, in Fig. 5.21, the leopard spots are clearly visible, although the animal does have a rather prominent mane also (in heraldry, older versions of the leopard – an animal believed to have been created in legend by the mating of a lion *leo* with a *pard,* often represent a leopard with some characteristics of the lion, including the possession of both a mane and spots!). This plate is from the John Andrews Collection at Oriel Plas Glyn-y-Weddw, Llanbedrog, Abersoch, and has been illustrated in Gambon (*Porslen Abertawe a Nantgarw*, 2016, 25/S20) and further examples are provided in Jones & Joseph (*Swansea Porcelain,* 1988. 119/4) and in John (*Swansea Porcelain,* 1958, Plate 81C). As yet, this crest still remains unidentified but see our later discussion. Another example illustrated in Phillips (*Heraldic Symbols on Swansea,* 2006, 18/48) is also more difficult to discern but an added feature is the identical border gilding pattern to that shown in Fig. 5.22, which indicates that it is probably originating from the same Swansea armorial service.

5.11 Rampant Lion Crest

A cup (London shape pattern with a triple ogee curved handle, as shown in Jones & Joseph, *Swansea Porcelain; Shapes and Decoration,* 1988) and saucer in duck-egg porcelain bear the crest of a lion or perhaps a leopard rampant with forepaws raised and dexter rear leg raised standing on the sinister rear leg on a gilt torse. The tail is curved back onto the body but not with the pronounced curvature adopted by the demi-leopard shown in Fig. 5.22; here, Fig. 5.23 shows the cup and the crest prominently displayed. Enlargement of the crest (Fig. 5.24) indicates that it is actually a

Fig. 5.22 The Swansea plate shown here is again of a Paris fluting type in duck-egg porcelain, but the gilding at the edge is now more detailed and prolific: the crest shows a gilt demi-leopard rampant with both forepaws raised. The leopard spots are visible, although the animal does have a rather prominent mane also and a pronounced curvature in the tail. This plate is from the John Andrews Collection at Oriel Plas Glyn-y-Weddw, Llanbedrog, Abersoch, and has been illustrated in Gambon (*Porslen Abertawe a Nantgarw*, 2016, 25/S20) and further examples are provided in Jones & Joseph (*Swansea Porcelain: Set Patterns and Decoration*, 1988. 119/4) and in John (*Swansea Porcelain*, 1958, Plate 84C). ***Reproduced with permission from Gwyn Jones, Director of the John Andrews Charitable Trust Collection, Oriel Plas Glyn-y-Weddw, Llanbedrog, Abersoch, Gwynedd***

rampant lion with a prominent mane, seen to be devoid of spots, and subtly different from the smoother lines of the leopard shown in the demi-leopard crest previously illustrated in Fig. 5.22, which is enlarged similarly in Fig. 5.25 for comparison purposes. This is the only example found of this crest on Swansea porcelain and it is illustrated in Phillips (*Heraldic Symbols on Swansea*, 2006, 18/45); this exemplar is in the Andrews Collection at Plas Glyn-y-Weddw, Llanbedrog, Abersoch, Gwynedd. Again, further work is needed to unequivocally identify and assign this crest.

Fig. 5.23 A cup (London-shaped pattern, with triple curved ogee handle) and saucer in duck-egg porcelain bearing the crest of a lion rampant with forepaws raised and dexter rear leg raised standing on its sinister rear leg on a gilt torse. The tail is curved back onto the body but not with the same pronounced curvature shown by the demi-leopard shown in Fig. 5.22. (*Reproduced with permission from Gwyn Jones, Director of the John Andrews Charitable Trust Collection, Oriel Plas Glyn-y-Weddw, Llanbedrog, Abersoch, Gwynedd*)

Fig. 5.24 Enlargement of the rampant lion crest shown on the Swansea London-shaped cup in Fig. 5.23 indicates clearly the prominent mane and is subtly different from the smoother lines of the leopard shown in the demi-leopard crest in Fig. 5.22. (*Reproduced with permission from Gwyn Jones, Director of the John Andrews Charitable Trust Collection, Oriel Plas Glyn-y-Weddw, Llanbedrog, Abersoch, Gwynedd*)

Fig. 5.25 Enlargement of the demi-leopard crest for comparison purposes with Fig. 5.24. (*Reproduced with permission from Gwyn Jones, Director of the John Andrews Charitable Trust Collection, Oriel Plas Glyn-y-Weddw, Llanbedrog, Abersoch, Gwynedd*)

References

Sir J. Bernard Burke, *Burke's General Armory—The General Armory of England, Scotland, Ireland and Wales: Comprising a Registry of Armorial Bearings from the Earliest to the Present Time* (Harrison & Sons, Pall Mall, London, 1884)

H.G.M. Edwards, *18th and 19th Century Porcelain Analysis: A Forensic Provenancing Assessment* (Springer -Nature, Dordrecht, 2020)

H.G.M. Edwards, *Porcelain Analysis and Its Role in the Forensic Attribution of Ceramic Specimens* (Springer -Nature, Dordrecht, 2021)

J. Fairbairn, *Crests of the Families of Great Britain and Ireland and Their Kindred in Other Lands*, (Volumes I and II T.C. and E.C. Jack, Edinburgh and London, 1905 and New Orchard Publishing, London, 1986)

O. Fairclough, The use of set pattern decoration at the Swansea China Works, in *Welsh Ceramics in Context*, ed. by J. Gray, vol. II, (The Royal Institution of South Wales, Swansea, 2005), pp. 192–208

A.C. Fox-Davies, *Armorial Families: A Complete Peerage, Baronetage and Knightage, and a Directory of some Gentlemen Coat -Armour and Being the First Attempt to Show which Arms Are in Use at the Moment Are Borne by Some Legal Authority*, vol I and II (T.C. and E.C. Jack, Grange Publishing Works, Edinburgh, 1895)

F. Gambon, *Porslen Abertawe a Nantgarw (Swansea and Nantgarw Porcelain)*, Oriel Plas Glyn-y-Weddw, Llanbedrog, Abersoch, Gwynedd, (2016)

W.D. John, *Swansea Porcelain* (Ceramic Book Company, Newport, Gwent, 1958)

A.E. Jones, S.L. Joseph, *Swansea Porcelain: Shapes and Decoration* (D. Brown, Cowbridge, UK, 1988)

K.S. Meager, *Swansea and Nantgarw Potteries: Catalogue of the Collection of Welsh Pottery and Porcelain on Exhibition at the Glynn Vivian Art Gallery, Swansea* (Swansea Corporation, Swansea, 1949)

E. Morton Nance, *The Pottery and Porcelain of Swansea and Nantgarw* (B.T. Batsford & Son, London, 1942)

D. Phillips, Heraldic Symbols on Swansea. *Nantgarw & Swansea Newsletter*, 18 (2006)

The Brownings Correspondence, Letters to and from Robert and Elizabeth Barrett Browning, eds. P. Kelley and R. Hudson, Volumes 1 and 2 (1809–1831) of 27 Volumes, (Wedgestone Press, Winfield, Kansas, 1984)

Chapter 6
Resume of the Current Situation with Nantgarw and Swansea Porcelain Armorials

Malum Consilium Quod Mutari Non Potest
It is a Bad Plan that Admits no Modifications
Publilius Syrus, 85-43 BCE, "Sententiae"
(Bickford-Smith, Publilius Syrii: Sententiae, 1895)
You see but you do not observe. The distinction is clear.
Sherlock Holmes to Dr John Watson, A Scandal in Bohemia.

Sir Arthur Conan Doyle, 1891.

Abstract A brief resume of the current situation thus far in which all Nantgarw armorials have been identified and 4 crests remain from the Swansea armorials. The results are displayed in tabular form in three tables for ease of comparison with each other.

Keywords Tabular displays of heraldic data · Swansea armorials · Nantgarw armorials

A summary of the current research that has been undertaken into Nantgarw and Swansea porcelain armorials recorded in the literature as achieved thus far can now be undertaken: the combined data relating to Nantgarw and Swansea armorial porcelains described in Chaps. 4 and 5 is summarised in Table 6.1 and the data relating to the crests and partial coats-of-arms for the Nantgarw and Swansea armorial porcelains described in Chaps. 4 and 5 are summarised in Tables 6.2 and 6.3, respectively.

Table 6.1 lists the identified arms thus far for the Nantgarw and Swansea China Works porcelains and the result is as follows: Nantgarw 7 from 7, Swansea 7 from 11, with four crests suggested as possible assignments tentatively thus far and awaiting further research and possible documentation in confirmation. This gives a total of 14 armorials identified positively from the whole collective research database of 18 armorials, i.e. giving a 78% success for definitive attributions of Welsh porcelain armorials, which increases to 100% when the possible and reasonable attributions for the four Swansea crests are included. In Table 6.2, the thirteen known crests, comprising 4 from Nantgarw and 9 from Swansea porcelains, have been listed. All of these have now been identified for Nantgarw (100%) but only 5 from the 9 (56%)

H. G. M. Edwards, *Welsh Armorial Porcelain*,
https://doi.org/10.1007/978-3-030-97439-8_6

Table 6.1 Nantgarw and Swansea armorials: an analytical summary

Manufactory	Armorial	Bearer of arms	Identified[a]	Possible	Literature amendment needed
Nantgarw	Full COA[b]	Lord Henry Thynne, Visc. Weymouth	Yes	–	Yes
	Partial COA	Henry Wyndham, Earl Dunraven	Yes	–	No
	Crest+motto (1)	Ramsay of Straloch	Yes	–	No
	Crest+motto (2)	Ramsay of Straloch	Yes	–	No
	Crest	Baron Phipps of Normanby	Yes	–	No
	Crest	Homfray of Penllyne	Yes	–	No
	Escutcheon	Jeffreys of Brecon Priory	Yes	–	Yes
Swansea	Partial COA	Clarke/Parkinson	Yes	–	Yes
	Crest+motto	Lloyd of Bronwydd	Yes	–	No
	Crest/Boar's Head	–	No	Yes	Yes
	Crest/Dolphin	Orme of Northampton	Yes	–	Yes
	Crest/Paschal Lamb	Wyndham Lewis	Yes	–	Yes
	Crest/Escallop shell	Clarke of Hereford / Newcastle	Yes	–	Yes
	Crest/Spreadeagle	Turbervill of Ewenny	Yes	–	Yes
	Crest/ Bird+Treestump	–	No	Yes	Yes
	Crest/Rampant Lion	–	No	Yes	Yes
	Crest/ Demi-Leopard	–	No	Yes	Yes
	Partial COA	Lockwood / Manners-Sutton	Yes	–	Yes

Analysis: Nantgarw 7 from 7 ---------- 100%
Swansea 7 from 11 (not including four tentative possible assignments)------ 64%
Total: 14 from 18 -- 78% definite for Welsh porcelain armorials
[a]Identified – a "Yes" entry gives the conclusion supported by current research. The "Literature Amendment Needed" Column reflects the need to amend the literature to account for an incorrect existing attribution or none made previously
[b]*COA* Coat-of-Arms

have been identified unequivocally for the Swansea exemplars, four crests having been logically assigned to potential arms bearers and these must await further research and perhaps supportive documentary confirmation to upgrade them from their tentative status. In Table 6.3, the full or partial coats-of-arms on the armorial porcelains are listed: Nantgarw is the only manufactory to display a full coat-of-arms on an armorial artefact in extant exemplars of Welsh porcelains and that has been re-assigned here to the arms of Lord Thomas Thynne, Viscount Weymouth at

Table 6.2 Summary of crests on armorial Nantgarw and Swansea Porcelains

Factory	No.	Description of crest	Identified bearer of arms	Crest in Fairbairn[a]	Motto provided	Result(Y/N)[b]
Nantgarw	1,2	Eagle rising regardant, membered, armed	Ramsay of Straloch	77.4	Migro et Respicio	Y
	3	Otter wounded by spear	Homfray of Penllyne	9.5	–	Y
	4	Demi-lion rampant with palm branch in both forepaws	Baron Phipps of Normanby	61.4	–	Y
Swansea	5	Dolphin naiant	Orme of Northampton	140.5	–	Y
	6	Boar chained to holly tree	Lloyd of Bronwydd	57.12	I Dduw B' Or Diolch	Y
	7	Escallop shell	Clarke of co. Hereford	141.1	–	Y
	8	Paschal lamb with pennon of St George	Wyndham Lewis of Greenmeadow	131.14	Duw ar fy Rhan	Y
	9	Eagle displayed, armed	Turbervill of Ewenny	48.11	–	Y
	10	Bird on sprouting tree stump	Lockwood	86.11	–	N
	11	Demi-leopard rampant	Hill	299.8	–	N
	12	Lion rampant	Edwards, Paget, Percy, Lloyd of Glasnevin, Villiers	67.5	–	N
	13	Boar's head erased/couped	Vernon, Seaton	42.2	–	N

[a]Referring to J. Fairbairn, *Crests of the Families of Great Britain and Ireland and Their Kindred in Other Lands,* Volumes I and II, T.C. and E.C. Jack, Edinburgh and London, 1905, and New Orchard Publishing, London, 1986. Volume 2 gives two-tone plates of selected crests, with 35.8 for example, being Plate 35 Crest No.8. The crests are reproduced in Fig. 36a (*a* Boar's head 42.2; *b* Bird on sprouting tree stump 86.11; *c* Demi-Leopard 299.8), Fig. 36b (*a* Dolphin naiant 140.5; *b* Otter 9.5; *c* Paschal lamb with pennon 131.14), Fig. 36c (*a* Eagle displayed armed; *b* Eagle rising regardant; *c* Escallop shell 141.1), Fig. 36d (*a* Demi-lion rampant with palm 61.4; *b* Boar chained to holly tree 57.12; *c* Lion rampant 67.5)

[b]*Y* Yes, *N* No for the unequivocal positive identification of the crest and family attribution: N therefore signifies a potential attribution without the necessary supporting confirmation being provided as yet

the time of the existence of the Nantgarw China Works, which has the escutcheon, supporters, crest, torse, and motto all supporting the assignment. However, it has also been noted in Chap. 4 that the arms portrayal of Viscount Weymouth which have been previously thought to be unequivocal in attribution is not straightforward at all and there are some discrepancies that require an explanation as to the message

Table 6.3 Coats-of-arms (full & partial) depicted on Nantgarw and Swansea Porcelains

Factory	Escutcheon	Supporters	Crest	Fairbairn plate[a]	Helm&Mantling	Torse	Coronet	Motto	Attribution
Nantgarw	Yes	Reindeer&Lion Reindeer statant	12/8		Yes	Yes	Viscount	Yes	Lord Henry Thynne, Viscount Weymouth
	Yes	No	Lion couped	90/14	No	Yes	No	Yes	Henry Wyndham-Quin Earl of Dunraven
	Yes	No	No	–	No	No	No	No	Jeffreys of Brecon
Swansea	Yes	No	Yes	141/14	No	Yes	No	Yes	Clarke of Co. Hereford Impaling Parkinson
	Yes	No	No	–	No	No	No	No	Lockwood Impaling Manners-Sutton

[a]The Plate and Number as given in Fairbairn, *Crests of the Families of Great Britain and Ireland and Their Kindred in Other Lands,* Volumes I and II, T.C. and E.C. Jack, Edinburgh and London, 1905, and New Orchard Publishing, London, 1986. Volume 2 gives two-tone plates of the selected crests which are described appropriately in the text

being revealed, assuming that these errors cannot simply be ascribed to carelessness in execution by the unknown enameller or atelier concerned, which is deemed to be highly unlikely. The identity of Viscount Weymouth as Lord Henry Frederick Thynne as portrayed in the literature is certainly incorrect as he did not succeed to this title until 1837, some 17 years after the closure of the Nantgarw China Works .The two partial coats-of-arms on porcelain artefacts have also been identified from the Nantgarw China Works, one with its crest and motto and the other with an escutcheon only and without crest or motto. The former is clearly identified with Henry Wyndham-Quin, the Earl of Dunraven, which is unequivocal; in contrast, the other exemplar has been labelled as a realistic identification in the absence of other supporting documentary information and without the presence of further supporting armorial devices to confirm the attribution.

At Swansea, two partial coats-of-arms have now been identified positively, namely those identified incorrectly in the literature as Clarke of Hereford impaling Parkinson and a hitherto unidentified partial coat-of-arms which has now been identified as Lockwood impaling Manners-Sutton. The former coat-of-arms and crest now should be re-assigned to Graham-Clarke of Newcastle-upon-Tyne impaling Parkinson of Kinnersley Castle, Herefordshire and this is supported by independent historical documentation. The second example of an escutcheon without any supporting heraldic devices only appeared in a low-resolution monochrome photograph initially in one literature citation and did not provide sufficient detail for its elucidation until a specimen from the same service; this has now been matched and has been attributed properly. So, for both the Swansea China Works and the Nantgarw China Works, the partial coats-of-arms which appear on each armorial service exemplar can be definitively attributed. Hence, for the category of partial coats-of-arms on armorial porcelains Nantgarw comes in at 3 from 3 definitively attributed and Swansea at 2 from 2, giving a 100% success rate for the attribution of armorials in this category for both Welsh porcelain manufactories.

The underlying message here is that for the interpretation of armorial arms on porcelain and their definitive attribution from the literature armories it is necessary to have at least two heraldic devices shown; in cases where only one has been displayed, the potential for a misattribution increases because of the commonality of arms possession between several ennobled arms-bearing families and their sharing of some heraldic achievements. However, it is still not an easy task to properly define the attribution to an arms-bearer unequivocally and resorting to independent historical documentation and familial genealogical evidence where that is available is very helpful and appropriate.

The quotation at the start of this Chapter is most appropriate here as we now have to address the changes that need to be made to the established assignments of several of the Welsh porcelain armorials, particularly relating to armorial artefacts from both the Nantgarw and the Swansea China Works, that have perhaps arisen in the literature either from a mis-interpretation of the heraldic symbols or from the adoption of apparently unwarranted assumptions from hearsay evidence. In the decipherment and matching of elusive armorial crests to a family bearer of arms, sometimes another method must be found to deliver the end-product, in this case novel information about arms-bearing families who could have commissioned armorial services in porcelain – and genealogical research then comes to the fore! In this respect, an added value is given because the dates of accession to a title, marriage, the birth of a child, a coming of age or an inheritance of new estates will afford a potential reason for the commissioning of a service. It should be remembered that a porcelain service of high quality was always an expensive item, especially for the gilding work, and its commission would not normally be undertaken lightly.

For the four crested armorial service exemplars from the Nantgarw China Works, a 100% success in identification is achieved: these comprise the two *Ramsay of Straloch* services, each bearing a crest and the motto "*Migro et respice*" which serves to aid the attribution unequivocally, the *Homfray of Penllyne Castle* with its unique crest of a speared otter imbrued, and the *Baron Phipps of Normanby* service which also made use of supporting genealogical and historical documentation to confirm its assignment , although here too the crest of a demi-lion is unique because of the palm branch held in both front paws. These attributions are deemed to be correct and do not require any modification or amendment to the existing literature. The situation with the crested china exemplars is seen to be particularly difficult with the nine armorials in this category from the Swansea China Works; here, only one has been found to be correctly attributed in the literature, namely that of Thomas Lloyd of Bronwydd, Ceredigion, which in addition to its unique crest and motto in Welsh of "*I Dduw B'Or Diolch*", also benefits from genealogical information that confirms it was commissioned as s service to celebrate the marriage of Thomas Lloyd and Ann Davies. The only other crested Swansea armorial that had been "positively" identified in the literature was that bearing a dolphin, which had been variously attributed to Abel Gower of Clunderwen or Garnett-Orme of Abbeytown, co. Mayo, both of which are incorrect and this now needs to be revised in attribution to Orme of Northampton – the key to this revision being the minor details possessed by the dolphin *naiant* crest. Of the remaining seven crests, all of which had been labelled as "unidentified" in the literature to date, three have now been positively identified and associated with Graham-Clarke of Newcastle-upon-Tyne (escallop shell crest quartered in gules and or), Wyndham Lewis of Greenmeadow, Llantrisant (Paschal lamb crest with halo plus a Welsh motto ("*Duw ar Fy Rhan*") and Sir Watcyn William-Wynn of Wynnstay (spread eagle armed). The other four crests were perceived to be multi-familial and a definitive attribution was not immediately possible without employing a novel approach which involved the examination of documentation relating to known local supporters of the Swansea and Nantgarw China Works and also, more widely, the major landowners in Glamorganshire in the first two decades of the nineteenth century. As a result, it has been found possible to make a logical suggestion for the attribution of these four crests, which involve a bird on a sprouting (fructed) tree stump, a demi-leopard rampant, a lion rampant and a boar's head couped, erased; it naturally requires supporting documentation preferentially along with familial genealogical evidence to move these tentative crest assignments to a more definitive attribution. Even so, the assignment of 5 from the

9 Swansea China Works armorial crested exemplars still gives a 55% success rate for this category and removes the classification of these armorials in the literature as "most remain unidentified".

Currently, the results of the present research into Welsh porcelain armorials has exposed several inconsistencies in the previous attributions in the literature which can now be corrected for accuracy in comprising and establishing a future armorials database: in total, six new definitive attributions have emerged from this study so far, but four further crests that appear on Swansea porcelain have been suggested as having potentially realistic and credible assignments, but still must realistically await definitive confirmation of their attribution. This is a significant step forward but now we need to examine other avenues to see if there is another route to the provision of novel information which can assist in the attribution of these remaining and partial armorial crests. It so happens that the unattributed crests all occur on Swansea porcelain and a novel approach that has not been attempted hitherto is to see if any commonality exists between the crested earthenwares and creamwares of the Cambrian Pottery and the porcelain of the Swansea China Works. It must be remembered that there could be other extant Swansea or Nantgarw armorial porcelain exemplars as yet unrecorded in the literature and even if this exercise does not yield evidence for any positive matches at this time, then groundwork will have been done for some potential links to be forged which will assist potentially in the better identification in the future of an unknown armorial exemplar which does not appear in our study of those that do appear currently in the literature.

6.1 The Input of Sherlock Holmes' Methodology to the Definitive Attribution of Armorial Porcelains

The great fictional detective, Sherlock Holmes, captured the imagination of an avid Victorian readership who were thirsting for examples as to how the new "forensic science" was being used to solve crimes and several Holmesian quotations have since emerged into our modern usage to illustrate his thinking and approach to the solution of otherwise seemingly insoluble mysteries. The full literary canon of Conan Doyle's Sherlock Holmes encompasses 56 short stories, commencing with *A Scandal in Bohemia* (1891) and finishing with *The Adventure of Shoscombe Old Place* (1927), and four full length novels, namely *A Study in Scarlet* (1888), *The Sign of Four* (1890), *The Hound of the Baskervilles* (1902) and *The Valley of Fear* (1915). Four Sherlock Holmes plays were also commissioned in this period, namely *The Angels of Darkness* (1889), *Sherlock Holmes: A Drama in Four Acts* (1899), *The Speckled Band* (1902) and *The Crown Diamond: An Evening with Mr Sherlock Holmes* (1921). It is interesting that Sir Arthur Conan Doyle maintained that his favourite Sherlock Holmes tale was *The Speckled Band*, which featured the evil Dr. Grimesby Roylott. Since then, several other Sherlock Holmes tales have been written by modern authors in the style of Conan Doyle and have enlarged the accessible canon of literature but all of these can be classified as pastiches and do not therefore properly belong to the original literary canon, although they all attempt to apply the original Sherlockian methodology to the solution of crimes. The basis of the Sherlock Holmes method of detection has some clearly enunciated principles of engagement, which involve the following stages:

1. The clear definition of the problem that requires a solution.
2. The assimilation of the relevant factual data relating to that problem.
3. The development of several working hypotheses which fit these facts.
4. The elimination of the least likely hypotheses to eventually arrive at the most likely hypothesis which leads to a solution.
5. The critical assessment and evaluation of the solution afforded from the identification and the adoption of the most likely hypothesis.
6. The final question as to whether this hypothesis provides the unique solution or possibly if a feasible alternative applies?

What is perhaps not generally appreciated by readers of the Conan Doyle Sherlock Holmes stories is that a rather novel approach by his central character has been adopted by Sir Arthur Conan Doyle which involves "inductive reasoning", often paraphrased as the "bottom-up" logical approach, in which the observed data set is extrapolated backwards to arrive at conclusions about how that final event came about or how that situation materialised. This methodology is diametrically in contrast with the "deductive reasoning" approach, where the detective works forwards from the event or situation. Most readers of the Conan Doyle canon have rather erroneously assumed that Sherlock Holmes' detective powers relied upon deduction rather than induction; the latter better describes the majority of his cases but he did use both methods in several cases to reach his final solution to the problem!

In comparison, therefore, we have here a direct analogy with the analytical theme of the current text, wherein we already have presented before us the armorial bearings in the form of full or partial coats-of-arms, which may normally include an escutcheon, a motto or a crest or a combination of these, and we then have to work backwards to find the "solution" which is of course the identity of the family member whose arms are being portrayed. This methodology involves "inductive reasoning" and has been clearly enunciated by Sherlock Holmes himself in *A Study in Scarlet*, which first appeared in *Beeton's Christmas Annual* (December 1887) and was summed up in a statement by Sherlock Holmes in Chap. 7 therein as follows:

> *In solving a problem of this sort, the grand thing is to be able to reason backwards .... This power is what I mean when I talk of reasoning backwards or analytically.*

The methodology of the analytical approach adopted in this text is precisely that stated above by Sherlock Holmes: the use of inductive "backwards" reasoning based on the Holmesian six-point plan outlined above, which can be further elucidated and illustrated by some direct quotations and exemplars from its use in tales from the Holmes literary canon. Finally, a quotation from Dr. Philip Hodgson (Travis and Hodgson, *Think Like a UX Researcher*, 2019) is clearly based on

Sherlock Holmes' deductive and inductive reasoning powers; this statement could well have been believed to have been a pronouncement from the great detective himself and poses a warning to be heeded in our quest for the accuracy of attribution of an armorial porcelain artefact:

> *Never, ever, ever, act on assumptions. Search out the facts and act on them!*

Firstly, one needs to establish the facts of the case or situation presented for our case studies: in our theme of armorial porcelains this means that we need to isolate the armorial bearings, be these partial coats-of-arms, crests and mottoes, and then to proceed with the investigation using the armories and heraldic listings available in the reference literature to seek out the distinctive features which will enable us to work backwards inductively to isolate and identify an individual from a family of origin. It is at this stage that the identification of a descriptively unique fact associated with the arms can facilitate the process of rejection of the least likely hypotheses which can then be invoked to narrow down the number of possibilities and arrive at the most reasonable solution. Sherlock Holmes has made several pertinent statements in his cases upon the use of this very procedure:

> *I never guess; it is a shocking habit and destructive to the logical faculty*
>
> Sherlock Holmes, The Sign of Four, 1890.

> *It is a capital mistake to theorise before one has the data. Insensibly one begins to twist the facts to suit the theories instead of theories to suit the facts.*
>
> Sherlock Holmes, A Scandal in Bohemia, 1891.

> *"Data, data, data", he cried impatiently "I cannot make bricks without clay".*
>
> Sherlock Holmes, The Adventure of The Copper Beeches, 1892.

To enlarge upon this process, the researcher needs to identify any unusual or distinctive feature possessed in the arms which may prove to be definitive for its attribution or rejection as stated above. A good example of this maxim which has emerged here in this current research using the inductive approach is provided by the Paschal lamb armorial crest on Swansea porcelain, which had hitherto been labelled as "unidentified" in the literature and which has now been identified and attributed here for the first time. The armorial crest displayed on the Swansea porcelain artefact features the lamb facing dexter with head turned to the front (affronted), surmounted by a halo, with its dexter front leg lifted and supporting a pennon which bears the Cross of St George. This proves to be unique to the assigned family arms-bearing member as other similar examples for consideration have a different attitude adopted by the lamb, which may also be minus its halo and may be bearing a different cross (for example, a saltire, or Cross of St Andrew) and often having an additional feature borne upon the head of the pennon, such as a crown, pheon or a fleur-de-lys finial. In some cases, the dexter raised leg of the lamb also stands upon a mound of earth or foliage. The attribution of this crest to Wyndham Lewis of Greenmeadow, Llantrisant, has been effected here based upon the application of this method to interpret the heraldic devices used in the crest. Another example that has been discussed fully here relates to the Pitt service, the earliest extant artefacts of an English armorial porcelain service in Chinese hard paste porcelain and dating from 1705, where slight differences between the crest of the stork depicted on the porcelain and other similar crests reproduced in the armories are sufficient to determine unequivocally the correct family bearer of these arms from a list which initially numbered 15 possible candidates for attribution from the wider Pitt family of arms bearers. In Sherlockian parlance the following quotations are therefore considered to be most apt to this situation relating to the precise idiosyncrasies possessed by the crests that are depicted on armorial porcelains:

You know my method: it is founded on the observation of trifles"

Sherlock Holmes, The Adventure of the Boscombe Valley Mystery, 1891.

It is, of course, a trifle, but there is nothing so important as trifles.

Sherlock Holmes, The Adventure of The Man with the Twisted Lip, 1891.

Each new discovery furnishes a step which leads on to the complete truth.

Sherlock Holmes, The Adventure of the Engineer's Thumb, 1892.

It has long been an axiom of mine that the little things are the most important... I have also trained myself to see what others have overlooked."

Sherlock Holmes, A Case of Identity, 1891.

Finally, even when the research into the portrayed arms seems not to have resulted in a definitive and unequivocal attribution as perhaps one would have expected and it is appreciated that some additional heraldic or genealogical familial information is required to arrive at a unique solution, the following Sherlockian quotations also address this eventuality:

Any truth is better than an indefinite doubt."

Sherlock Holmes, The Adventure of the Yellow Face, 1893.

It is an old maxim of mine that when you have excluded the impossible, whatever remains, however improbable, must be the truth.

Sherlock Holmes, The Adventure of the Beryl Coronet, 1892.

Never trust to general impressions but concentrate yourself upon details.

Sherlock Holmes, The Adventure of the Blue Carbuncle, 1892.

Before we start to investigate, let us try to realise what we do know, so as to make the most of it, and to separate the essential from the accidental".

Sherlock Holmes, The Adventure of the Priory School, 1904.

Elementary....It is one of those instances where the reasoner can produce an effect which seems remarkable to his neighbour because the latter has missed the one little point which is the basis of the deduction".

Sherlock Holmes, The Adventure of the Crooked Man, 1893.

It is clearly manifest that the adoption of inductive reasoning and the observation of "trifles" that are associated with the crests portrayed on the armorial porcelain exemplars under study should eventually result in, if not an unequivocal, then a most likely and credible hypothesis for the candidature of matching a family arms bearer to the arms or heraldic device being displayed on the porcelain artefact. The application of inductive reasoning to analytical science was actually first proposed by Sir Francis Bacon (1561–1626), Lord Verulam and first Viscount St Albans, Lord Chancellor of England under King James I between 1617 and 1621, who advocated the development of the scientific method and initiated the scientific revolution in the late sixteenth century and early seventeenth century. Bacon argued for the generation of scientific knowledge through inductive reasoning and accurate observation, just as Sherlock Holmes did some three hundred years later (Bacon, *The Twoo Bookes of Francis Bacon: Of the Proficience and Advancement of Learning, Divine and Human*, 1605). Sir Francis Bacon is often credited historically with the use of Publilius Syrus' opening quotation given here as his motto but independent verification of this was not achieved upon cross-checking in Fairbairn (*Crests of the Families of Great Britain and Ireland and Their Kindred in Other Lands,* 1905) and Burke (*Burke's General Armory—The General Armory of England, Scotland, Ireland and Wales: Comprising a Registry of Armorial Bearings from the Earliest to the Present Time*, 1884). It seems that the association of Bacon with Syrius is an apocryphal one. The actual motto of Sir Francis Bacon given in the armories is "*Mediocria Firma", meaning "Mediocrity is Stable"*, which accompanies his coat-of-arms (which comprise two estoiles sable on an argent field with fess gules, quartered with a barry of six or and azure for his heiress wife, these being the Quaplode arms). The motto and crest of a boar regardant erminois for Sir Francis Bacon are also adopted from those of the Quaplode family – several historians have commented hitherto that the boar adopted as his crest may be a pun on bacon! Again, as has been found on several occasions in this research, the legend and the facts discerned upon checking do not seem to agree in the detail. Bacon is also credited with adopting "*Scientia Potentia Est"*, meaning "*Knowledge is Power*" as a motto, which is an attractive proposition as it certainly conforms to his scientific beliefs (unlike mis motto which gives credibility to mediocrity!) but the motto as given in the armorial literature does not support this allegation.

References

Sir F. Bacon, *The Twoo Bookes of Francis Bacon: Of the Proficience and Advancement of Learning, Divine and Human* (Henrie Tomes, Grays Inn Gate, Holborn, London, 1605)

R.A.H. Bickford-Smith, *Publilii Syrii: Sententiae* (C.J. Clay/Cambridge University Press, London, 1895)

Sir J. Bernard Burke, *Burke's General Armory – The General Armory of England, Scotland, Ireland and Wales: Comprising a Registry of Armorial Bearings from the Earliest to the Present Time* (Harrison, London, 1884)

J. Fairbairn, *Crests of the Families of Great Britain and Ireland and Their Kindred in Other Lands,* vols I and II (T.C. and E.C. Jack, Edinburgh/London, 1905, New Orchard Publishing, London, 1986)

D. Travis, P. Hodgson, *Think Like a UX Researcher* (Routledge /Taylor & Francis, London/New York, 2019)

Chapter 7
A New Approach

Aut Viam Inveniam Aut Facium
I Shall Either Find a Way or Make One
Hannibal Barca, 218 BCE: in response to his generals' questioning the wisdom of his plan to use elephants crossing the Pyrenees and the Alps to attack Rome.
Do not go where the path may lead, go instead where there is no path and leave a trail.
Ralph Waldo Emerson (1803–1882).
The greater danger for most of us lies not in setting our aim too high and falling short but in setting our aim too low and achieving our mark.
Michelangelo Buonarroti (1475–1564).

Abstract A new approach adopted in an endeavour to determine the missing heraldic data for Welsh armorials: a survey of the landowning gentry and aristocracy according to the land records – the support offered to the Swansea and Nantgarw China Works by local benefactors, and their relationship to those who had already commissioned armorial services from the Cambrian Pottery. William Weston Young's activities as land and estate surveyor and his identification of Nantgarw supporters from his travels in South Wales. Young's "Ten true men of Glamorgan", each putting money into the establishment of the Nantgarw China Works.

Keywords Financial support for Swansea and Nantgarw China Works · William Weston Young · Titled landowners · "The ten true men of Glamorgan"

The premier historian of the Swansea and Nantgarw factories and their porcelains in the second half of the twentieth century was undoubtedly Dr. William John of Newport, who in his four seminal texts (John, *Nantgarw Porcelain*, 1948; John, *Swansea Porcelain*, 1958; John, *William Billingsley*, 1968; John et al., *The Nantgarw Porcelain Album*, 1975) covered many aspects of the output from these

H. G. M. Edwards, *Welsh Armorial Porcelain*,
https://doi.org/10.1007/978-3-030-97439-8_7

manufactories, the stylistic designs of the porcelain that was produced and the personalities of the proprietors and decorators involved in the Nantgarw and Swansea China Works; these books are amply illustrated with coloured and monochrome plates of the porcelain under discussion. Although Ernest Morton Nance's earlier single work (*The Pottery and Porcelain of Swansea and Nantgarw*, 1942) is still regarded as the major source book for Nantgarw and Swansea porcelains, encompassing as it does the results of some 40 years' research on the topic, it is truly a veritable mine of information but is consequently much less sympathetic in content for the general reader who is not looking for the special details and minutiae which can frequently obscure the main storylines. Nevertheless, the accuracy and wealth of detailed information in Morton Nance's book has been respected by generations of successive ceramic historians who are interested in the manufacturing operations of the Nantgarw and Swansea manufactories. Despite this extensive coverage, armorial porcelains of Nantgarw and Swansea hardly receive a mention and the reader will surely now appreciate the dearth of information that is available on this topic for Welsh porcelains in comparison with their contemporary Chinese and English porcelain analogues. Clearly, as addressed by the above quotation of Hannibal Barca, it will be necessary to discover some other route or lateral approach to deliver novel information about armigerous families and armorial porcelains when the normal consultation of the armorial literature has arrived at an apparent impasse regarding the definitive attribution of an armorial ceramic artefact.

Two statements made by Dr. William John in his books are worthy of analytical comment in the context of Welsh porcelains: firstly, he claimed to have seen almost 5000 pieces of surviving Nantgarw porcelain in his lifetime and in this regard he confirmed that in his opinion there was just one Nantgarw body, echoing the earlier visual observation of Herbert Eccles that all the Nantgarw pieces submitted for the Nantgarw and Swansea Centennial Exhibition at the Glynn Vivian Art Gallery in Swansea in 1914 were all of the one body type, namely a soft paste phosphatic porcelain. It may not be generally appreciated too that Morton Nance devotes approximately half of his seminal text to the history and products of the two manufactories of earthenwares at Swansea, namely, the Cambrian Pottery and the Glamorgan Pottery, both of which were contemporary with the shorter-lived porcelain manufactory at the Swansea China Works and in fact outlived its demise by many years. The Cambrian Pottery had actually been in existence for almost 50 years before porcelain was made on the adjacent site at Swansea and the Glamorgan Pottery was founded in 1814, the same year that William Billingsley and Samuel Walker joined forces with Lewis Weston Dillwyn to make porcelain at Swansea. In contrast with Morton Nance (*The Pottery and Porcelain of Swansea and Nantgarw,* 1942), the first dedicated coverage of ceramics at Swansea and Nantgarw by William Turner, published some 45 years before his own book in 1897, does not even mention the presence of armorials in either the Welsh porcelain or the earthenwares (Turner: *The Ceramics of Swansea and Nantgarw. A History of The Factories with Biographical Notices of the Artists and Others, Notes on the Merits of the Porcelain, the Marks Thereon Etc.,* 1897). William Turner, in the preparation of his book on the ceramics of Swansea and Nantgarw in the late nineteenth century, was able to interview people such as Henry Morris, who had worked at the Swansea China Works, and who was able to recount some information about his experiences, the personnel who worked there, what was made there and who supported the activities of the manufactory. It is quite remarkable therefore that Henry Morris never mentioned to William Turner, or to Colonel Grant Francis (who had interviewed him earlier in 1850), the provision of armorial services at the Swansea China works, especially since he was himself responsible for enamelling the Clarke of Hereford /Graham-Clarke of Newcastle impaling Parkinson armorial Swansea service that has been described in Chap. 5. In contrast, Morris did mention other services that he had decorated along with his colleagues in the enamelling workshop at Swansea, including William Billingsley, David Evans, William Pollard, Thomas Baxter and George Beddow.

7.1 Local Landowners and Landed Gentry

Secondly, a particularly interesting comment that has much relevance here was the statement made by Dr. John that many of the Nantgarw pieces that he saw in his extensive personal research for his books were in the local country houses of the gentry and local Welsh landowning families. Dr. John believed that many of these pieces were locally decorated and therefore would have been ordered or commissioned directly through the Nantgarw China Works and not through their London outlet of John Mortlock's in 47, Oxford Street. Ernest Morton Nance also maintained that it was through William Weston Young's personal contacts as a surveyor and land agent with the local gentry and landowners and in his travels through South Wales that personal appeals would have been made by him to these local patrons for support of the Nantgarw China Works which had just started up in Phase II in 1817, and of which Young was a founder partner and major financier in the enterprise. A record of William Weston Young's travels in South Wales and of the people he visited is gained through reading his Diaries (W.W. Young, *The Diaries of William Weston* Young (1776–1847), *1802–1843,* 30 Volumes, West Glamorgan Archives Service, Swansea, SA13SN), which have been transcribed in part by some authors and historians, such as Morton Nance (*The Pottery and Porcelain of Swansea and Nantgarw,* 1942) and Edwards (*Porcelain to Silica Bricks: the Extreme Ceramics of William Weston Young, 1776–1847*, 2019). What is clear from his Diary entries is that William Weston Young had a large circle of local landowners whom he visited and advised professionally and that he did not spend much time at all in Nantgarw, where the mechanics of the porcelain production would be left effectively to William Billingsley and Samuel Walker who were in residence there. This does not imply that he had little interest in the success of the Nantgarw China Works or in its products as indeed he had invested considerable amounts of money into the business, both personally and through the encouragement of his clients to make some similar investment. This investment at Nantgarw totalled £2100, more than half being put up by Young personally, which would have been equivalent to approximately £250,000 today ($300,000).

The contribution of William Weston Young to the successful Nantgarw operation as the financial backer and partner to the china production activities of William Billingsley and Samuel Walker in the period 1817–1820 was exemplary and materially catalytic towards the success of the Nantgarw enterprise initially. It is realised, of course, that when William Weston Young and Thomas Pardoe eventually had the responsibility for the decoration and disposal of the remnant stocks of Nantgarw porcelain in the post-Billingsley/Walker period, 1820–1823, much if not all of their production would have been deployed locally either through the Nantgarw China Works itself at Tyla Gwyn or through the two Aubrey auction sales held at Cardiff and Cowbridge in 1820 and 1822. Historically, William Weston Young and Thomas Pardoe whom it would transpire have such a major input to the Nantgarw China works after William Billingsley and Samuel Walker had departed in early 1820 for Coalport had actually developed their personal friendship when both were employed as decorators and enamellers by William Dillwyn and Lewis Weston Dillwyn at the Cambrian Pottery, Swansea, between 1802 and 1810 (Edwards, *Porcelain to Silica Bricks: The Two Ceramic Extremes of William Weston Young, 1776–1847*, 2019).

It would certainly be appropriate, therefore, to undertake here a review of the local gentry and major landowning families to try and assess what their response might have been to William Weston Young's surveying business activities and potential later marketing opportunities for the Nantgarw China Works which had been ongoing in South Wales for at least a decade before the Nantgarw China Works set up for its second phase of operations in 1817. In this respect, it is reassuring to read the entries in Young's Diaries (William Weston Young, *The Diaries of William Weston Young (1776–1847), 1802–1843,* 30 Volumes, West Glamorgan Archives Service, Swansea, SA13SN; Edwards, *Porcelain to Silica Bricks: The Two Ceramic Extremes of William Weston Young, 1776–1847*, 2019) which reinforce the idea that he was indeed travelling extensively around Glamorganshire, Brecknock and Monmouthshire and visiting influential landowning people, several of whom are named specifically therein. In many cases he seemed to have been a regular visitor at their country residences, for example the Mansel-Talbots at Margam Park, where he received a commission to construct a memorial to Thomas Mansel-Talbot using local stone, such as Mumbles marble and Penrice sandstone – an effort which necessitated his visiting Margam regularly over a period of some 5–6 years from 1814 from his home in Newton Nottage near Porthcawl (Edwards, *Porcelain to Silica Bricks: The Extreme Ceramics of William Weston Young, 1776–1847*, 2019). In fact, the first meeting between Billingsley, Walker and Dillwyn to discuss their proposed involvement with the fledgling Swansea China Works was recorded by William Weston Young as taking place at Margam Park, which resulted in Dillwyn then visiting Nantgarw and immediately engaging the services of Billingsley and Walker to assist him in his ambition to realise the manufacture of porcelain at Swansea. It is not stated in his Diaries but it is certainly implicit that Young's influence was material is setting up this meeting as he knew well Lewis Dillwyn on the one hand and William Billingsley and Samuel Walker on the other. It would only have taken one or two persons to order Nantgarw services locally for their personal use and dinner entertaining and the word would soon be passed around other visitors to dinner or tea parties who would then see for themselves this beautiful ceramic which everyone, especially in London society, was desirous of acquiring from John Mortlock who as Nantgarw sole agent wanted more porcelain than Billingsley and Walker could manufacture to satisfy his demanding clientele! For the first time, therefore, a new approach is to be adopted here for the potential identification of armorial Nantgarw and Swansea china: namely, a survey to be undertaken of the main landowners, gentry and aristocracy in the region who may have been able to acquire porcelains from the two Welsh factories between about 1814 and 1823. These armigerous families would have additionally been able to commission armorial porcelains from both Nantgarw and Swansea.

From the records of the Glamorgan Land Registry (Glamorgan Land Registry Records, Archives and Land Registry GB214D19, Glamorgan Records Office, Cardiff) it is possible to construct, firstly, a list of the major landowners, several of whom owned land in the County of Glamorgan but who lived outside its borders – and to compare these names with those whom it is known had commissioned porcelain services from Swansea and Nantgarw. Secondly, to identify names of landowners who appear in William Weston Young's Diaries as people whom he recorded that he had visited perhaps on a regular basis as part of his land surveying business. Table 7.1 gives a list of the selected major landowners in Glamorganshire in the early nineteenth century, who might therefore fit into our consideration of the two criteria of "eligibility" and "being mentioned" in the Diaries as we have described above. Many of these would have been entitled to bear crests or coats-of-arms on their china and indeed, several are members of the peerage and aristocracy: what is doubly interesting is the cross-correlation of the names on this list and a comparison of those who do not have a record of formally owning Swansea or Nantgarw china – as this information will then surely place them in focus for being primary candidates for the potential attribution of their crests for matching with hitherto unidentified armorial porcelains? In Table 7.1, therefore, we can summarise the information presented as follows:

Of the 49 major landowners listed in Table 7.1, 20 have been recorded as already owning Nantgarw and/or Swansea porcelain services, i.e. 41%. This can hardly be a coincidence and lends credibility to the main thesis, first commented on by Dr. John from his own research experience, that titled persons, landowners and local gentry of the district would indeed have been a major clientele for the purchase of china from the Nantgarw and Swansea manufactories. This list comprises the major landowners, several of whom were wealthy business people such as the industrialists and Merthyr Tydfil ironmasters, the Crawshays, Homfrays, Guests and Hill, who were acknowledged nationally as some of the wealthiest industrialists of their time. This is no accident of circumstance, as Dr. Samuel Johnson iterated upon his accompanying his friend, the physician Dr. Butter, to the Derby China Works on the 20th September 1777 and being urged by his friend to purchase a service of china whilst there: Dr. Johnson commented to Boswell that he could not afford Derby china as the cost was too prohibitive –

Table 7.1 Prominent landowners in South Wales in the early nineteenth century (from the Glamorgan Archives and Land Registry Records, GB214D19)

Henry Paget, Marquess of Anglesey[a], Llanddaniel, Anglesey
Crawshay Bailey[b], Maindiff Court, Abergavenny, Monmouthshire
Sir Joseph Russell-Bailey, Glanusk Park, Brecon, Brecknock
Walter Thomas Baskerville, Clyro Court, Radnorshire
Arthur Davies - Bennington, Pantygoetre, Monmouthshire
John Crichton-Stuart, Marquess of Bute[a, b], Cardiff Castle, Glamorganshire
John Jeffreys Pratt, Marquess of Camden, Wherwell House, Andover, Hampshire
Nicholas Jeffreys, The Priory, Brecon, Brecknock
John Nicholl-Carne, St Donat's Castle and Nash Manor, Glamorganshire
John Campbell, Earl Cawdor, Castell Martin, Pembrokeshire
George Talbot Rice, Lord Dynevor[a], Dinefwr House, Llandilo, Carmarthenshire,
Thomas Wyndham, Earl of Dunraven[a, b], Dunraven Castle, Glamorganshire
John Bruce-Pryce, Dyffryn House, St Nicholas[a], Glamorganshire
William Booth-Grey, Dyffryn St Nicholas and Monknash[b], Glamorganshire
John Traherne, Coedarhydyglyn and Castellau Fawr[a], Glamorganshire
Lord Talbot, Baron Talbot of Hensol Castle[a], Glamorganshire
John Rhys-Williams, Miskin Manor, Glamorganshire
Sir Josiah J. Guest, Dowlais House, Merthyr[a], Glamorganshire
Howell Gwyn and the Moore-Gwyns, Dyffryn Clydach, Neath[a], Glamorganshire
Edward Villiers, Earl of Jersey, Eaglesbush House, Briton Ferry, Neath[a], Glamorganshire
Wyndham Lewis, Greenmeadow, Llantrisant[a], Glamorganshire
John Dillwyn-Llewellyn, Penllergaer House, Swansea[a, b], Glamorganshire
Sir John Nicholl, Merthyrmawr House, Glamorganshire[a]
Christopher Mansel -Talbot, Margam Park, Taibach, Port Talbot, Glamorganshire
Sir Charles Morgan, Lord Tredegar, Tredegar House, Monmouthshire
T.P. Turbervill, Ewenny Priory, Glamorganshire[a]
Charles Kemeys-Tynte, Cefn Mably House, Glamorganshire
Archer Windsor-Clive, Lord Windsor[b], St Fagan's Castle, Glamorganshire
Morgan Williams, Aberpergwm House, Vale of Neath[a], Glamorganshire
Sir Watcyn Williams-Wynn[a], Wynnstay, Ruabon, Denbighshire
John Edwards-Vaughan, Rheola House, Resolven, Vale of Neath, Glamorganshire
William Ormsby-Gore, Lord Harlech, Porkington and Clennenau, Shropshire and Caernarvonshire
John Rolls, Lord Llangattock[a], The Hendre, Monmouthshire
Capel Hanbury, Pontypool Park, Monmouthshire
William Herbert-Jones, Clytha Park, Raglan, Monmouthshire
Robert Lascelles Jenner, Wenvoe Castle, Glamorganshire
Sir Lloyd Kenyon, Gredington, later Lord Kenyon[a], Flintshire
Charles Mansell Lewis, Stradey House, Llanelli, Carmarthenshire
Augusta Hall, Lady Llanover, Ty Uchaf, Monmouthshire
Sir Thomas Dennis Lloyd, Bronwydd, Cardiganshire[a]
Lewis Mathias, Lamphey Court, Pembrokeshire

(continued)

Table 7.1 (continued)

Sir Edward Lloyd, later Lord Mostyn, Mostyn House, Llandudno, Caernarvonshire
Thomas Penrice, Kilvrough Manor, Swansea, Glamorganshire
John Phillips, Picton Castle, Carmarthenshire
Sir John Stepney, Llanelli House, Llanelli, Carmarthenshire
George Charles Venables-Vernon, Lord Vernon, Sudbury Hall, Derbyshire[a]
Richard Hoare Jenkins, Llanharan House, Llanharan[b], Glamorganshire
Henry Seymour Allen, Preseli House, Pembrokeshire
Henry Somerset, Duke of Beaufort and Earl of Glamorgan, Badminton House, Gloucestershire

[a]Recorded owners and their descendants who possessed Swansea and/or Nantgarw porcelain services or artefacts
[b]Mentioned in William Weston Young's Diaries as active supporters of the Nantgarw China Works

in fact, he noted that the cost of purchasing fine porcelain was equivalent to what it would have cost him in solid silver at that time. In a letter to Mrs. Hester Thrale, Dr. Johnson said:

- *The Derby china is very pretty but I think that the gilding is all superficial and the finer pieces so dear that perhaps silver vessels of the same capacity may be bought at the same price*
- Two classic contemporary examples to illustrate this statement are, firstly, a 150 piece, 12-place setting, breakfast service of armorial porcelain that was ordered by Admiral Lord Horatio Nelson from Chamberlain's Worcester manufactory in August 1802, which was delivered to Mrs. Emma Hamilton at their country home of Merton Place in Surrey, where some pieces were used and the remainder taken on board *HMS Victory* for Admiral Nelson's use there at sea. This breakfast service was known as the *Horatia* service after Lord Nelson and Emma Hamilton's baby daughter. A plate from this armorial service is illustrated in Fig. 7.1: readers should note the wealth of information that is available for checking the attribution of the heraldic achievements on this plate compared with that of the simple escutcheon depicted on the Nantgarw plate shown in Fig. 4.8 (and its enlargement in Fig. 4.9). Mrs. Hamilton eventually received the invoice for this service from Chamberlain's Worcester factory, which was rather tactlessly presented to her during the week of Nelson's State funeral following his death at the Battle of Trafalgar (21st October 1805) in January 1806; the amount requested was £120 10s 6d, which today would be equivalent to approximately £12,000. Secondly, in 1831, Queen Adelaide approved the final specimens prepared by the Royal Rockingham China Works for an armorial service for King William IV – a sumptuous dessert service of 200 pieces with several special central table pieces, including three-tier fruit comports and ice pails, costing in total £3000 guineas, equivalent to approximately £250,000 today. This service was not delivered from the Rockingham China Works during the King's lifetime and was first used at Queen Victoria's Coronation banquet in Windsor Castle in 1837 (Cox & Cox, *Rockingham Porcelain, 1745–1842*, 2001).

Fig. 7.1 Plate from the Chamberlain's Worcester armorial breakfast service of 150 pieces commissioned in August 1802 by Admiral Lord Nelson, the so-called *Horatia* service, displaying his full coat-of-arms comprising the escutcheon, supporters, crest, torse, mantling, helm and motto in full colour which assist in the unequivocal identification of the family arms bearer. *Reproduced by courtesy of Bonhams Auctioneers and Valuers, London*

- Several of the titled persons in the list, although not recorded as specifically owning Nantgarw or Swansea china services, did commission porcelain from other earlier or contemporary china works, for example, John Jeffreys Pratt, the Marquess of Camden, John Campbell, Earl Cawdor, Henry Somerset, the Duke of Beaufort, and Sir John Stepney of Llanelli. It is perhaps not unreasonable to suggest a hypothesis that they might also have furnished their South Wales estates with local high quality and highly esteemed china services?
- A comparison of the names in the titled landowners list in Table 7.1 with the list of 74 major collectors of Welsh porcelain consulted by William Turner in the last decades of the nineteenth century for the preparation of his book on Swansea and Nantgarw porcelains (Turner, *The Ceramics of Swansea and Nantgarw*, 1897) indicate that there are some 20 identities between the two lists, which confirms the hypothesis that the major landowners were also supporters and collectors of Welsh porcelains even after the closure of the manufactories some 60–70 years hitherto. Again, this cannot be a coincidental occurrence. Several of these noted collectors listed by Turner, such as William Bemrose, were based outside Wales.
- Seven of the major landowners listed in Table 7.1 were resident outside Glamorganshire, representing 15% of the total. There would be many other minor landowners, some of whom were entitled to bear arms and with seats in South Wales, who presumably might have supported the Swansea and Nantgarw manufactories with porcelain service commissions. Examples would include the Mackworths of Gnoll House, Neath, the Vivians of Swansea (antecedents of

Lord Swansea), Francis Gwynne of Llansannor, Cowbridge and the Youngs of Preswylfa, Neath. In this context, a search of John Bateman's book on the landowners of Britain and Ireland was thought to be worthwhile. This book (Bateman, *The Great Landowners of Great Britain and Ireland: A List of All Owners of Three Thousand Acres and Upwards Worth Eight Thousand Pounds a Year,* 1883) resulted from the undertaking of the first national survey of landowners accomplished since the Domesday Book in 1873 and consists of several thousand entries, citing the acreage and annual income derived from the land. Many of the names cited in the Glamorgan Land Registry Records also appear in this list as expected, especially for the major landowners such as Sir Watkin Williams-Wynn of Wynnstay (146,000 acres), Earl Cawdor (101,700 acres), the Marquess of Bute (117,000 acres) and the Duke of Beaufort (51,000 acres). Other landowners who are resident in South Wales are also named with extensive holdings that do not appear in the Glamorgan Land Registry, such as Sir James William-Drummond (Edwinsford, Llandilo, 9800 acres), Arthur Picton Saunders Davies (Pentre, 8000 acres), Henry Miers (Ynyspentwrch, Swansea, 5000 acres) and Mrs. Wood (Stouthall, Swansea, 5800 acres) and this can probably be attributed to that fact that their land ownership was outside the Glamorgan county boundary as Lady Llanover (Ty Uchaf, Abergavenny, 6500 acres) has a similar acreage of land ownership yet she appears in the Glamorgan Land Registry and also in our Table 7.1. Bateman comments that he found it difficult to generate precise data for Welsh land ownership because of a "desperate simplicity of names" and this resulted in some 60% of Wales being owned by families named Jones, Davies, Evans and Williams. This is in itself an interesting statistic as there is only one Williams in our list (namely, Morgan Williams of Aberpergwm House, Glyn Neath) and others have defining surnames such as Herbert-Jones, Davies-Bennington, Rhys-Williams and Williams-Wynn, which armorially make them quite specific and identifiable. Where possible, each entry in Bateman was submitted to families for checking the accuracy of the data – and several were found to be inaccurate, for example, the Duke of Wellington stated punctiliously that his own entry was "very seriously inaccurate". In an appendix, Bateman included the land ownership details for each designated county and Glamorganshire realised the following analytical data: 6 peers (85,000 acres), 19 great landowners (50,000 acres), 36 squires and 300 yeomen with an additional 8000 small property owners for a total county land ownership of 475,000 acres. It is clear then that, within Glamorganshire, landowners alone would be a considerable number of potential clients for Welsh porcelains locally.

It seems then, surprisingly perhaps, that there would have been significant local support for the Swansea and Nantgarw porcelain products which in one respect belies their comparative rarity amongst the surviving locally decorated specimens today. In some cases, these might involve a simple armorial decoration with perhaps just a gilt crest to indicate the arms-bearing family ownership, as vouchshafed on the Swansea crested porcelains reviewed above. It is interesting that a visual comparison of the Nantgarw and Swansea armorials reveals their simplicity in comparison with some of their contemporary Chinese export porcelains and certainly in comparison with the armorial *Pendock-Barry (Barry-Barry)* Derby service, an example of which is shown in Fig. 1.11, and the Welsh porcelain versions are perhaps then more akin to the Lord and Lady Cremorne service of Derby porcelain shown in Fig. 1.17. The reason for this situation may not reside entirely with decorating costs as Nantgarw and Swansea were strictly the last of the classical soft paste phosphatic porcelain manufactories of the Georgian era, when the revived rococo and highly baroque decoration was already starting to emerge and usher in the Victorian era of opulence and a characteristic perhaps almost "over-the-top" prevalence towards sumptuous decoration and gilding. The *Barry-Barry* (Fig. 1.13) and *Lord Ongley* (Fig. 1.17) Derby porcelain services are classic examples of these sumptuous and perhaps almost over-decorated porcelains of this Regency and immediate post-Regency era where hardly any of the porcelain substrate can be seen through the overlying enamels – in contrast with the *Baron Phipps* Nantgarw armorial service shown in Fig. 4.4, which emphasises the wonderful clarity and beauty of the translucent porcelain substrate so well. John Crichton-Stuart, the Marquess of Bute and Lord Dumfries, an avid supporter of the Nantgarw manufactory with several accredited commissions recognised from the Nantgarw China Works for use in his seat at Cardiff Castle and estate at Mount Stewart in Scotland, ordered a service of Nantgarw porcelain of the highest quality, which was to be glazed but left entirely undecorated and ungilded to demonstrate the beauty and perfection of the porcelain of which it was composed. Examples of this unique service unfortunately have not survived knowingly to the present time but one can appreciate the concept and idea when looking at the *Baron Phipps* service, a plate from which is shown in Fig. 4.4, the nearest equivalent that we have today perhaps to the completely undecorated and ungilded *Marquess of Bute* service: it should be pointed out to readers that at least two other *Marquess of Bute* Nantgarw services exist which have been decorated in the Sevres style with bright ground colours of blue and emerald green at the verge and extensive French-style Empire gilding destined for his use in his home at Cardiff Castle. It is an interesting but relevant digression in this context to consider some of the paintings of Henri-Fantin-Latour (1836–1904), a French Realism artist and friend of James McNeill Whistler and Edouard Manet, who in 1864 painted a simple and even perhaps rather stark white porcelain cup and saucer against a black background; this painting, illustrated in Fig. 7.2, was acquired from the bequest of Herbert Thompson by the Fitzwilliam Museum, Cambridge, whose Director, Luke Syson, describes it as an enigma and as a modest Holy Grail in its simplicity – there is engendered within it a directness of encounter and there are present "*no extras to detract us from its beauty*" (Syson, 2021). This surely, was precisely the feeling of the Marquess of Bute when he commissioned his undecorated Nantgarw service, which would be appreciated by porcelain connoisseurs for its pure simplicity and untrammelled beauty as envisaged by William Billingsley when he first set out to manufacture "*the finest porcelain in the world*" at the end of the eighteenth century. Occasionally, pieces of glazed but undecorated Nantgarw porcelain do surface, such as the example illustrated in Fig. 7.3 here, a magnificent oval dish with its characteristic Nantgarw moulding of scrolls, ribbons, flowers and foliage from which it is

Fig. 7.2 Painting of a "White Cup and Saucer", oil on canvas, height 19.4 cm, width 28.9 cm, by Henri Fantin-Latour (1836–1904), 1864. Bequest of Herbert Thompson, 1920, accession number 1016. *Reproduced by permission of the Syndics of The Fitzwilliam Museum, Cambridge*

Fig. 7.3 Nantgarw porcelain large oval dish, with characteristic moulding, undecorated but glazed, demonstrating fully the beauty of the exquisite porcelain. *Private Collection*

possible to derive an idea of the beauty of such a service as that which would have been ordered by the Marquess of Bute; sadly, these pieces do not generate the interest at auction as do their decorated, enamelled and gilded analogues. The origin and provenance of such pieces is now lost but it is reasonable to speculate that they were acquired from the final Nantgarw sale of remnant porcelains in 1822, and especially after Thomas Pardoe's death in 1823, prior to which he was known to be still actively decorating porcelain left in the white at Nantgarw, it is thought both glazed and unglazed, some of which would still have been available for disposal locally afterwards.

In contrast to the *Baron Phipps* service in Fig. 4.4 and the undecorated, glazed Nantgarw dish shown in Figs. 7.3 and 7.4 shows a Nantgarw plate, impressed NANT-GARW C.W., which has been fully enamelled in a blue background pigment completely covering the white porcelain substrate –the purpose of this type of decoration is unknown, but apparently this specimen also has an applied red enamel stencil mark SWANSEA underneath, so it is an intriguing possibility to speculate that this could be an example of Nantgarw porcelains offered in the final Swansea China Works sale of 1826 which had been purchased at a Nantgarw sale hitherto in the white and decorated afterwards by a Swansea artist, thereby occasioning the unusual presence of both factory marks?

Fig. 7.4 Nantgarw China Works plate decorated completely in a blue background, embellished with floral groups of pansies centrally and at the verge, impressed NANT-GARW C.W. and also carrying a red stencilled SWANSEA mark. Perhaps one of the Nantgarw china plates offered by the Bevingtons in the Swansea China Works closing sale and decorated by local Swansea artists. Note that the translucency of the porcelain is completely masked by the enamelling coverage. *Private Collection*

7.2 A Test Case

To assess the value of this current approach towards the identification of an unknown crest or a partial coats-of-arms on Swansea or Nantgarw porcelain, it is appropriate to now examine the Nantgarw plate which exhibits a simple gilt escutcheon in the verge, as described in Sect. 4.6. above and illustrated in Fig. 4.8; it seems at first that we can derive little information from such a simple representation, which is diametrically opposite to that provided by the Viscount Weymouth armorial Nantgarw plate or the Lord Nelson armorial Chamberlain's Worcester plate, both of which have displayed in full their coats-of-arms and heraldic achievements provided thereon, so making their identification a relatively straightforward procedure. An enlargement of the escutcheon on the unknown Nantgarw armorial is shown in Fig. 7.5, from which we can discern a chevron set between three inverted vees (^), all applied in gilt and surrounded by a stylised foliage of fern fronds, bows and ribbons, also in gilt. Reference to our selected listing of landowning entitled families in Table 7.1 reveals that this simple coat-of-arms appears on the heraldic achievements of two local families: the first of these being Bruce-Pryce of Dyffryn St Nicholas, whose son became Lord Aberdare, a well-known collector and connoisseur of Welsh porcelain and the second appears in the arms of Nicholas Jeffreys of

Fig. 7.5 Enlargement of the escutcheon on the armorial Nantgarw plate shown in Fig. 4.8 and decorated locally by Thomas Pardoe at the Nantgarw China Works, *ca.* 1820-1823. Although very basically painted, the escutcheon clearly shows a chevron between three inverted vees (^) which are believed to represent spearheads (pheons) – and research now affirms that these arms are those of Nicholas Jeffreys of The Priory, Brecon, whose daughter Elizabeth married the 1st Marquess of Camden. The Jeffreys coat-of-arms comprises a chevron between three arrowheads (pheons) argent, the points embrued, which are represented in the Camden arms, 2nd and 4th quarters. *Reproduced by courtesy of Bonhams Auctioneers and Valuers, London*

The Priory, Brecon. In the latter case, this shield also appears in the quartered arms of the Marquess of Camden, elder son of the Earl Camden, Sir Charles Pratt, who was Lord Chancellor between 1766 and 1770 and who was raised to an Earldom in 1786: a plate from his esteemed Derby service and beautifully decorated by William Billingsley is shown in Fig. 1.16. He married Elizabeth Jeffreys, daughter of Nicholas Jeffreys of The Priory, Brecon, and their son, John Jeffreys Pratt, was created Earl of Brecknock and Marquess Camden in 1812. As heiress of the Brecon Priory estates, Elizabeth Jeffreys (1724–1779) and her courtship by John Pratt was the topic of a book by Sally Holloway (Holloway, *The Game of Love in Georgian England: Courtship, Customs and Material Culture,* 2019). John Jeffreys Pratt became the Lord Lieutenant of Ireland and the 2nd Earl Camden, with the family seat located at Wherwell House, Andover, Hampshire. The arms of the Marquess of Camden are – quarterly 1st and 4th: sable, on a fess argent between three elephant's heads erased argent, three mullets sable (Pratt); 2nd, sable, a chevron between three spear heads argent the points embrued (Jeffreys); 3rd, gules, an inescutcheon vair, between eight cross-crosslets or (Molesworth). Crests: dexter an elephant's head erased or (Pratt); sinister, a dragon's head erased vert and holding in the mouth a sinister hand couped at the wrist gules and about the neck a chain and pendant therefrom a portcullis or (Jeffreys). Supporters: dexter, a griffin sable, sinister, a lion or armed and langued gules gorged with a collar argent and charged with three mullets sable. Motto: *"Judicium parium aut lex terrae"* which translates as *"The judgement of my peers or the law of the land"*. Clearly, the Marquess of Camden, in his coat-of-arms was accentuating his links with his maternal family, the Jeffreys of Brecon. For the other possibility, we have John Knight Bruce-Pryce (1784–1872), of Dyffryn St Nicholas, and originally of Llanblethian in the Vale of Glamorgan. As John Knight Bruce he succeeded to his uncle's estate in Aberdare in 1805 and thereafter in 1837 to the estate of Dyffryn St Nicholas through Frances Anne Grey, daughter of William Booth Grey, who was one of the original financial supporters of the Nantgarw China Works listed as "one of the ten true men of Glamorgan" by William Weston Young in 1817. Upon succeeding to the Dyffryn St Nicholas estate he adopted the additional surname of Pryce and became John Knight Bruce-Pryce. His eldest son, Henry Austin Bruce (1815–1895) became Home Secretary in 1863 and was raised to the peerage as Lord Aberdare in 1873. Hence, at the time of the commissioning of this Nantgarw service in *ca.* 1820, the family had not yet adopted the Pryce surname, so chronologically it is apparent that the coat-of-arms would not have reflected this in the heraldic achievement in the period between 1820 and 1823. Likewise, the Bruce coat-of-arms as depicted in Lord Aberdare's heraldic achievement is totally different, as is the motto and crest – and we can therefore most surely discount the Bruce-Pryce connection to Nantgarw at this time, although Lord Aberdare did possess a superb collection of Welsh porcelain later in the nineteenth century. Therefore, we are left with the very reasonably deduced and unique possibility that the Nantgarw armorial plate decorated by Thomas Pardoe exhibiting the escutcheon shown in Fig. 4.8 is in fact that of Nicholas Jeffreys of Brecon Priory.

So, from being completely unknown and unidentified in heraldic origin we have now achieved a distinct and realistic possibility for the correct attribution of this particular piece of armorial porcelain: in confirmation, it would be a task for a future research project to examine the family records and extant archives of the Jeffreys at The Priory, Brecon, and perhaps elsewhere, to look for evidence of the commissioning and use of this service and also through local auction sales records from the family estates for a Nantgarw service and its description which would then confirm our hypothesis. This approach, *viz.* the appraisal of a section of a limited range of local landowning families who are entitled to bear arms, is vindicated through this particular case study: of course, we still are faced with some missing vital information in the heraldic achievements in the form of a motto or perhaps a crest which would have potentially enabled a further narrowing down of the possibilities and pre-selection to have been achieved, and in this particular case then providing a definitive confirmation of the potential attribution of this heraldic escutcheon.

7.3 Local Support and Sponsorship for the Swansea and Nantgarw China Works

Previous historical accounts of the commissions received by the Swansea and Nantgarw China Works have detailed to a greater or a lesser extent the support and sponsorship of local businessmen, the aristocracy and the gentry for each undertaking. It may be reasonably expected, therefore, that some of these persons who had the right to bear arms would have manifestly demonstrated their support very obviously through the commissioning of crested armorial services. The absence of factory records both at Swansea and at Nantgarw makes it difficult to assess quantitatively in this respect the contributions of local landowners and benefactors but the following list gives an idea of the extent of this practice which evolved through their ordering of porcelain services, most of which are recognised now as being decorated in the normal "anonymous" fashion and vogue:

7.3.1 Nantgarw

Morton Nance (*The Pottery and Porcelain of Swansea and Nantgarw*, 1942) has considered this aspect carefully regarding financial support for the Nantgarw China Works and has compiled two lists of local influential businessmen and gentry, which he has differentiated into men of business or "traders" and "county gentlemen". The first of these categories involves names that appear in William Weston Young's Diaries (Young, *The Diaries of William Weston Young (1776–1847), 1802–1843,* 30 Volumes, West Glamorgan Archives Service, Swansea, SA13SN) whom he recorded as being providers of financial support for the Nantgarw manufactory, namely:

> Walter Coffin of Llandaff Court; John Bruce of Duffryn, Aberdare; William Crawshay of Cyfarthfa Castle, Merthyr Tydfil; Crawshay Bailey of Merthyr Tydfil; Richard Hill,

Plymouth Lodge, Merthyr Tydfil;? (Author: Anthony) Bacon of Aberaman; John Edwards of Rheola House, Glyn Neath, Vale of Neath;? (Author: William) Mitchell of Aberdare; Richard Forman of Penydarren, Merthyr Tydfil; Richard Blakemore of Velindre, Melingriffith; Josiah Guest of Dowlais, Ebbw Vale.

It should be noted that these names include some of the giants of the Industrial Revolution that had started to sweep through Britain in the early 1800s (see Appendix IV): the four leading ironmasters in the country at that time were Crawshay, Guest, Homfray and Hill, all of whom are recorded as having contributed financially to the Nantgarw China Works through William Weston Young in some form or other. Lady Charlotte Guest, wife of Sir Josiah Guest of Dowlais, later became a noteworthy connoisseur of porcelain and amassed a large collection which as Lady Charlotte Schreiber, as she became known after her second marriage to William Schreiber, was donated to the Victoria & Albert Museum in South Kensington, London, in 1884. She was famed for her tenacity in seeking out specimens of rare porcelains, being known as *La Grande Dame du Chasse* in this respect (Edwards, *Porcelain Analysis and Its Role in the Forensic Attribution of Ceramic Specimens,* 2021; Jacob-Hanson, *Charlotte Schreiber – The Unforgotten Grand Dame of the Chasse,* 2012; Rackham, *The Catalogue of The Schreiber Collection of English Porcelain, Pottery, Glass and Enamels collected by Charles Schreiber MP and the Lady Charlotte Elizabeth Schreiber and Presented to The South Kensington Museum in 1884, Volume I: Porcelain, Volume II: Earthenwares, Volume III: Enamels and Glass*, 1915; Schreiber, *The Lady Charlotte Schreiber Journals: Confidences of a Collector of Ceramics and Antiques Through Britain, France, Holland, Belgium, Spain, Portugal, Turkey, Austria and Germany from the Year 1869–1885*, 1911; Owens, *Lady Charlotte Guest: The Exceptional Life of a Female Industrialist*, 2020).

The second list of "county gentlemen" identified by Morton Nance and William Weston Young, which includes the "ten true men of Glamorgan" who each put up the sum of £100 (which at today's value would be approximately £11,000) at the invitation of William Weston Young to ensure the commencement of porcelain manufacture at Nantgarw in the second phase of operations there in 1817, is reproduced below:

William Booth-Grey of Dyffryn St Nicholas (son of the Earl of Stamford); John Crichton-Stuart, Marquess of Bute and Lord Dumfries of Cardiff Castle; Sir John Nicholls of Merthyrmawr, Vale of Glamorgan; Robert Jones of Fonmon Castle; Wyndham Lewis of Greenmeadow, Llantrisant; Dr. Whitlock Nicholl of Cowbridge and Adamsdown, Cardiff; Thomas Mansel-Talbot of Margam; Griffith Llewellyn of Baglan Hall; Rev. William Perkins Lisle, Prebendary of Llandaff and Rector of St Fagan's; Lord Windsor of St Fagan's; R. Franklin Esq.; Thomas Wyndham of Dunraven Castle, Southerndown, Vale of Glamorgan. It is curious that in William Weston Young's list of "the ten true men of Glamorgan" here there are actually 12 men cited (but see the further discussion on this point in Appendix V)!

Added to this list we have the local aristocracy and gentry who are known to have purchased porcelain from the Nantgarw China Works between 1817 and 1820 or who met with Billingsley, Walker and Young in the buildup to their production there and whose names are recorded in William Weston Young's Diaries (Young, *The*

Diaries of William Weston Young (1776–1847), *1802–1843,* 30 Volumes, West Glamorgan Archives Service, Swansea, SA13SN) and these include:

> The Talbot-Rices of Penrice Castle, the Turbervills of Ewenny Priory, William Williams of Aberpergwm House, Vale of Neath, John Traherne of St Hilary, Vale of Glamorgan, William Vaughan of Llantrisant, Richard Jenkins of Llanharan House, Thomas Edmondes of Cowbridge, Baron Talbot of Hensol Castle, Vale of Glamorgan, Richard Rickards of Usk Priory and the Aubreys of Llantrithyd.

In Table 7.2 a list of landowners and named supporters is presented with the results of a diligent search of Fairbairn (Fairbairn, *Crests of the Families of Great Britain and Ireland and Their Kindred in Other Lands,* Volumes I and II, 1905) and Fox-Davies (Fox-Davies, *Armorial Families: A Complete Peerage, Baronetage and Knightage, and a Directory of Some Gentlemen Coat -Armour and being the First Attempt to Show Which Arms are in Use at the Moment are Borne by Some Legal Authority*, Volumes I and II, 1895) to establish their right to bear heraldic crests and, if so, whether or not these appear in some form on Swansea porcelain as hitherto unidentified crests.

7.3.2 Swansea

Lewis Weston Dillwyn, the owner of the Cambrian Pottery, Swansea, and the proprietor of the Swansea China Works established when he had the idea of making porcelain there on site in 1812, was himself a member of the local gentry, with his seat at Sketty Hall, Swansea. He therefore moved in circles which already included landed gentlemen and the aristocracy, such as the Vivians of Singleton Park, later Lord Swansea, and Lord Dynevor of Dinfewr Castle, Llandyfeisant, Carmarthenshire. Unlike Billingsley, Walker and Young at Nantgarw he had no need to acquire independent investors to secure his production operations at Swansea as the original purchase of the Cambrian Pottery and the future site of the Swansea China Works had been made by his father, William Dillwyn, in 1802, but it should not be assumed from this that he was a dilettante dabbling in the manufacture of porcelain as a hobby as he was an esteemed scientist, a Fellow of the Royal Society and a confidante of renowned scientists such as Sir Joseph Banks and Sir Henry de la Beche. The handwritten notes that have survived from his experimental trials and observations of his porcelain body compositions, glazing and their kiln firings undertaken from 1815 until August 1817 with Samuel Walker as his dedicated kiln manager bear witness to his exactitude and scientific precision in undertaking experiments. His rather terse comments relating to the results of his experimental firing sequences of his trial porcelains and of their bodies and glazes are classic examples of the experimental notes and conclusions made *in situ* by a practising chemist (Dillwyn, *Notes of Recipes for Porcelain Bodies and Glazes 1815–1817,* Library of the Victoria & Albert Museum, South Kensington, London, deposited by John Campbell Esq. in 1920; reproduced in full in H. Eccles & B. Rackham, *Analyses of English*

Table 7.2 Named supporters of Nantgarw and Swansea China works and their armorial crests from Fairbairn[a]

Nantgarw Sponsor/ Client	Location	Crest[a]
Coffin	Llandaff Court, Glam.	Long cross sa. 128/19
Bruce	Duffryn, Aberdare, Glam.	Lion passant and trefoil 4/14
Crawshay	Cyfarthfa Castle, MT, Glam.	Greyhound 28/7
Crawshay Bentley	Merthyr Tydfil, Glam.	Demi-lion rampant 67/5
Hill	Plymouth Lodge, MT, Glam.	*Falcon beaked on trunk of tree 78/8*[b]
Bacon	*Aberaman, Glam.*	*Boar's head couped, griffin's head 2/7; Demi-leopard with spots 125/4*
Edwards	*Rheola House, VoN*	*Boar's head erased 16/14; Lion's head erased bend erminois 81/4*
Mitchell	*Aberdare, Glam.*	*Pelican, wings elevated 109/15*
Forman	Penydarren, MT, Glam.	Scimitar in hand 22/8
Blakemore	Velindre, Melingriffith	Moor's head profile sa. chaplet roses
Guest	Dowlais, Ebbw Vale, Glam.	Swan's head and neck erased 66/10
Booth-Grey	*Dyffryn St Nicholas, Glam.*	*Demi-leopard rampant 12/14*
Crichton-Stuart	Cardiff Castle, Glam.	Demi-lion rampant 67/10; Wyvern wings endorsed fire 51/2
Nicholls	Merthyrmawr, Glam.	Cornish chough on coronet 84/3
Jones	*Fonmon Castle, Glam.*	*Boar's head erased 21/7*
Lewis	Greenmeadow, Llantrisant, Glam.	Cornish chough ppr. 100/13
Nicholl	Cowbridge, Glam.	Wolf's head erased 14/6
Mansel-Talbot	Margam Abbey, Port Talbot, Glam.	Chapeau, flame fire erminois 71/11
Llewellyn	*Baglan Hall, Glam.*	*Paschal lamb 48/13*
Perkins Lisle	Llandaff, Cardiff, Glam.	Stag lodged ppr. 67/2
Windsor	St Fagans, Glam.	Buck's head collared 111/13
Franklin	?	Dolphin's head erased branches140/12
Wyndham	*Dunraven Castle, Ogmore, Glam.*	*Lion's head erased fetterwork chain 90/14*
Talbot-Rice	Penrice Castle, Glam.	Crow 23/8
Turbervill	*Ewenny Priory, Bridgend, Glam.*	*Eagle displayed 48/11*
Williams	*Aberpergwm House, VoN*	*Buck statant collared 88/9; Lion rampant regardant 10/15*
Traherne	St Hilary, Glam.	Goat's head on ducal coronet 72/2
Vaughan	*Llantrisant, Glam.*	*Lion rampant 67/5*

(continued)

Table 7.2 (continued)

Nantgarw Sponsor/ Client	Location	Crest[a]
Jenkins	Llanharan House, Glam.	Battleaxe handle or 14/8; Lion's paw erased with bezant 97/10
Edmondes	Cowbridge, Glam.	Two hands holding heart 43/15
Talbot	Hensol Castle, Glam.	Lion statant tail extended 107/1; Talbot 120/8
Aubrey	Llantrithyd, Glam.	Eagle's head erased 20/7
Swansea		
Vivian	Singleton Park, Swansea	Lion's head erased 7/10
Dynevor	Dinefwr Castle, Llandeilo, Carms.	Crow 23/8
Mackworth	Gnoll House, Neath	Cock 67/11
Gwynne	Llansannor, Cowbridge	Lion rampant regardant with boar's head 10/15
Young	Preswylfa, Neath	Demi-lion collared 18/13
Exeter/Cecil	Burghley House, Stamford	Chapeau with two lions 36/8
Acheson	Gosford Castle, Armagh	Cock gules, trumpet or 91/6
Coutts	London	Demi-centaur 70/3; man from middle with bow and arrow and quiver 247/11
Dillwyn	Sketty Hall, Swansea	Stag's head couped 91/14
Williams-Wynn	*Wynnstay, Denbigh*	*Eagle displayed 48/11*
Lewis	*Greenmeadow, Llantrisant, Glam.*	*Paschal lamb 131/14*
Somerset	Badminton House, Glos.	Portcullis with chains pendant 178/3
Biddulph	Makerstoun House, Kelso	Wolf rampant, guttee de sang 28/5
Lysaght (Lisle)	Monksnorth, Cork	Dexter arm embowed in armour, holding a sword 195/2; arm in armour, dagger 120/11

MT Merthyr Tydfil, *VoN* Vale of Neath; *Glos.* Gloucestershire, *Glam.* Glamorganshire, *Carms.* Carmarthenshire

A? signifies that the location of the family seat is not specified, so cross-checking of the name of the arms -bearer against a potential local supporter is not possible

[a]The numbers cited after each crest refer to the Plate/Crest numbers in Fairbairn's *Crests of the Families of Great Britain and Ireland and Their Kindred in Other Lands,* Volumes I and II, T.C. and E.C. Jack, Edinburgh and London, 1905 and New Orchard Publishing, London, 1986

[b]The italicised script signifies that this armorial crest has been identified already on Swansea and Nantgarw porcelains

Porcelain in the Victoria and Albert Museum Collection, Victoria & Albert Museum Publications, South Kensington, London, 1922). Unfortunately, his experimental attempts to increase the robustness of his delicate duck-egg porcelain body by the introduction of a low phosphatic and highly steatitic magnesian body were doomed to failure when this new "trident" body, which was an otherwise robust porcelain variant on his delicate duck-egg porcelain body, was shunned by his clientele who refused to purchase what they considered to be a visibly inferior china and he faced bankruptcy proceedings. He then signed over the lease of the operations of the Swansea China Works to Timothy and John Bevington, who really wanted the site for expansion of their existing earthenware and pottery business. It is believed that the Bevingtons never made any porcelain at Swansea thereafter, although this is by no means certain, but they employed their resident enamellers in the workshop to decorate the residual stocks of porcelain remaining after Lewis Weston Dillwyn's departure. This would include remnants of both the esteemed phosphatic duck-egg porcelain and the publicly less acceptable magnesian bodied trident wares. Thus, we might expect to see both of these bodies represented in the limited range of Swansea armorial porcelains under consideration here, as indeed we do observe. For example, the Clarke of co. Hereford/Graham-Clarke of Newcastle-upon-Tyne impaling Parkinson armorial service is of the trident magnesian porcelain and the boar's head and Venn crested armorial services are examples of the duck-egg porcelain, the former with the desirable Paris fluting moulding and the latter with the equally desirable Nantgarw-type moulded ribbons and foliage edge moulding.

7.3.3 Analysis of Table 7.2 Supporters

Some points can be observed from an analysis of the supporters for the Swansea and Nantgarw China Works who are listed in Table 7.2:

- Firstly, for the 46 names of supporters identified in the Table 7.2, all appear in Fairbairn (Fairbairn, *Crests of the Families of Great Britain and Ireland and Their Kindred in Other Lands,* 1905 and 1986) and in Burke's *Armory* (Burke, *Burke's General Armory – The General Armory of England, Scotland, Ireland and Wales: Comprising a Registry of Armorial Bearings from the Earliest to the Present Time*, 1884). Hence it can be concluded that these names would indeed have been in the privileged position to commission armorial services in porcelain as they all had the pre-requisite right to bear arms.
- References to Fairbairn (Fairbairn, *Crests of the Families of Great Britain and Ireland and Their Kindred in Other Lands,* 1905 and 1986) reveals that 11 of the 46 entitled to bear arms have crests which are identical or very similar in type to those that appear on Nantgarw and Swansea porcelains, and also on the associated earthenware armorials from the Cambrian Pottery, Swansea. This represents some 25% of the names on the list, which is by no means exclusive, and again such an incidence surely cannot be coincidental.
- Of the 46 nominated supporters of the Swansea and Nantgarw China works listed in Table 7.2 no fewer than 39 armigerous families were local to Glamorganshire and Carmarthenshire and it is surely inconceivable that a significant proportion of these would not have commissioned porcelain from the Welsh factories, especially when their porcelain artefacts were making such an impact

on the more distant London society geographically in the second decade of the nineteenth century.

Some of the crests are "shared" in type or common to arms-bearing individuals from several families, with similar basic depictions of crests, and the presence of additional heraldic devices on the crest then becomes important, such as the wearing of a collar or a coronet, or the beast holding another device, such as a cross or a branch. The importance of matching the crest exactly with the registered armorial in the armories can be seen when we refer back to that assigned to Wyndham Lewis of Greenmeadow, Llantrisant: namely, a Paschal lamb bearing a pennon with the Cross of St George and accompanied by a halo (Fig. 5.12). In the list in Table 7.2 the crest of Griffith Llewellyn of Baglan Hall also displays a Paschal lamb but the portrayal of this crest in Fairbairn (Fairbairn, *Crests of the Families of Great Britain and Ireland and Their Kindred in Other Lands,* Volumes I and II, T.C. and E.C. Jack, Edinburgh and London, 1905 and 1986, Plate 48/crest 13), shows a subtly different picture and the crest of Llewellyn shows a Paschal lamb bearing a pennon with the cross of St Andrew (a saltire) and minus its halo, and additionally carrying a spear point (a pheon) at the top of the pennon as a finial, as can be seen later in Fig. 8.6.

7.4 Comparison with Armorial and Crested China from the Cambrian Pottery, Swansea

In a second novel approach that has not been reported hitherto the author decided to investigate the occurrence of armorial and crested china that was commissioned through the Cambrian Pottery, Swansea, to ascertain if there were any potential clues that might emerge from a cross-referencing of the heraldic achievements appearing there that might inform the assignment of those crests that are still uncertainly attributed for the Swansea porcelain armorial services (Chap. 3). This exercise, whilst revealing and confirming that the occurrence of armorial services and artefacts from the Cambrian Pottery were, surprisingly perhaps, not as profuse as they were previously imagined to be from citations in the literature, nevertheless did not prescribe many direct cross-matching possibilities for their porcelain analogues, except for one successful incidence of the armed and spread eagle, which is now quite reasonably re-assignable either to the Turbervill family of Ewenny Priory or alternatively to Sir Watcyn Williams-Wynn of Wynnstay, who has a similar armed spread eagle crest. However, several other assigned crests on the Cambrian Pottery earthenwares, notably those of a bull's head horned (Galton), portcullis (Duke of Beaufort), cockerel beaked (Alcock) and pelican (Richardson) were all identified and are now incorporated in the database for potentially matching with any similar crest that may surface in the future on analogous porcelain artefacts.

The role of the ironmasters and their dynasties in Merthyr Tydfil and their impact upon particularly the Nantgarw manufactory and an amplified discussion of the part played by the "Ten True Men of Glamorgan" cited by William Weston Young in his

Diaries (*The Diaries of William Weston* Young (1776–1847), *1802–1843,* 30 Volumes, West Glamorgan Archives Service, Swansea, SA1 3SN) are given in Appendix IV and Appendix V, respectively. These were accredited as some of the wealthiest "men in trade" then in Britain and several of them, aside from being inveterate gamblers, were also local benefactors and appeared as such in Young's Diaries. It is frankly inconceivable that some of these would not have acquired high quality porcelain services for their entertaining and socialising in their South Wales estates. Of course, it is possible that they did use porcelain or creamwares at the dinner table and it may be that these were purchased instead from Staffordshire, Coalport, Derby or Worcester but there is little in the extant records to indicate this having happened.

In this respect, therefore, the approach adopted here has been successful and can be considered to have been a worthwhile piece of background research. In the next and concluding Chapter, the crests for local landowning families will be compared with those on the armorials that have yet to be identified on Swansea porcelain.

References

J. Bateman, *The Great Landowners of Great Britain and Ireland: A List of all Owners of Three Thousand Acres and Upwards Worth Eight Thousand Pounds a Year. Also, One Thousand Three Hundred Owners of Two Thousand Acres and Upwards in England, Scotland, Ireland and Wales, their Acreage and Income from Land, Culled from the Modern Domesday Book (1873), Corrected in the Vast Majority of Cases by the Owners Themselves*, 4th edn. (Harrison, London, 1883)

Sir J. Bernard Burke, *Burke's General Armory – The General Armory of England, Scotland, Ireland and Wales: Comprising a Registry of Armorial Bearings from the Earliest to the Present Time* (Harrison, London, 1884)

A. Cox, A. Cox, *Rockingham Porcelain, 1745–1842* (The Antique Collectors Club, Woodbridge, 2001)

L.W. Dillwyn, *Notes of Recipes for Porcelain Bodies and Glazes 1815–1817,* Library of the Victoria & Albert Museum, South Kensington, London, deposited by John Campbell Esq. in 1920

H. Eccles, B. Rackham, *Analyses of English Porcelain in the Victoria and Albert Museum Collection* (Victoria & Albert Museum Publications, London, 1922)

H.G.M. Edwards, *Porcelain to Silica Bricks: The Extreme Ceramics of William Weston Young, 1776–1847* (Springer, Dordrecht, 2019)

H.G.M. Edwards, *Porcelain Analysis and its Role in the Forensic Attribution of Ceramic Specimens* (Springer, Dordrecht, 2021)

J. Fairbairn, *Crests of the Families of Great Britain and Ireland and Their Kindred in Other Lands,* vols I and II (T.C. and E.C. Jack, Edinburgh/London, 1905, New Orchard Publishing, London, 1986)

A.C. Fox-Davies, *Armorial Families: A Complete Peerage, Baronetage and Knightage, and a Directory of Some Gentlemen Coat -Armour and being the First Attempt to Show Which Arms are in Use at the Moment are Borne by Some Legal Authority*, vols I and II (T.C. and E.C. Jack, Grange Publishing Works, Edinburgh, 1895)

Glamorgan Land Registry Records., Archives and Land Registry GB214D19, Glamorgan Records Office, Cardiff

S. Holloway, *The Game of Love in Georgian England: Courtship, Customs and Material Culture* (Oxford University Press, Oxford, 2019)

C. Jacob-Hanson, Charlotte Schreiber – the unforgotten Grand Dame of the "*Chasse*". Northern Ceramics Newsletter, No. 166, June 2012, pp. 50–61 (2012)
W.D. John, *Nantgarw Porcelain* (Ceramic Book, Newport, 1948)
W.D. John, *Swansea Porcelain* (Ceramic Book, Newport, 1958)
W.D. John, *William Billingsley* (Ceramic Book, Newport, 1968)
W.D. John, K. Coombes, G.J. Coombes, *Nantgarw Porcelain Album* (Ceramic Book, Newport, 1975)
E. Morton Nance, *The Pottery and Porcelain of Swansea and Nantgarw* (B.T. Batsford, London, 1942)
V. Owens, *Lady Charlotte Guest: The Exceptional Life of a Female Industrialist* (Pen & Sword Books, Barnsley, 2020)
B. Rackham, *The Catalogue of the Schreiber Collection of English Porcelain, Pottery, Glass and Enamels Collected by Charles Schreiber MP and the Lady Charlotte Elizabeth Schreiber and Presented to the South Kensington Museum in 1884, Volume I: Porcelain, Volume II: Earthenwares, Volume III: Enamels and Glass* (Victoria & Albert Museum Publications, London, 1915)
L.C. Schreiber , *The Lady Charlotte Schreiber Journals: Confidences of a Collector of Ceramics and Antiques Through Britain, France, Holland, Belgium, Spain, Portugal, Turkey, Austria and Germany from the Year 1869–1885*, edited by Montague John Guest (1839–1909), John Lane, London/New York, 1911
L. Syson, Just a plain cup and saucer, CAM (Cambridge Alumni Magazine), Issue 93 (Easter Term 2021, The University of Cambridge Development and Alumni Relations, 2021), p. 43
W. Turner, *The Ceramics of Swansea and Nantgarw. A History of the Factories with Biographical Notices of the Artists and Others, Notes on the Merits of the Porcelain, the Marks Thereon Etc* (Bemrose/The Old Bailey, London, 1897)
W.W. Young, *The Diaries of William Weston Young (1776–1847), 1802–1843,* 30 vols, West Glamorgan Archives Service, Swansea, SA1 3SN. https://archiveshub.jisc.sc.uk/data/gb216-d/dxch/ddxch/i/hub

Chapter 8
Summary and Conclusions

Search for the truth is the noblest occupation of man: its publication a duty.
Anne-Louise Germaine de Stael – Holstein (1766–1817).
All truth passes through three phases. First it is ridiculed. Second it is violently opposed. Third it is accepted as being self-evident.
Arthur Schopenhauer (1788–1860).
Facile Fere Factum.
What is Easy has Usually Already Been Done.
(Written on a Blackboard in the Clarendon Physics Research Laboratory, University of Oxford, neighbour to the "Abbot's Kitchen", the first purpose-built Chemistry Laboratory in Britain).
Festina lente.
Make haste slowly.
Octavius Augustus Caesar, 63 BCE-14 CE.
Sat celeriter fieri quidquid fiat satis bene.
That which has been done well has been done quickly enough.
Octavius Augustus Caesar, 63 BCE-14 CE.
A clever person solves a problem. A wise person avoids it.
Albert Einstein *(1879–1955), Nobel Laureate in Physics 1921.*

Abstract Summary and final results: a comparison of the crests of local landowners and supporters with the unidentified heraldic crests on the Swansea porcelain exemplars affords a match for the four hitherto unidentified crests, but the escutcheon still remains unidentified. Although the matching of the unidentified crests is not an unequivocal statement, it is a great step forward and a vindication of the approach taken. By this means a database has now been established for the first time which will facilitate the rapid identification of new armorial porcelain exemplars which may surface in the future for the Swansea and Nantgarw China Works.

H. G. M. Edwards, *Welsh Armorial Porcelain*,
https://doi.org/10.1007/978-3-030-97439-8_8

Keywords Summary of results · Compilation of preliminary database of local landowners and supporters and their crests

The study of armorial porcelains and earthenwares, demanding as it is in time and resources, is indeed an absorbing exercise for ceramics research, occupying the interdisciplinary region at the arts/science interface between analytical science, historical documentation, heraldry and genealogical research. The two quotations cited above from Octavius Augustus Caesar, the son of Julius Caesar, were made by him in response to what he perceived as impetuous decisions made by his generals but they do resonate very well with our armorial research here and have been paraphrased by historians as "Take your time and do it right first time". A classic example of the importance of these studies in establishing the provenance of a porcelain artefact is provided by the detailed genealogical and heraldic research which, together with the analytical chemical science, has revealed novel information about the practice of the Derby China Works in the early days of the proprietorship of Robert Bloor through the conclusions drawn from the construction and composition of the armorial bearings on a *Barry-Barry* (*Pendock-Barry*) service dinner plate (Burke, *Burke's General Armory—The General Armory of England, Scotland, Ireland and Wales: Comprising a Registry of Armorial Bearings from the Earliest to the Present Time*, 1884). This has facilitated the attribution of the painter and the date of commissioning of this service to a reasonably precise date of 1806 when no specific factory records or other documentation were otherwise available to assist in this endeavour and has resulted in a significant amendment being made to historical precedent and the satisfactory resolution of an existing literature controversy on the subject (Edwards, *Porcelain Analysis and Its Role in the Forensic Attribution of Ceramic Specimens,* 2021; Twitchett, *Derby Porcelain: An Illustrated Guide, 1748–1848*, 2002; Denyer, Denyer & Edwards, *The Pennock-Barry Derby Porcelain Service: A Reappraisal,* 2021). The reason for its commissioning from the Derby China Works by its family arms bearer, Pendock Barry-Barry, has also been established and provides an interesting facet of its historical background: the genealogical research underpinning this particular study has taken several years in completion and leading down several pathways before its final outcome was achieved, in accordance with Octavius Augustus Caesar's recommendation given above.

In this current text, an analogous study on Welsh porcelains from the Nantgarw and Swansea China Works, the focus has been centred on the heraldic research underpinning the designation of the unidentified heraldic crests and escutcheons which appear on the armorial porcelain exemplars of the Nantgarw and Swansea porcelain manufactories between 1815 and 1822. In this respect, the Latin quotation that concludes this Chapter, namely *Finis Coronat Opus (The End Crowns the Work),* is particularly appropriate as the coronet features so strongly in the interpretations of the heraldic symbolism of crests and escutcheons on armorial china.

A summary of the results of this current research into the Nantgarw and Swansea armorial porcelains has been made in Table 6.2, where the thirteen heraldic crests that have been discovered thus far, representing twelve distinct and different crests, are enumerated and compared with their analogues in Fairbairn (*Crests of the*

Families of Great Britain and Ireland and Their Kindred in Other Lands, Volumes I and II, 1905). These twelve crests are actually distributed over thirteen different porcelain armorial exemplars since the *Ramsay of Straloch and Barra* crest appears on two distinctly different and identifiable Nantgarw services that were commissioned at different periods of the Nantgarw manufactory production. Likewise, in Table 6.3 are summarised the results for the coats-of-arms and partial coats-of-arms that exist outside of the purely crested china exemplars for the Nantgarw and Swansea manufactories: included in this Table is the depiction of the "Garter Star" motif which we have considered in detail here and have now excluded from its belonging to the genuine armorial category on the reasonable presumption that it is in fact the adopted signature motif of the decorative work of Thomas Pardoe. The crests considered in Table 6.2 are reproduced in Figs. 8.1, 8.2, 8.3, and 8.4 as they appear in the relevant Plates in Fairbairn (Fairbairn, *Crests of the Families of Great Britain and Ireland and Their Kindred in Other Lands,* Volumes I and II, 1905); similarly, those referenced in the coats-of-arms cited in Table 6.3 are shown in Fig. 8.5.

The following points can now be made with reference to the summary listings of the armorials cited in Tables 6.2 and 6.3:

- Hitherto, at the start of this research into Welsh porcelain armorials there was a dearth of information about the origin of many of these as was clearly stated by Fergus Gambon (*Porslen Nantgarw y Abertawe,* 2016) in his survey of the Welsh porcelain in the superb Andrews Collection at Plas Glyn-y-Weddw and this could be quantified as follows: for Nantgarw, one full coat-of-arms and one partial coat-of-arms comprising an escutcheon and motto were both already known and identified in the literature (even though the attribution of the former has now been shown to be incorrect), whereas one partial coat-of-arms in the form of an escutcheon was unidentified, and four crested pieces were known, all of which had been correctly assigned hitherto, giving an overall success rate of 86% for

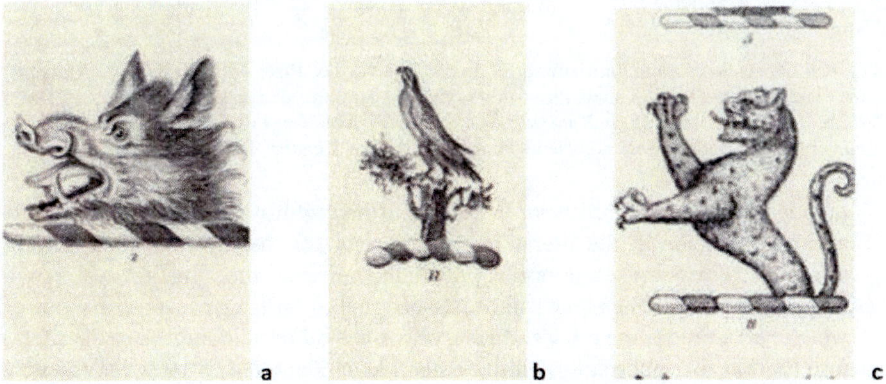

Fig. 8.1 Crests depicted in Fairbairn: *a* Boar's head erased Plate 42/2; *b* Bird on sprouting (fructed) tree stump Plate 86/ 11; *c* Demi-leopard rampant Plate 299/8. (From Fairbairn, *Crests of the Families of Great Britain and Ireland and Their Kindred in Other Lands,* Volumes I and II, T.C. and E.C. Jack, Edinburgh and London, 1905 and New Orchard Publishing, London, 1986)

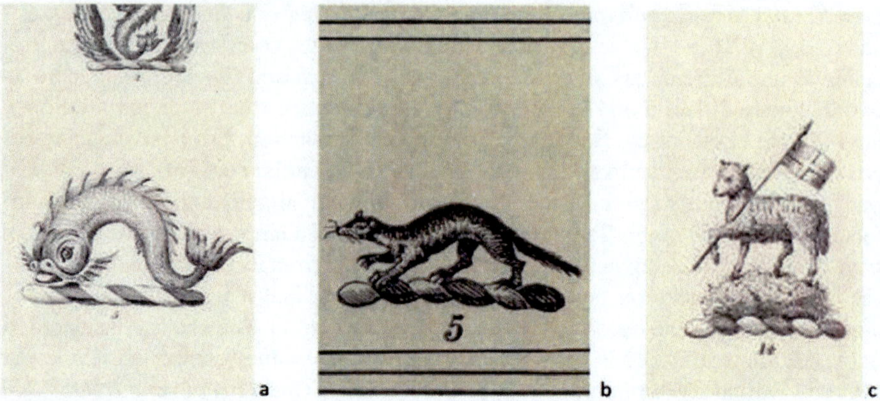

Fig. 8.2 Crests depicted in Fairbairn: *a* Dolphin naiant Plate140/5; *b* Otter Plate 9/5; *c* Paschal lamb with pennon of St George 131.14. (From Fairbairn, ***Crests of the Families of Great Britain and Ireland and Their Kindred in Other Lands,*** Volumes I and II, T.C. and E.C. Jack, Edinburgh and London, 1905 and New Orchard Publishing, London, 1986)

Fig. 8.3 Crests depicted in Fairbairn: *a* Eagle displayed armed Plate 48/11; *b* Eagle rising regardant Plate 35/8; *c* Escallop shell Plate 141/1. (From Fairbairn, ***Crests of the Families of Great Britain and Ireland and Their Kindred in Other Lands,*** Volumes I and II, T.C. and E.C. Jack, Edinburgh and London, 1905 and New Orchard Publishing, London, 1986)

the identification up until now. It was confirmed in this work that not all of the attributions made in the literature were correct for these Nantgarw porcelain armorials. For Swansea armorial porcelain, however, the situation was much more dire in that of the two partial coats-of-arms comprising escutcheons (one of which had a motto, the other did not), only one had been identified in the literature (but, again, incorrectly) an the other was unidentified, whereas of the nine heraldic crests known, only one had been identified and attributed correctly (or perhaps only partially so!) and the other eight remained unidentified (including another which had had been incorrectly attributed), giving an overall conservative success rate of only 18% at best for the Swansea armorials.

Fig. 8.4 Crests depicted in Fairbairn: *a* Demi-lion rampant with palm Plate 61/4; *b* Boar chained to a holly tree Plate 57/12; *c* Lion rampant Plate 67/5. (From Fairbairn, ***Crests of the Families of Great Britain and Ireland and Their Kindred in Other Lands,*** Volumes I and II, T.C. and E.C. Jack, Edinburgh and London, 1905 and New Orchard Publishing, London, 1986)

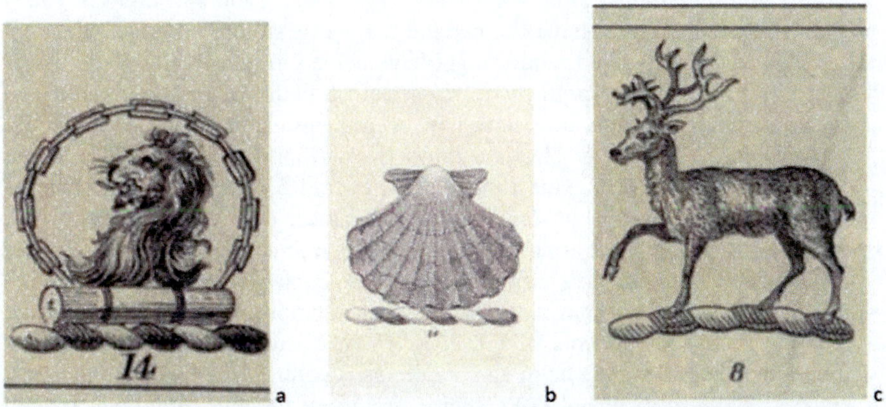

Fig. 8.5 Crests depicted in Fairbairn: *a* Reindeer statant Plate 12/8; *b* Lion erased /couped within fetterlock and chain Plate 90/14; *c* Escallop shell Plate 141/14. (From Fairbairn, ***Crests of the Families of Great Britain and Ireland and Their Kindred in Other Lands,*** Volumes I and II, T.C. and E.C. Jack, Edinburgh and London, 1905 and New Orchard Publishing, London, 1986)

- The thirteen items of crested china in Table 6.2, comprising 12 distinctly different crests, can be subdivided into the individual factory origins of 4 Nantgarw and 9 Swansea: the Nantgarw list includes the two variant but nevertheless distinct armorial services of *Ramsay of Straloch and Barra*, which both have the same crest and motto depicted but which are accompanied by different gilding and decorative patterns on the porcelain artefact which registers them as separate services chronologically. In the *Gartre'n Ol Exhibition of Nantgarw Porcelain* held at the Nantgarw China Works Museum in 2019, only one armorial piece of

Nantgarw porcelain was represented out of more than seventy pieces of the finest Nantgarw porcelain and that was one of the two versions of the *Ramsay of Straloch* crested porcelain, namely that shown in Fig. 4.2. The only piece of armorial Nantgarw porcelain which had not been identified hitherto was the plate decorated locally by Thomas Pardoe and bearing a simple gilt escutcheon displaying three pheons and a chevron (Fig. 4.8) and this has now been assigned here to Nicholas Jeffreys of Brecon Priory. All the Nantgarw armorial and crested porcelain has therefore now been identified successfully here, with a 100% success rate – bearing in mind that the full armorial of Viscount Weymouth had nevertheless been incorrectly ascribed in the literature.

- It has been a rather less satisfactory outcome for the Swansea armorials in that of the 9 items of crested Swansea porcelain, 5 have been positively identified unequivocally as a result of the current study but four remain still unidentified uniquely, but with a number of potential arms-bearing families now logically being put in the frame as seriously possible contenders. All of these have the proper crest and associated torse depicted on the porcelain artefacts but, unfortunately, this is still not sufficient to arrive at a unique attribution of arms-bearer unequivocally and definitively using Fairbairn (*Crests of the Families of Great Britain and Ireland and Their Kindred in Other Lands,* Volumes I and II, 1905) because several families have claimed the usage of a similar or almost identical crest. Here, an associated motto might have helped to narrow down the list of possibilities somewhat: for the two Swansea crests in the armorials list which are accompanied by a motto, both have now been identified and attributed securely (C.N. Elvin, *A Handbook of Mottoes Borne by the Nobility, Gentry, Cities, Public Companies, Etc.,* 1971). This results in an overall 55% success rate thus far achieved for the heraldic attribution of Swansea crested porcelain.
- In Table 6.3, five items of armorial china depicting escutcheons, are split between 3 Nantgarw and 2 Swansea exemplars, and these contain potentially more information for an attribution in theory as they each comprise more than two of the seven defining components of the coats-of arms: namely, the escutcheon, a motto, the supporters, the helm and mantlings, a coronet, the torse and the crest (*Burke's General Armory—The General Armory of England, Scotland, Ireland and Wales: Comprising a Registry of Armorial Bearings from the Earliest to the Present Time*, 1884).
- Of the three Nantgarw partial or complete coats-of-arms, two have been positively and unequivocally identified (one requiring an amendment to the existing literature) and the third, which comprises a rather simply portrayed gilt escutcheon only, has been identified as a very strong candidate here for Jeffreys of Brecon Priory. Here the lack of supporting information, as found with an exemplary solitary crest and/or torse, is still nevertheless technically insufficient for an absolutely unequivocal attribution to be arrived at but the confidence in this current attribution remains reasonably high. This gives a conservative 67% success rate for the identification of the Nantgarw armorials and a realistic 100%. For the Swansea analogues, it is possible to identify both of the two listed partial coats-of-arms, namely that of Clarke of co. Hereford impaling Parkinson of Kinnersley

(but see the later comment that confirms this should now be amended to refer to Graham-Clarke of Newcastle-upon-Tyne and Parkinson of Kinnersley— as John Graham-Clarke was married to Mary Elizabeth Parkinson of Kinnersley Castle); the other was illustrated first in the work of Jones and Joseph (*Swansea Porcelain: Shapes and Decoration*, 1988), and revealed there as an unattributed example in their text only through a monochrome picture from which the escutcheon is seen but ill-defined and for which no supporting descriptive information was provided additionally in the text. Its source provenance was not specified in the text which meant that it was impossible to backtrack and research it further for the clarification of the armorial details from just this single but incomplete literature citation. It may be assumed, perhaps, that it was in the personal collection of Sir Leslie Joseph, which was certainly one of the largest private collections of Welsh porcelain if not the largest to have been made in the twentieth Century, and it may, therefore, have been dispersed in the sale of that collection some years ago? The Sir Leslie Joseph Collection from Coedargraig, Porthcawl, was auctioned at Margam Park by Sotheby's between 14th and 16th May 1992 after his death: it comprised 1118 lots and over 2000 pieces of Swansea and Nantgarw porcelain, along with some pieces from other factories which had been decorated by artists such as Thomas Baxter. The whole collection realised over £1.1 Million at auction. A personal communication has been made to the author relating to a polychrome armorial plate in the National Museum of Wales Amgueddfa Cymru Collection (Renton 2021), which is seen to be identical to the same service that is illustrated and unattributed in the Jones & Joseph book (*Swansea Porcelain: Shapes and Decoration*, 1988) and has now firmly identified this particular armorial depiction as that of Lockwood of Dews Hall, Essex, impaling Manners-Sutton. This means that there is now, as a result, a 100% success rate for the Swansea items comprising partial coats-of-arms in this category. The genealogical research underlying this assignment revealed that all three Manners-Sutton daughters at this time married Lockwoods, and indicated how vital it is to consider the fine detail when making the attribution.

- During the course of this research, the hitherto unidentified crest of a Paschal lamb on Swansea porcelain bearing a pennon with the cross of St George and a Welsh motto "*Duw ar fy Rhan*" has now been attributed to Wyndham Lewis of Greenmeadow, Llantrisant, who had already commissioned a Nantgarw service in a *rose pink du Barry* ground colour to match the fabled Sevres colour palette. This Welsh porcelain crested service with its motto is noteworthy in that it provides only the second known example of an armorial service in Swansea porcelain with a Welsh motto, the other being the Thomas Lloyd of Bronwydd service commissioned on the occasion of his marriage in 1819, which bore the unique crest of a boar chained to a holly tree and carrying the motto "*I Dduw B'Or Diolch*" (Fairclough 2005). It is interesting that both of these mottoes carry an exhortation to God (in Welsh, *Duw*, or in mutated form following a vowel, *Dduw*). Several other Welsh landowning families have adopted Welsh mottoes for their crests, such as Sir Watcyn Williams-Wynn and Henry Somerset, the

Duke of Beaufort, but as yet these have not featured on any porcelain or earthenware armorial artefact made at the Cambrian Pottery or the Swansea China Works.

- The precise structure and composition of the crest depicted on the porcelain artefact is of critical importance to its rigid acceptance for unequivocal identification and attribution from a comparison with the heraldically correct versions displayed in Fairbairn (*Crests of the Families of Great Britain and Ireland and Their Kindred in Other Lands,* Volumes I and II, 1905). A typical example is the Paschal lamb referred to earlier, which also appears as a Paschal lamb derivative bearing a pennon with the Cross of St Andrew and not that of St George and also is minus its halo (Fig. 8.6) as possessed by Griffith Llewellyn of Baglan Hall. Another is the dolphin *naiant* belonging to the family Orme, which appears on a Swansea porcelain armorial service exemplar and which had been incorrectly assigned in the literature hitherto to a branch of the Orme family, namely Garnett-Orme, from co. Mayo, Ireland, whereas in fact it is more correctly described and attributed to that of Orme of Northampton, as the Irish family dolphin *naiant* also displays a poleaxe spear through its body (Fig. 8.6), which the Northampton Orme crest and the porcelain plate depiction does not have! In this context, it is important to emphasise that care must be exercised in the interpretation of a monochrome or gilded crest on the porcelain artefact as normally any colour tinctorial heraldic restrictions would not thereby be registered; for example, in the dolphin crest of Orme of Northampton discussed above, the fins and teeth are stated in Fairbairn's armory to be coloured (heraldically, tinctured) gold (heraldic *or*) on a black body (heraldic *sable*), which of course is not faithfully

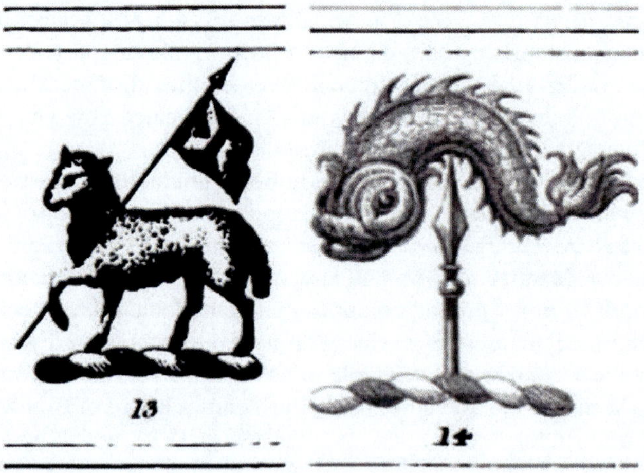

a b

Fig. 8.6 Crests depicted in Fairbairn: *a* Paschal lamb with a pennon bearing a St Andrew's cross Plate 48 / 13 *b* Dolphin naiant speared Plate 49/14. (From Fairbairn, *Crests of the Families of Great Britain and Ireland and Their Kindred in Other Lands,* Volumes I and II, T.C. and E.C. Jack, Edinburgh and London, 1905 and New Orchard Publishing, London, 1986) representable on a purely gilded armorial device such as that displayed on a porcelain artefact which has formed the basis of our discussion here.

- Furthermore, the generic "*bird on a sprouting tree stump*" crest which also appears on a Swansea porcelain artefact proved especially rather difficult to precisely identify because the equivalent crests in Fairbairn (*Crests of the Families of Great Britain and Ireland and Their Kindred in Other Lands,* Volumes I and II, 1905) are very specific in the type of bird employed and its attitude: noted examples of a bird in this respect are an eagle, pheasant, chough, dove, raven and crow and each may also have with it an associated heraldic device such as a spear, arrow, snake or a conger eel. In the Swansea porcelain artefact representation of this crest the bird is identifiable visually as potentially perhaps as a dove, or a better a martlet, but this is not absolutely certain so this can give rise to an element of doubt in its assignment. It has been noted that, occasionally, the painting of an animal or bird on china can be a little relaxed in detail, especially if the animal is more exotic and had not been encountered hitherto by the painter, such as a tiger, elephant or a whale. The most reasonable assignment of this crest perhaps in the current study is that of Lockwood of Dews Hall, Essex, who already has an armorial service in Swansea porcelain bearing an escutcheon in which his arms are impaling those of Manners-Sutton. In this instance the bird resting on the fructed tree stump (specifically, an oak tree in the case of Lockwood) is identified as a martlet and the depiction of this crest on the particular Swansea armorial piece under investigation bears out this identification from the appearance of the feet of the bird shown in the crest.
- Finally, there remains one further item of Swansea china which needs to be considered here and that is the Lliw (Llwchwr) Corporation punchbowl, which was presented by John Bevington in 1825 to the Llwchwr council and decorated by William Pollard at the Swansea China Works. This bowl has been already described and considered here in Chap. 3 under the Cambrian Pottery exemplars, since hitherto it has been itemised as such. This punchbowl has recently been acquired by the National Museum of Wales Amgueddfa Cymru and now resides in their Welsh Ceramics Collection and a personal communication from Andrew Renton to the author reveals that this had been catalogued by Sotheby's auctioneers prior to its sale as Swansea porcelain! This means that, if in fact this new classification as porcelain is substantiated, that there is now another item of Swansea armorial porcelain to complement the existing list, and furthermore, it will then undoubtedly be the largest in size of all known examples of Welsh armorial porcelain in existence. It appears that although the Llwchwr bowl can now confidently be re-assigned as porcelain rather than earthenware, despite its rather poor translucency, it cannot be firmly classified as one of the three Swansea porcelain types, namely glassy, duck-egg or trident made by Lewis Weston Dillwyn, and the conclusion is that it is either a rare exemplar of Bevington's Swansea porcelain or perhaps that it is an "interloper" that was bought in from Staffordshire for decoration at Swansea. The only way forward here to resolve this impasse would be to undertake a chemical analysis of the punchbowl body and compare the data with other porcelains of the period and especially of the

three known Swansea variants; another approach would be to estimate quantitatively the percentage transmission of the porcelain to visible radiation and thereby differentiate from other contemporary factories. Some examples (it is believed about five) of possible Bevington Swansea porcelain artefacts exist in the form of the *Bevington-Gibbins* service (Table 8.1) which was decorated in a blue and white willow pattern and was said to be of a poor translucency: perhaps this was derived similarly to the Llwchwr punchbowl and a combination of chemical analysis and transmission tests would be useful in establishing a common original source between them, or not.

It is manifestly clear that the greatest success in this exercise has been achieved in identifying the armorial bearings on a piece of porcelain when the maximum amount of information is accessible from the arms displayed there as the data in Tables 6.1 and 6.2 demonstrate. This, of course, is not unique to Nantgarw and Swansea porcelains and the absence of factory records and correspondence with clients does not materially help the researcher in the identification of armorial commissions for these two factories. However, even where manufactory records and historical accounts do exist for other factories they are occasionally brought into question by conflicting contemporary or later analytical evidence; a classic example of this occurs for the Derby China Works, where apparently comprehensive and exhaustive listings of porcelain commissions were preserved in the Derby China Works archive. Over 50% of the porcelain production at Derby was in the form of figures, almost 400 models of which apparently were meticulously recorded in the Derby lists, yet the lists of Derby figures from the eighteenth Century compiled directly from the factory records and cited by Haslem and Bemrose do not agree and there are several omissions, which are now being redressed or amended by modern research. John Haslem was an employee in the decorating workshop at the Derby China Works in the 1820s and 1830s until its closure in 1848 and he had access to the appropriate factory archival documentation first-hand from which he could compile his list (Haslem, *The Old Derby China Factory,* 1878). Bemrose's list on the other hand was derived later from an affidavit that was sworn in 1819 for the Duesbury *versus* Kean litigation over the affairs of the Derby China Works (Bemrose, *Bow, Chelsea and Derby Porcelain: Being Further Information Relating to the Factories Obtained from Original Documents Not Hitherto Published,* 1898). Furthermore, recent work by the late Dr. Peter Bradshaw (Bradshaw, *Derby Porcelain Figures 1770–1848*, 1990), a well-respected connoisseur of eighteenth Century and later porcelain figures, has revealed even more discrepancies arise when the actual descriptions of some of the surviving numbered Derby figures are compared with their entries in the official lists. This has been highlighted by more recent research carried out by Roger Massey and reported in the *Transactions of the English Ceramic Circle* in 2008 on a hitherto unknown Derby figure he has located with the incised number "371" on its base which does not even exist in any of these Derby listings (Massey, 2008)! Another example of an unlisted, very rare and perhaps unique, Derby biscuit figure is that numbered "390" in incised numeric script underneath, along with the Derby crown and crossed batons and the triangle mark

Table 8.1 Nantgarw and Swansea Porcelain Named Services and Their Origins

Service Name	Original Commission /Association	Type	References
Nantgarw			
Viscount Weymouth	Lord Thomas Thynne, Longleat House, Bath[a]	Dinner-dessert	John et al. (1975)
Wyndham	Henry Windham-Quin, Dunraven Castle[a]	Dinner-dessert	Gambon (2016)
Ramsay of Straloch 1	Col. John Ramsay of Straloch & Barra[a]	Dinner-dessert	Morton Nance (1942)
Ramsay of Straloch 2	Col. John Ramsay of Straloch & Barra[a]	Dinner-dessert	Bonhams[b]
Jeffreys of Brecon	Nicholas Jeffreys, Brecon Priory[a]	Dinner-dessert	Gambon (2016)
Homfray of Penllyne	John Homfray, Penllyne Castle[a]	Tea & coffee	Morton Nance (1942)
Baron Phipps of Normanby	Constantine Henry Phipps, first Earl of Mulgrave, Whitby, Yorkshire[a]	Dinner-dessert	Edwards (2017)
Wyndham Lewis 1	Wyndham Lewis MP, Greenmeadow, Llantrisant	Dinner-dessert	John (1948)
Wyndham Lewis 2	Wyndham Lewis MP, Greenmeadow, Llantrisant	Dinner-dessert	John (1948)
Mackintosh 1	E. Priest Richards/Harriet Diana Richards / Plas Newydd, Cardiff / Alfred, The 28th Mackintosh of Mackintosh of Moy	Dessert	John (1948) and John et al. (1975)
Mackintosh 2	?	Dessert	John (1948)
Mackintosh 3	William Legge, fourth Earl of Dartmouth, Blakelea House, Marsden, Yorkshire	Tea & coffee	John (1948)
Wilde	W.T. Wilde MP	Tea & coffee	Edwards (2017) and John (1948)
Misses Booker	Southerndown[c]	Dessert	John (1948)
Edmunds 1	Edward Edmunds/ Edwards, Nantgarw	Tea & coffee	John (1948)
Edmunds 2	Edward Edmunds/ Twyning, Nantgarw	Tea & coffee	John (1948) and Hillis (2005)
Edmunds 3	Edward Edmunds/ Duncombe, Nantgarw	Tea & coffee	John (1948) and John et al. (1975)
Edmunds 4	Edward Edmunds, Nantgarw	Tea & coffee	Renton (2003) and John (1948)
Sir John Williams	Sir John & Lady Williams, Aberystwyth	Dessert	Edwards (2017)
Brace	William Brace MP, Risca	Dinner-dessert	Edwards (2017)
Farnley Hall	Walter Ramsden Hawksworth Fawkes, Otley	Dinner-dessert	Edwards (2017)

(continued)

Table 8.1 (continued)

Service Name	Original Commission /Association	Type	References
Cardiff Castle	John Crichton Stuart, Marquess of Bute, Cardiff Castle	Dinner-dessert	John (1948) and John et al. (1975)
Spence -Thomas	?	Breakfast	John et al. (1975)
Lady Seaton	?	Dinner-dessert	Edwards (2017)
George, Prince Regent	Ernest Augustus, Duke of Cumberland	Tea & coffee	John (1948), Jewitt (1878) and Turner (1897)
Duke of Cambridge	Prince Regent/Adolphus, Duke of Cambridge	Dinner-dessert	John et al. (1975) and Edwards (2017)
Duke of Gloucester	Prince Regent/ William, Duke of Gloucester	Tea & coffee	John et al. (1975) and Edwards (2017)
Lord Kenyon 1	George, second Baron Kenyon of Gredington	Tea & coffee	Edwards (2017) and John et al. (1975)
Lord Kenyon 2	George, second Baron Kenyon of Gredington	Dinner-dessert	John (1948), Bonhams[b]
Theodore Ellis	Theodore & Mary Ellis, Worcester, USA	Dinner-dessert	John et al. (1975)
Duchess of Richmond	Charles Gordon Lennox, fifth Duke of Richmond, Goodwood House, Chichester	Dessert	Edwards (2017)
Marquess of Bute 1	John Crichton Stuart, Marquess of Bute, Cardiff Castle	Dessert	Morton Nance (1942)
Marquess of Bute 2	John Crichton Stuart, Marquess of Bute, Cardiff Castle	Dinner-dessert	Morton Nance (1942)
Marquess of Bute 3	John Crichton Stuart, Marquess of Bute Cardiff Castle	Dinner-dessert	Morton Nance (1942) and Turner (1897)
Duke of Norfolk 1	Henry Charles Howard, 13th Duke of Norfolk, Earl of Surrey, Arundel Castle	Dinner-dessert	Morton Nance (1942)
Duke of Norfolk 2	Henry Charles Howard, 13th Duke of Norfolk, Earl of Surrey, Arundel Castle	Dinner-dessert	Morton Nance (1942)
Duke of Newcastle	Henry Pelham-Clinton, fourth Duke of Newcastle -under Lyme, Clumber Park, Nottingham	Dinner-dessert	Morton Nance (1942)
Earl Spencer	George Spencer-Churchill, fifth Duke of Marlborough, Althorp House, Northants	Dinner-dessert	John et al. (1975)
Ferguson	?	Dinner-dessert	Edwards (2017), John et al. (1975) and Renton (2003)
Princess Charlotte	Charlotte, Princess Royal, London	Cabaret	Morton Nance (1942)
Hensol Castle	William Talbot, first Earl Talbot/ Benjamin Hall	Breakfast	Morton Nance (1942)
Usk Priory	Richard Rickards	Dessert	John et al. (1975)

(continued)

Table 8.1 (continued)

Service Name	Original Commission /Association	Type	References
Ewenny Priory 1	Richard Turbervill Picton / Turbervill	Dessert	Edwards (2017)
Ewenny Priory 2	Richard Turbervill Picton / Turbervill	Dessert	Edwards (2017)
Young	William Weston Young, Preswylfa, Neath	Tea & coffee	John et al. (1975)
Marquess of Anglesey	Henry Paget, Marquess of Anglesey	Tea & coffee	John et al. (1975)
Earl of Jersey	George Villiers, fifth Earl of Jersey, Briton Ferry	Dessert	John (1948) and John et al. (1975)
Mortlock 1	King George IV/ John Mortlock, London	Dinner-dessert	Morton Nance (1942)
Mortlock 2	John Mortlock, London	Dinner-dessert	Turner (1897) and Morton Nance (1942)
Llangattock	John Rolls, The Hendre, Monmouth	Tea & coffee	Edwards (2017)
Aberpergwm 1	William Williams, Aberpergwm House, Neath	Dessert	John (1948)
Aberpergwm 2	Marie Jane Williams, Aberpergwm House, Neath	Dessert	John (1948)
Swansea			
Clarke/Parkinson	Graham-Clarke of Newcastle/ Parkinson, Kinnersley[a]	Dinner-dessert	Gambon (2016)
Lockwood/Manners-	Sutton Lockwood/Manners-Sutton, Dews Hall [a]	Dinner-dessert	Renton (2021)
Lloyd of Bronwydd	Sir Thomas Lloyd, Bronwydd, Cardigan[a]	Dessert	Fairclough (2005)
Dolphin crest	Orme of Northampton[a]	Dessert	Gambon (2016)
Escallop crest	Clarke of Hereford/ Kinnersley[a]	Dessert	Gambon (2016)
Paschal lamb	Wyndham Lewis MP, Llantrisant[a]	Dessert	Gambon (2016)
Spread eagle	Sir Watcyn Williams-Wynn, Wynnstay[a]	Dessert	Morton Nance (1942)
Boar's head	Edwards, Rheola House, Glyn Neath [a]	Breakfast	Gambon (2016)
Bird on tree stump	Lockwood, Dews Hall[a]	Dessert	Gambon (2016)
Demi-lion rampant	Vaughan, Llantrisant [a]	Tea & coffee	Gambon (2016)
Demi-leopard rampant	Richard Hill of Penydarren[a]	Tea & coffee	Gambon (2016)
Burdett-Coutts	Sir Thomas Coutts, London	Dinner-dessert	John (1958)
Lysaght	John Lysaght, Baron Lisle of Mountnorton / William Lysaght, Swansea	Dinner-dessert	John (1958)
Nicholl	Sir John Nicholl/Judy Nicholl, Merthyrmawr	Tea & coffee	Morton Nance (1942)

(continued)

Table 8.1 (continued)

Service Name	Original Commission /Association	Type	References
Biddulph	John Tregenna Biddulph, Lord Biddulph Ledbury Park	Dinner-dessert	Bonhams[b], Edwards (2017) and Edmundson (2003)
Lady Seaton	Arthur Jones, Bryn-Newydd, Swansea	Dinner-dessert	Morton Nance (1942) and Edwards (2017)
Marquess of Anglesey	Henry Paget, Marquess of Anglesey	Tea & coffee	John (1958) and John et al. (1975)
Garden Scenery	Lewis Dillwyn, Sketty Hall/Sir Michael Dillwyn-Venables-Llewellyn, Llysdinam	Dessert	John (1958)
Marino Ballroom	John Vivian, Marino, Singleton Park	Dinner-dessert	Jones and Joseph (1988)
Bosahan	?	Dessert	Jones and Joseph (1988) and Morton Nance (1942)
Venn	?[a]	Dinner-dessert	Gambon 2016
Ivor Vachell	?	Dessert	Morton Nance (1942)
Bevington-Gibbins	Joseph Gibbins, Swansea	Dinner-dessert	Fairclough (2005) and Morton Nance (1942)
Vivian 1	John Vivian, Singleton Park, Swansea	Dinner-Dessert	Jones and Joseph (1988) and Morton Nance (1942)
Vivian 1	John Vivian, Singleton Park, Swansea	Dessert	Morton Nance (1942)
Lord Dynevor	George Rice, third Baron Dynevor, Dinefwr Castle, Llandyfeisant	Tea & coffee, Dessert	John (1958)
Gosford Castle	Archibald Acheson, second Earl of Gosford, Markethill, Armagh, Ireland	Dessert	John (1958)
Marquess of Exeter	Brownlow Cecil, second Marquess of Exeter, Burghley House, Stamford, Lincolnshire	Dessert	Edwards (2017)
Sayers	E. Sayers, Swansea	Breakfast	Jenkins (1970) and Morton Nance (1942)
Lord Swansea	Henry Hussey Vivian, first Baron Swansea, Singleton Abbey, Swansea	Dessert	Morton Nance (1942)
William Weston Young	WWY Young of Waun-Ceirch, Neath (nephew)	Dessert	Turner (1897) and Morton Nance (1942)

The assignment of the origin of the commission of the service listed here is based upon the designated recipient or family: for the most part the originator is not precisely known in the absence of manufactory records and most named services, therefore, have acquired their nomenclature from the dispersal of family effects upon an inheritance or their sale afterwards. Very few services still remain in their original family ownership: rare examples of these are the *Farnley Hall* dinner-dessert service of Nantgarw porcelain at Farnley Hall, Otley, Yorkshire (G. Horton -Fawkes Esq.), and the *Garden Scenery* dessert service of Swansea porcelain at Llysdinam, Powys (Sir Michael Dillwyn-Venables-Llewellyn).

(continued)

Table 8.1 (continued)

There are 83 named services collectively distributed here between Nantgarw and Swansea porcelains: 52 for Nantgarw (63%) and 31 for Swansea (37%). The 18 armorials, therefore, researched in this text represent 21% of these named services in Welsh porcelain. Of these named services 22, comprising 11 from Nantgarw and 11 from Swansea, have no basis in documentary history for ascribing their original attribution and they exist in this list because previous ceramic historians have mentioned them in the literature or because they have been recorded at some time later in the possession of their named owners at that time; of these, the *Mackintosh 2, Brace, Spence-Thomas, Lady Seaton(Nantgarw), Kenyon, Ferguson* and *Venn* services are well-known from their characteristic patterns. The *Bosahan, Vivian* and *Vachell* services are mentioned only *en passant* by Ernest Morton Nance (1942) and have not been identified or correlated further with any particular patterns and decoration, although they were seen by him personally when researching the preparation of his book (*The Pottery and Porcelain of Swansea and Nantgarw*, 1942) in the 1930s. A further service, known as the *Sayers* service,which is mentioned by Morton Nance (1942), refers to a complete and very large breakfast, tea and coffee service in Swansea duck-egg porcelain with simple gilding, estimated to be in excess of two hundred pieces of Paris fluted moulding which was purchased by a Mr. E. Sayers complete in the early twentieth Century and noted by Ernest Morton Nance. Its original commission of purchase is now lost, but its existence was advertised in the Swansea China Works porcelain sale of Timothy and John Bevington in 1826 as a "large breakfast service in Paris fluting with broad gold bands". An example of a tea cup and saucer of this *Sayers* service pattern is shown in Fig. 5.15. One should be aware, therefore, that any armorial or heraldic emblem borne by a named service might not actually be truly representative of that arms-bearing family unless it has passed through a formal inheritance, otherwise the armorial may not be truly representative of the nominated owner who may have acquired the service later and it has now become technically then an associated named service

[a]This superscript signifies an armorial service which displays one or more heraldic symbols enabling the identification of the individual arms-bearing member of the family who commissioned the service originally to be accomplished potentially more rigidly

[b]According to the background information supplied at sale by Bonhams Auctioneers and Valuers, New Bond Street, London (courtesy of Fergus Gambon, Director of British Ceramics)

[c]A rather mysterious entry in Dr. John's book (*Nantgarw Porcelain*, 1948) refers to the Misses Booker of Southerndown who possessed "*several Nantgarw services*". A diligent search by the author has failed to uncover the identity of the Misses Booker, although they are reported as being active in local community affairs in Southerndown in the early part of the twentieth Century. However, Richard Blakemore (1775–1855) was cited by William Weston Young as a supporter of the Nantgarw China Works in his Diaries (WW Young, *Diaries,* 1830; Edwards 2019). Initially, Francis Homfray developed a water/corn mill at Melingriffith in 1749 and started the manufacture of tinplate there with Messrs. Harford & Partridge. Richard Blakemore acquired the Melingriffith Ironworks on the Glamorgan Canal at Taffs Well, just some three miles from Nantgarw, in 1810 and it prospered rapidly under his direction into the largest tinplate works in the world, producing some 100,000 tons per annum, and he diversified into sheet iron production at Pentyrch Forge, building tramways between the two sites. His home was Velindre House, overlooking the Canal and the Nantgarw China Works and also The Leys in Monmouthshire. He became High Sheriff of Glamorgan in 1826. His sister, Ann, married the Rev. Luke Booker, Chaplain to the Prince Regent, and he adopted his nephew, Thomas Booker (1801–1858), who inherited the Melingriffith estate until his death in 1858. Thomas Booker's son, also Thomas, acquired the estate in the name of Thomas Booker-Blakemore and in 1872, there were 12 tinplate and sheet iron mills operating there, trading as T.W. Booker and Co. It is clear that Richard Blakemore would have had ample opportunity to acquire some of the Nantgarw porcelain being manufactured literally in his sight from Velindre and that this would probably have been passed on to the Bookers – it seems therefore that the Nantgarw services possessed by the Misses Booker would most probably have originated with their ancestor Richard Blakemore, a confirmed Nantgarw China Works supporter, and it is sad that these have now seemingly lost their provenance and have become anonymous

Fig. 8.7 William Duesbury II Derby China Works biscuit porcelain, *ca.* 1795, showing an elegant young woman dressed as *Flora* in a classical dance pose wearing a long dress and typical shoes of the Georgian period with buckles and heels, carrying a garland of flowers. Height 12 inches (30 cm). Marked on the base with the Duesbury Derby mark of incised crown, crossed batons and six dots, a number "390" and the incised triangle mark of the "repairer" Joseph Hill, who constructed the figure. Modelled by Jean-Jacques Spangler after a character of Angelika Kaufmann and entitled "*A Paris Opera Girl in the Role of Flora*". *Private Collection*

of the "repairer" Joseph "Jockey" Hill (Fig. 8.7); the figure was modelled by Jean-Jacques Spangler, *ca.* 1795, after a character drawn by Angelika Kaufmann entitled "*A Paris Opera Girl in the Style of Flora*". Dr. Peter Bradshaw researched this figure and found that it was indeed another hitherto unrecorded example of eighteenth Century biscuit porcelain from the Derby China Works and he consequently deemed it to be very rare.

A recent paper by Stephen Hanscombe in the *Transactions of the English Ceramic Circle* on armorial porcelain, which mainly comprised Chinese and Worcester porcelain decorated by James Giles in his atelier in London between about 1755 and 1770, refers to several well-executed partial armorial bearings which have all failed to have been identified for precisely the same reasons that have emerged here, namely, the number of family arms-bearers entitled to use the arms and crests depicted on the china is prohibitive for an unequivocal and unique attribution to be made from the strictly limited amount of information available and depicted on the china artefacts (Hanscombe, 2008). This is perhaps only to be expected yet may still be surprising in view of the large amount of subsidiary detailed historical information that is available currently for the activities of both James Giles and for especially Worcester armorial porcelains (*Armorial Worcester Porcelain of the First Period: Specimens from the Marshall Collection in the Ashmolean Museum and Other Sources*, 1964; Sandon, *The Ewers-Tyne Collection of Worcester Porcelain at Cheekwood,* 2007).

Since the Swansea China Works in particular sold to many local and regional clients, who it can be presumed purchased their porcelain directly from the manufactory, unlike Nantgarw where 90% of its output was negotiated for sale by John Mortlock through his Oxford Street, London, shop the author considered that it may be useful to see if local families had the appropriate arms which were then reproduced on Swansea porcelain which may be instrumental in narrowing down the field of possible contenders for their attribution. This geographical difference in client base is manifest in that much of the surviving Nantgarw porcelain is London-decorated and the local production was enamelled by only a small workforce which would have included William Billingsley himself possibly William Pegg and perhaps several unrecognised young workers between 1817 and 1820, and later by William Weston Young and Thomas Pardoe between 1820 and 1823 at the end of the Nantgarw second phase after Billingsley and Walker had departed for Coalport. In contrast, Swansea had several locally based accomplished artists such as David Evans, Henry Morris, Thomas Baxter, George Beddow, William Pollard and the mysterious "de Junic" (Edwards, *Porcelain Analysis and Its Role in the Forensic Attribution of Ceramic Specimens,* 2021) who were based there at various times between 1814 and 1823. Having said that, surviving examples of Nantgarw armorials that were locally decorated by William Billingsley comprise items from the *Homfray* tea and coffee service, with its crest of a speared otter, and marginally, perhaps entering this armorial category on a wider remit, several small porcelain beakers with gilt Masonic emblems (such as the square and compasses etc.) which have been noted (Morton Nance, *The Pottery and Porcelain of Swansea and Nantgarw*, 1942) and for which the original invoice still exists in the Nantgarw China Works archive in Billingsley's handwriting as being supplied to a Mr. Thomas Johnes of Cardiff in December 2019. An example of one of this set of Masonic beakers is shown in Fig. 2.6 and discussed in Chap. 3. No armorial porcelain artefacts have been identified thus far as having been painted by William Billingsley during his time at Swansea. It is also possible that the armorial *Baron Phipps* service (Fig. 4.4) was gilded locally at Nantgarw by William Billingsley although the general feeling is that it was London decorated.

In this context, therefore, it was thought possible that the gentry and armigerous families who had purchased decorated services of Swansea porcelain might have also acquired services of simpler crested porcelain with their minimal accompaniment of decoration, perhaps even just accompanied by a simple gilding pattern. Certainly, many of the Swansea crested pieces surviving today attest to a simplified decoration that would be compatible with a more local commissioning and it is appropriate here to therefore consider families who had already supported the

Swansea China Works, or even Nantgarw, with the purchase of other porcelain services. Although both the Swansea and Nantgarw China Works have associated with them some prestigious "named" services (as listed in Table 8.1) one must be careful to differentiate between the services that were actually bought by the family directly from the factory or an established retailer acting upon their behalf, such as Mortlock's, and those which were purchased later at the sales and dispersal of family effects in country houses (Edwards, *Nantgarw and Swansea Porcelain: A Scientific Reappraisal*, 2017). Good examples of such Swansea and Nantgarw named services that were acquired sometime after their initial manufacture and have hence then adopted or acquired the names of their later purchasers or more distant family inheritors are the *Lady Seaton, Burdett-Coutts, Brace, Garden Scenery, Sir John and Lady Williams, Mackintosh* and *Duncombe* services. In contrast, other named Nantgarw and Swansea services which are labelled according to their first owners and commissioners are exemplified by the *Marino, Hensol Castle, Gosford Castle, Marquess of Exeter, Marquess of Anglesey, Marquess of Bute, Prince of Wales, Duke of Cambridge, Duke of Gloucester, Baron Phipps, Farnley Hall, Greenmeadow, Thomas Lloyd* and *Ramsay of Straloch and Barra* services.

To cite one example to illustrate this point of nomenclature of a service by its later acquisition which has now come down as its definitive labelling, the *Lady Seaton* service in Swansea porcelain was ordered originally by Arthur Jones of Bryn-Newydd House, Swansea, *ca.* 1817–1820. This was purchased as a "*handsome Swansea china dinner service of 145 pieces*" for the sum of £13 on behalf of Sir Arthur Pendarves Vivian on March 8th, 1864, from the sale of his uncle's estate, namely the Arthur Jones mentioned above (as cited by Ernest Morton Nance, *The Pottery and Porcelain of Swansea and Nantgarw*, p. 343, footnote 1, 1942). Then it passed by inheritance to Lady Seaton of Bosahan House, Cornwall, after the death of Sir Arthur Vivian in 1926, whence it acquired its current name of the *Lady Seaton* service. To complicate matters further, another "*Lady Seaton*" service also exists in Nantgarw porcelain, which is London-decorated, and hand painted in cobalt blue enamel and accompanied by the characteristic London gilt dentil edging. Lady Seaton is recorded as having a discussion about her Swansea service with Ernest Morton Nance, probably in the early 1930s, as she died in 1937, in which the Nantgarw service named after her was not referred to at all: it is apparent that neither of these services were seemingly commissioned by Lady Seaton herself - and the Swansea service was locally decorated in blue transfer (designed and drawn by Henry Morris) whereas the Nantgarw service was hand-enamelled and decorated in London! The Nantgarw exemplar, therefore, appears to have acquired its nomenclature from its superficial similarity stylistically to its Swansea congener and in reality has nothing to do with its actual ownership by Lady Seaton, unlike its Swansea counterpart – in fact, as stated earlier, Morton Nance does not even mention a Nantgarw *Lady Seaton* service in his interview with Lady Seaton. Examples of the Nantgarw and Swansea *Lady Seaton* services are shown in Figs. 8.8 and 8.9, respectively. The so-called "Lady Seaton" service in Nantgarw porcelain is illustrated in Morton Nance's book but it has no reference or appellation there to any connection

Fig. 8.8 Dinner plate from the *Lady Seaton* service of Nantgarw porcelain, 1817–1819. The association of this service with Lady Seaton of Bosahan Manor, Cornwall, is unknown. Marked impressed NANT-GARW C.W., underglaze blue hand-enamelled pattern, London decorated through John Mortlock's, Oxford Street, with the characteristic dentil gold edging. *Private Collection*

Fig. 8.9 Salad bowl, square with indented corners, from the Swansea porcelain *Lady Seaton* service decorated locally in a pale blue transfer pattern from woodcut designs executed by Henry Morris. Note that the blue tone is paler than the Nantgarw analogue and it is devoid of gilding. The shape is rather unusual for the Swansea China Works and there is some accredited belief that the salad bowl is of another porcelain which has been acquired and decorated *en suite* to complete this service at Swansea. This is a piece of the Swansea service commissioned by Arthur Jones of Bryn-Newydd, Swansea, and acquired by Lady Seaton through her relative Sir Arthur Pendarves Vivian. *(Reproduced by courtesy of Bonhams Auctioneers and Valuers, London)*

with Lady Seaton, and only the Swansea service is so named after her, which suggests that its appellation and nomenclature is a more recent acquisition and is based upon a pattern similarity between the two services.

Specimens of the *Lady Seaton* Nantgarw service are relatively more readily encountered nowadays than their *Lady Seaton* Swansea analogues (a rather surprising statement in view of the large size of the *Lady Seaton* Swansea service mentioned earlier – some 145 pieces!), although it must be said that the latter are generally considered by connoisseurs to be rather "inferior" versions of the former visually, in that they have an applied transfer decoration from woodcuts executed by Henry Morris at Swansea, rather than being hand-painted in blue enamel, and they also have a much less striking and paler blue decoration than their Nantgarw analogues. The porcelain of the Swansea *Lady Seaton* service is also visibly inferior in translucency to that of its Nantgarw version. There are only three citations with illustrations of the *Lady Seaton* Swansea service in the literature, namely, Morton Nance (*The Pottery and Porcelain of Swansea and Nantgarw*, 1942, Plate CLIV F and G, facing page 344, where a muffin dish, a soup plate and a salad bowl are all illustrated in monochrome), Jones & Joseph (*Swansea Porcelain: Shapes and Decoration*, 1988, page 214, a soup plate illustrated in Plate 2 in monochrome) and in Fairclough (*Welsh Ceramics in Context*, Vol II, 2005, page 206, illustration 10.24 in colour) of a salad bowl which seems to be identical with that shown in Morton Nance's illustration). This salad bowl is of a rather unusual shape for Swansea china and it has been suggested that perhaps it was bought in from another factory in the white to complete the service commission requirements locally and possibly during the tenure of the Bevingtons at a time when other Swansea transfer and Imari type patterns were being exploited commercially after Lewis Dillwyn had departed from the Swansea China Works. The *Bevington-Gibbins* Swansea service mentioned earlier is another case of a blue and white transfer printed willow pattern porcelain service from the period of tenure of the Bevingtons at the Swansea China Works and similarly has a poorer translucency.

Genealogically, this story of Sir Arthur Pendarves Vivian and Lady Seaton impinges upon the theme of this text and a closer examination can perhaps shed some light on the matter of the attribution of the *Lady Seaton* Swansea service. Sir Arthur Pendarves Vivian (1834–1926) was the third son of John Henry Vivian (1785–1855), the Swansea industrialist, copper works owner and supporter of Swansea China Works. His elder brother was Henry Vivian, who became the first Baron Swansea, and his uncle the first Baron Hussey Vivian. He left Trinity College, Cambridge, prior to his graduation to take over the family copper smelting business and colliery in Port Talbot upon the death of his father in 1855. In 1867 he married Lady Augusta Emily Wyndham-Quin, daughter of the third Earl of Dunraven, who died in 1877; he married for the second time Lady Jane Dalrymple, daughter of the tenth Earl of Stair, who died in 1914. He bought the estate of Bosahan Manor in Manaccan, Helston, Cornwall, in 1882, thereby vacating his former estate at Glendorgal, St Columb Major, also in Cornwall. Upon his death in 1926, his estate passed to Baroness Seaton, Elizabeth Beatrice Fuller-Elliott-Drake, of Buckland Abbey, a relative of Sir Francis Drake. The Lady Seaton (1862–1937) of our eponymous service was the wife of the third Baron Seaton and was known as Baroness Seaton of Seaton. It was this Lady Seaton with whom Ernest Morton Nance had discussions and who gave her name to the Swansea and Nantgarw services. Technically, therefore, the *Lady Seaton* Swansea porcelain service, a blue transfer- printed service with a twelve lobed plate design of a rather poor translucency, should be re-catalogued as the *Arthur Jones* service! In contrast, the *Lady Seaton* Nantgarw service is much admired as it goes against the trend operating at that time of having very heavily decorated and gilded pieces and undoubtedly displays the whiteness and translucency of Nantgarw porcelain at its finest. The reason for its acquisition and nomenclature as the *Lady Seaton* Nantgarw service is unknown and there is no record extant of it ever belonging to Lady Seaton; the author can only surmise that their superficial similarity in design has given rise to a relatively recent assimilation of the name by the Nantgarw service. Both services would have been contemporary with each other in manufacture and are assignable to the period *ca.* 1817–1820.

8.1 Cross-Correlation with the Cambrian Pottery Armorial Earthenwares

The correlation between the armorial earthenwares and creamwares produced at the adjacent Cambrian Pottery with those at the Swansea China Works may seem a little conjectural and perhaps even pointless to undertake but it was deemed to be worthy of a wider research investigation here. It transpires that, contrary to expectations, the statements and predictions in the literature, the Cambrian Pottery was not that prolific in the production of true armorials: this deviation from the expected norm probably can be ascribed to the way that armorials have been defined hitherto as it is quite apparent that the Cambrian Pottery and its later foundation but contemporary neighbour, the Glamorgan Pottery, certainly produced a range of inscribed, commemorative and monogrammed earthenwares for a predominantly locally based clientele, and also especially addressing Cornish themes based upon tin mining and salted pilchards. These have probably been counted as "armorial" in concept and hence have distorted the true production situation of armorial china at the Cambrian Pottery.

Several crests are noted on Cambrian pottery earthenwares which also appear on their porcelain analogues: specific examples here include the armed spread eagle with wings extended, the demi-leopard and rampant lion. Another crest which is observed to appear on two different Cambrian Pottery earthenware armorials is that of a pelican – the first of these is associated with an escutcheon and a legend which clearly identifies it as belonging to John Richardson of Pantygwydir, Carmarthen, who was Mayor of Swansea in 1845, and the other appears as a solitary crest on the verge of an undated Cambrian Pottery creamware service. Are both of these, therefore, representative of Richardson commissions and can they be cross-correlated and identified as such if found on Swansea porcelain in the future? The attitude of the armed spread eagle with wings extended crest on a Swansea porcelain service needs to be examined further in this context too as it has been previously assigned to the Venn family, but for its meticulously correct forensic attribution the Venn crest as cited in the literature armories is completely different and comprises a cubit arm in armour holding an arrow. Perhaps, like the *Lady Seaton* service, this is an acquired nomenclature service .... in other words, a *named* service through a later possession and acquisition .... and the real and correct assignment of attribution for this armorial crest now needs to be made. Two possibilities are immediately apparent, namely, the Turbervills of Ewenny Priory and the Williams-Wynn of Wynnstay: both families claim crests that have an armed spread eagle with extended wings but the Turbervill crest shows the eagle with head facing the front (affronted) and talons displayed rampant whereas the Williams-Wynn eagle shows the beak and head extended dexter, as is depicted in the Venn service. It is tentatively suggested, therefore, that the so-called Venn service should be re-classified as the *Williams-Wynn of Wynnstay* service. Likewise, the demi-leopard shown on Cambrian Pottery earthenwares is very dependent upon the spots being displayed prominently, as in some heraldic achievements the spots are coloured definitively in black or red on a gold background (defined as *sable* and *gules* in the associated heraldic descriptors of tincture) and this often cannot be achieved when the crest is simply gilded as it normally would be on a porcelain artefact, where the spots can then be rather difficult to discern from the background gold field. Several of these crests are shown in Fig. 8.10, namely the pelican, the couped bull's head of Galton, the lion rampant, demi-lion, and the armed spread eagle with wings extended.

As an aside during this research into the Cambrian Pottery, Swansea earthenwares which bore armorial designs, the intriguing example of a potentially "foreign" armorial on a Cambrian Pottery plate has been mentioned only by Morton Nance in the literature (Morton Nance, *The Pottery and Porcelain of Swansea and Nantgarw*, 1942) and described therein as "foreign" because of its unusual pictorial composition which seemingly did not reflect a British heraldic armorial composition. In an unrelated comment elsewhere, Ernest Morton Nance alludes to the particular interest of Mrs. Hester Lynch Piozzi in the closure of the Cambrian Pottery warehouse in Fleet Street, London and to her approval of the agency for Swansea earthenwares being then taken on by Mortlock's of Oxford Street, some years before Swansea porcelain was being manufactured and presumably then later also being distributed to clients in London via Mortlock's agency in the capital. The author has followed up this lead and has discovered an intriguing history (which has been further described in detail in Chap. 3) relating to Mrs. Hester Lynch Piozzi, whose full name was Mrs. Hester Lynch Thrale Piozzi Salusbury, born in 1741 and an heiress of considerable wealth and entitled to bear arms through her father, Sir John Salusbury, a prominent landowner in North Wales. Her arms are cited in Burke's armory (Burke, *Burke's General Armory—The General Armory of England, Scotland, Ireland and Wales: Comprising a Registry of Armorial Bearings from the Earliest to the Present Time*, 1884) and her motto immediately struck a chord with the very unusual portrayal of the two lion supporters of the arms portrayed on the

Fig. 8.10 Crests depicted in Fairbairn: *a* Spread eagle wings extended, armed, Plate 48 / 11; *b* Pelican in piety Plate 33/17; *c* Rampant lion Plate 67/5; *d* Bull's head Plate 43/8; *e* Demi-lion Plate 67/10. (From Fairbairn, ***Crests of the Families of Great Britain and Ireland and Their Kindred in Other Lands,*** Volumes I and II, T.C. and E.C. Jack, Edinburgh and London, 1905 and New Orchard Publishing, London, 1986)

Cambrian Pottery plate. The author realised that the "arms" on this Cambrian Pottery earthenware plate would possibly be appropriate for attribution to this lady as she was well known in Georgian England as the doyenne of the anecdote and for her mastery of the language in the upper echelon literary circles in which she moved in London in the late eighteenth Century. Such a quirky type of armorial bearing would certainly not be out of character with her demeanour and a confident assignment of the arms to her origin can be made here – even though this plate is not porcelain and only comes within our more broadly defined remit as Swansea "china". Therefore, Ernest Morton Nance, and also Bernard Rackham, it is maintained by the author, were incorrect to assume a foreign nobility origin for this armorial plate along with the implications that this would have generated at that time in a Europe bedevilled with political upheaval and the Napoleonic Wars. It is believed, therefore, that this plate is not an overseas foreign commission but rather represents an unusual British "armorial" artefact which would have perhaps been commissioned by Mrs. Hester Lynch Thrale Piozzi Salusbury from Thomas Pardoe and decorated by him in Bristol, where they both lived as near neighbours around the middle of the second decade of the nineteenth Century, i.e. about 1815–1818.

8.2 Recapitulation of Analyses

Taking the summary of the crests and coats-of arms that appear on Nantgarw and Swansea porcelain artefacts that have been reviewed above, the analytical results are as follows:

- Full and Partial Coats-of-Arms: 3/3 Nantgarw have all been identified and 2/2 Swansea, the hitherto unidentified Swansea partial coat-of-arms comprising an escutcheon was only referred to once in the literature, namely by Jones & Joseph (*Swansea Porcelain: Shapes and Decoration*, 1988) and where the image was too indistinct to analyse initially – also, no reference had been cited as to its provenance or current placement, so this was initially shelved as necessarily remaining as a case study for future research. However, it appears that the National Museum of Wales holds an exemplar of this same armorial service, which had already been identified conclusively as the arms of Lockwood impaling Manners-Sutton and this then completed satisfactorily the identification of both of the known Swansea porcelain partial armorials.
- Crests: In summary, 4/4 for Nantgarw have all been identified and 5/9 from Swansea. The 4 remaining Swansea crests that have not been unequivocally identified in this current analytical research with possible contenders for their attribution are as follows:
- Bird on a Sprouting (Fructed) Tree Stump: Lockwood of Dews Hall, Essex or possibly Hill of the Plymouth Ironworks, Merthyr Tydfil.
- Demi-leopard: Booth-Grey of Dyffryn St Nicholas.
- Boar's Head Erased: Edwards of Rheola House, Glyn Neath, Vale of Neath.
- Lion Rampant: Vaughan of Llantrisant.

All of these contenders have the correct crest and entitlement for their potential assignment to the appropriate Swansea armorial service, they are all local landowners and have been mentioned in William Weston Young's Diaries as people whom he visited during his estate surveying duties, with the exception of Lockwood, who nevertheless had business and landowning interests in Swansea (some of which he rented to provide the site for the Glamorgan Pottery in 1814) and he had already commissioned an armorial Swansea porcelain service bearing his arms impaling Manners-Sutton (shown in Fig. 5.20). The coat-of-arms of Lockwood of Dews Hall is shown in Fig. 5.21 and clearly shows his crest of a bird (identified here as a martlet) sitting atop a fructing tree stump as appears on the crested Swansea porcelain we have identified above. It is not unreasonable to propose, therefore, that at some time these gentlemen did acquire some Swansea porcelain as did many of their social circle, several of whose names will appear in Tables 7.1 and 7.2. A brief thumbnail sketch of each of these four possible contenders now follows to illustrate their standing in the Glamorganshire community in which William Weston Young moved.

8.2.1 Lockwood of Dews Hall, Essex

From the partial coat-of-arms on a Swansea plate shown in Fig. 5.20, the arms have been identified as those of Lockwood impaling Manners-Sutton. The family of Lockwood of Dews Hall, Essex, has already featured earlier (Sect. 5.8.1) in the consideration of their crest of a martlet sitting on top of a fructed tree stump which Fairbairn describes as "*On the stump of a tree erased ppr. a martlet sable.* The Lockwoods, of Dews Hall, Lambourne, Essex, had an extinct ancestry linked with Wood, of Gayton Park, where the crest in deference to the name is "*an oak tree fructed vert*", with a preponderance of martlets and oak trees in evidence on the arms. Clearly the type of bird exhibited in Hill's crest, a falcon, is not manifest here and the consensus judgement must therefore be that this crest is much better attributed to Lockwood, of Dews Hall, Essex, with a martlet sitting on a fructed tree stump.

Lord Robert Manners with his assumption of the name of Sutton, became Manners-Sutton with a seat at Kelham Hall, near Newark in Nottinghamshire. Lord Thomas Manners-Sutton, the first Baron Manners (1756–1842) became Baron of the Exchequer and Lord High Chancellor of England with estates at Haddon Hall in Derbyshire and at Oakham Castle in Rutland. Lord George Manners-Sutton, the third son of the Duke of Rutland, had three daughters, Louisa Bridget, who married Edward Lockwood Perceval of Bishop's Hall, Essex, in 1790 and who died in 1800, Mary, who married the Rev., Richard Lockwood of Fifield, Essex in 1799 and died in 1829, and Charlotte Manners-Sutton, (born 1764) who married Thomas Lockwood of Dews Hall, Lambourne, in co. Essex and Gayton, Northampton, in June 1789 and died in 1827. It is interesting that all three Manners-Sutton daughters married Lockwoods from Essex! In 1812, Thomas Lockwood (born 1768, died 1836) leased land he owned in Swansea to Baker, Bevans and Irwin, who founded the Glamorgan Pottery on the site in 1814. The armorial bearings on the Swansea plate of the finest duck-egg porcelain match those of Thomas Lockwood impaling Manners-Sutton. What is striking is that the three daughters of Lord George Manners-Sutton all married into branches of the Lockwood family within the space of ten years. A potential reason for commissioning the service between 1815 and

1825 is not immediately obvious as it cannot be to celebrate a wedding, but perhaps a special birthday or birth of a child? Research into the genealogy of Thomas Lockwood and Lady Charlotte Manners-Sutton reveals that Thomas was born in 1768 so he would have attained the age of 50 in 1818, but their only daughter, Lady Horatia Charlotte Lockwood (born 1797, died 1838) married Viscount Hood, and their daughter Mary Isabella was born in 1818! So, 1818 would indeed have been a memorable year for celebration in the Lockwood household to occasion the commissioning of a fine Swansea porcelain service!

8.2.2 Richard Hill, Plymouth Ironworks, Merthyr Tydfil

Richard Hill was one of the four great, fabulously rich ironmasters of the Industrial Revolution, the others being Sir Josiah John Guest of Dowlais, Richard Crawshay of Cyfarthfa Castle and Samuel Homfray of Penydarren. Richard Hill acquired the Plymouth Ironworks in 1803 but dies three years later, leaving his son, also Richard Hill Jr. (often called Richard Hill II), to inherit with his brothers, Anthony and John. The Dowlais Ironworks of Josiah Guest soon became the largest producer of iron and steel in the world, but the largest blast furnace was operational at the Plymouth Ironworks. An historical challenge was made between Samuel Homfray and Richard Crawshay for the sum of 500 guineas each, overseen by Richard Hill as an adjudicator, that Richard Trevithick (1771–1833), a Cornish engineer, would not be able to construct a railway to run the 9 miles between Penydarren and Abercynon in the Rhondda Valley to haul coal to feed their blast furnaces hauled by a steam locomotive. Trevithick duly obliged and constructed his *"Puffing Devil"*, a fully functional high pressure steam engine which ran on tracks and hauled 10 tons of coal in 5 wagons with 70 passengers in four hours and came back again with the empty wagons via a temporary halt for steam pressure failure due to a severed bolt. This was undoubtedly the first ever train run on rails in the world, carrying goods and passengers and the journey was successfully made on the 21st February 1804, predating George Stephenson and his *Locomotion No.1* by over 20 years, surely making Trevithick the true Father of the Railways! It also demonstrates the acumen, wealth and business standing of these industrialists, all of whom appear in our lists as potential supporters of the Nantgarw and Swansea porcelain manufactories. A fuller account of the birth of steam-hauled freight and passengers, initiated by these four industrial supporters of the Nantgarw China Works is given in Appendix IV.

8.2.3 William Booth-Grey of Dyffryn St Nicholas

The second son of George Harry Grey, the fifth Earl of Stamford, of Dunham Massey, Cheshire, William married Frances Anne Pryce, daughter and heiress of Thomas Pryce of Dyffryn St Nicholas in 1802. He is noteworthy for his mention in

William Weston Young's Diaries as one of the "Ten True Men of Glamorgan" who are nominated as the main financial supporters of the Nantgarw China Works, Phase II, start-up operation in 1817. It is inconceivable surely that such a meritorious stand in support for this venture would not also have resulted in further practical support being given through his acquisition of some porcelain from the fledgling local enterprises of Swansea and Nantgarw.

8.2.4 John Edwards of Rheola House, Glyn Neath, Vale of Neath

A cousin of the architect John Nash, John Edwards commissioned Nash to build Rheola House between 1812 and 1814. As a Parliamentary lawyer he would have had access to London society as well as that in Glamorganshire: his next-door neighbour was William Williams of Aberpergwm House, Glyn Neath, who is recorded as having commissioned several important pieces of porcelain from the Nantgarw China Works, especially favouring the local painting of Thomas Pardoe. Again, it is surely inconceivable that John Edwards would not have sought to acquire some of the finest Welsh porcelain, whether locally or from the London rooms of John Mortlock or other ateliers as did his immediate neighbour, William Williams, with whom he may well have shared dining on occasion and thereby come across Welsh porcelain.

8.2.5 William Vaughan, Lanelay Hall, Llantrisant

An attorney-at-law, he was related to John Edwards of Rheola House, to whom he bequeathed his estate, after which John Edwards became known as John Edwards-Vaughan with an entitlement to display both heraldic crests! A near neighbour of William Vaughan of Lanelay Hall was Wyndham Lewis of Greenmeadow, a partner of Sir Josiah Guest in the Dowlais Ironworks and also MP for Maidenhead alongside Benjamin Disraeli. Wyndham Lewis was a supporter of the Nantgarw China Works, from whom he commissioned a fine porcelain service and he has here also been assigned an armorial service in Swansea porcelain, adopting his crest of a Paschal lamb with his Welsh motto. After his death in 1821, his widow, Mary, married Benjamin Disraeli, Lord Beaconsfield. Again, as a near neighbour of Wyndham Lewis, William Vaughan would surely have experienced some fine dining from Welsh porcelain dinner and dessert services in the Lewis household which may have reasonably resulted in his personal acquisition of a Welsh armorial service from Swansea or Nantgarw bearing his own crest?

8.3 Lost Services and Information Tracking

Researchers into individual family histories may uncover some historical documentation which will reveal their support of the local China Works and perhaps the purchase of an armorial service or possibly an inventory of family effects which was made prior to their sale and dispersal in the auction rooms? As an illustration of this, the author some years ago noted a passing reference in the literature, mentioned by Dr. William John in his seminal work *Nantgarw Porcelain* (John, *Nantgarw Porcelain*, 1948), to a large Nantgarw porcelain service that had been acquired for Farnley Hall; no location was specified, and it was assumed that this service was now lost. In his efforts to trace the whereabouts of this service, the author contacted Guy Horton-Fawkes Esq., of Farnley Hall, Otley, North Yorkshire, who kindly invited him to meet and discuss the possibility that this was the perhaps the Farnley Hall alluded to in the literature (Edwards, *Nantgarw and Swansea Porcelains: A Scientific Reappraisal,* 2017). The first thing the author noted upon entering the drawing room of Farnley Hall was a beautiful Swansea vase decorated by Henry Morris, which is shown here in Fig. 8.11! This very promising start led to a detailed

Fig. 8.11 Large Swansea porcelain vase, campana shaped with flared rim and supported on a round tapered pedestal base with fluted moulding, embellished with a beaded edge rim moulding and an overhanging fringe of acanthus leaves. The bowl is gadrooned and ornamented with two pairs of gilded satyr mask heads. Superbly decorated in London in the manner of Moses Webster at the workshop of Robins & Randall with a basket of flowers and random butterflies in the style of the *Burdett-Coutts* and *Lysaght* Swansea services and accompanied by the characteristic London dentil-edge gilding. Vase type shown in Jones & Joseph (*Swansea Porcelain: Shapes and Decorations*, 1988, page 122). *Courtesy of Guy Horton-Fawkes Esq., Farnley Hall, Otley, North Yorkshire*

inspection of the inventory of the cellars of Farnley Hall, which unearthed some 25 pieces of London-decorated Nantgarw porcelain stacked there on a stone shelf, long forgotten by its owners and mixed up with a rather splendid early nineteenth Century Bloor Derby dessert service with a caerulean blue ground decorated with animals and birds, probably by Richard Dodson. The 35 pieces of Nantgarw porcelain indeed proved to be the remains of the long-lost Nantgarw *Farnley Hall* service mentioned by Dr. John. It has resided there since its historical acquisition by Walter Ramsden Hawkesworth Fawkes MP in the latter years of the second decade of the nineteenth Century. Walter Ramsden Fawkes (1769–1825), the son of Walter Ramsden Beaumont Hawkesworth, took the name Fawkes in 1792 as heir to Francis Fawkes of Farnley Hall; he became MP for the North Riding of Yorkshire and its High Sheriff in 1823. He was a lifelong friend of the artist Joseph Mallord William Turner, who used to spend his summers over many years in Yorkshire travelling and painting local scenes whilst staying at Farnley Hall. A pair of tureen stands from the Farnley Hall service as rescued are shown in Fig. 8.12 and a very large meat platter in Fig. 8.13, which interestingly displays the bearded tulip, a characteristic of the mysterious "Swansea" artist "de Junic," who it is alleged moved to London and there decorated Nantgarw porcelain (Edwards, *Porcelain Analysis and Its Role in the Forensic Attribution of Ceramic Specimens,* 2021). A serendipitous tale, undoubtedly, which although perhaps not unique in the annals of porcelain research

Fig. 8.12 Pair of tureen stands, Nantgarw porcelain, impressed NANT-GARW C.W., from the *Farnley Hall* service, Nantgarw border moulding and decorated in London with vignettes of flowers and birds with dentil-edge gilding. *Courtesy of Guy Horton-Fawkes Esq., Farnley Hall, Otley, North Yorkshire*

Fig. 8.13 Large meat platter from the large *Farnley Hall* Nantgarw dinner-dessert porcelain service, impressed NANT-GARW C.W., with moulded Nantgarw border and dentil-edge gilding. Superbly painted with six vignettes of garden flowers and birds and a central floral group which contains the bearded tulip characteristic of the painter "de Junic" who worked in the London ateliers during the second half of the decade between 1810 and 1820, decorating Nantgarw and Swansea porcelain following his initial employment locally at the Swansea China Works. *Courtesy of Guy Horton-Fawkes Esq., Farnley Hall, Otley, North Yorkshire*

and discovery, is attractive in concept for the thought as to what else may be lurking hidden and waiting to be re-discovered in the attics and cellars of some of our great country houses.

A second tale has a similar ending: in 2015, at the disposal of the estate of the late Duchess of Richmond, chatelaine of Goodwood House in Surrey, the auctioneers discovered approximately twenty London-decorated Nantgarw plates and dishes of the desirable *Brace* service type genre at the back of a corner cupboard in pristine condition, apparently never having been used (Fig. 8.14) — there were no marks of utensils at all on them and they were obviously an inherited part dessert service that had been destined for display only after its purchase or for some other reason had never been used after its purchase!! This part Nantgarw dessert service (which was rather deprecatingly paraphrased in the local Richmond *Argus* newspaper as *"a lot of old crockery"*) was purchased for an estimated £20,000 by one delighted client. A family ancestor, Charlotte, Duchess of Richmond, organised *"the most famous ball in history"* (Longford, 1986) in Brussels on the eve of the Battle of Waterloo, June 15th 1815, at which the Duke of Wellington and his senior officers received news from Prince Willem of Orange of Napoleon's advance with his army into the kingdom of the Netherlands at Charleroi and of his impending battle

Fig. 8.14 Nantgarw porcelain half dessert service, unused, comprising twenty pieces: 5 square dishes, two oval dishes, 2 shell dishes, small tazza, large tazza, and nine dessert plates. London-decorated and of the generic *Brace* type service. Formerly in the possession of the Duchess of Richmond, Goodwood House, Surrey. The Duchess of Richmond hosted a grand ball in June 1815 in Brussels, where the Duke of Wellington in attendance first heard of the advance into the kingdom of the Netherlands by Napoleon on the eve of the Battle of Waterloo. (see Fig. 8.15). When informed of the invasion of Napoleon, the Duke of Wellington immediately ordered his staff officers to prepare for the battle, but took a supper first at 1.30 am. The Duke of Richmond commanded the rearguard British army in defence of Brussels should Napoleon's forces break through. A very historic Nantgarw service as the Richmonds did not return to London until 1818 and then the Duke died in Canada in 1819, where he was Governor-General, so the household was probably re-organising itself when the service would have been delivered in 1818/1819 and the service would possibly have been put into storage. *Private Collection*

with Field Marshal Blucher at Ligny, thereby annulling the impending British invasion of France (Fig. 8.15). Charles Lennox, the fourth Duke of Richmond, had been placed in command of the British defence of Brussels. Wellington calmly took supper at the ball at about 1.30 am and rode off with his staff to give battle with Napoleon at Quatre Bras and at Waterloo the following day (Miller, *The Duchess of Richmond's Ball*, 2004). It is intriguing to think that soon after their return to Britain in 1818 the Duke and Duchess of Richmond must have ordered their service of London-decorated Nantgarw porcelain, probably from John Mortlock in Oxford Street. The Duke of Richmond died in 1819 in Canada, where he had just been appointed Governor-General, so possibly his Nantgarw service was delivered in London to his widow and then forgotten? His widow, Charlotte, Duchess of Richmond, died in 1842, and sometime after that this Nantgarw service, which was

Fig. 8.15 "*The Duchess of Richmond's Ball*", oil on canvas, 44 x 60 cm, by Robert Alexander Hillingford, ca. 1870. Presently in Goodwood House, Surrey, seat of the Dukes of Richmond. *Public Domain*

apparently never used, must have been divided up. Truly, a Nantgarw service of historical importance.

The moral of these two tales is that services for these grand country houses could have been been sequestered many years ago and there could be undiscovered armorial porcelains still waiting to be unearthed and researched, which will potentially activate and extend the incipient armorial database for Nantgarw and Swansea porcelains that has been developed here.

8.4 Final Assessment of Unattributed Crests

Although the coats-of-arms and escutcheons have now been positively identified for Nantgarw and Swansea porcelains, including the monochrome escutcheon depicted on the Swansea plate illustrated in Jones & Joseph (*Swansea Porcelain: Shapes and Decoration,* 1988), which has now been rectified with the subsequent discovery of its polychrome illustrated exemplar, there remain four crests on Swansea porcelain that have yet to be more definitively attributed. Table 1 in Appendix V gives the results of an heraldic analysis of the "Ten True Men of Glamorgan" (who actually are twelve in number!) as cited by William Weston Young in his Diaries (W. W. Young, *The Diaries of William Weston Young, 1776–1847*(1802-1843), 30 volumes, West Glamorgan Archive Service, Swansea, SA1 3SN) and discussed by Ernest Morton Nance (*The Pottery and Porcelain of Swansea and Nantgarw*, 1942). No fewer than four crests that have been adopted by these gentlemen actually appear on Swansea porcelain, which certainly gives credence to the approach adopted for the investigation of the family heraldry of local supporters and landowners. It is now possible, therefore, to reasonably assign the four hitherto unattributed Swansea porcelain crests as follows:

- Demi-leopard rampant: William Booth-Grey of Dyffryn St Nicholas.
- Lion rampant: William Vaughan of Lanelay Hall, Llantrisant.
- Boar's head, erased and couped: John Edwards of Rheola House, Glyn Neath.
- Bird on fructed tree stump: Thomas Lockwood of Dews Hall, Essex, or possibly Richard Hill of the Plymouth Ironworks.

It is accepted that there could still be some conjecture about this assignment, which would require further research to be undertaken into individual family documentation and archives, but the balance of evaluated opinion would be in favour of such an assignment being made at least until evidence was presented which would contradict it or confirm its veracity. This means that all armorials in Welsh porcelain have now had a definitive or highly probable candidature for their attribution, giving an overall success rate of 100%.

8.5 Analytical Differentiation Between London Decorated and Locally Decorated Porcelains

As part of the holistic determination of the origins and provenancing of a porcelain artefact from the Nantgarw and Swansea factories the source of the decoration is a component detail of interest: we have seen earlier in this text that much information can be gleaned from an attribution of the decoration to a local enamelling or gilding or to a London sourced atelier. In some cases it has been possible to do this for the armorial exemplars considered here but in others it has not been so straightforward. In an earlier analytical scientific publication (Edwards, *Porcelain Analysis and Its Role in the Forensic Attribution of Ceramic Specimens,* 2021) it was revealed that through the use of non-invasive Raman spectroscopy the chemical molecular spectroscopic signatures of the glaze used for Nantgarw porcelains is essentially different for the *Nantgarw No.1.* glaze adopted by Billingsley and Walker between 1817 and 1820 and the *Nantgarw No.2.* glaze used by Young and Pardoe to finish the remnant porcelain stocks between 1820 and 1823. The glaze interrogation could be carried out non-destructively on perfect porcelain pieces without incurring any damage, so this means that in cases of doubt it is now perfectly possible to access this type of information, which may be critical in one's expert evaluation of an individual piece to assess whether it has been locally decorated at the Nantgarw China Works during the Billingsley/Walker period or during the Young/Pardoe period. Of course, experts maintain that they are able to distinguish between these two potential scenarios from the creamier colour and thinner texture of the Young/Pardoe glaze in contrast to the stark and glossy whiteness of the Billingsley glaze, but this may not always be as clear cut visually as it should be and then the scientific evidence will be mandatory for an objective conclusion to be made.

Secondly, the apparently didactic evidence offered by the presence of dentil edge gilding being indicative of a London decorated artefact has been examined and whereas this statement is absolutely true, the converse is not: this means that the absence of dentil-edge gilding does not then automatically confer the placing into the category of local decoration upon the piece concerned. As we have seen in our analysis here, it appears that some 75% of accepted and classified London-decorated pieces in the literature do not have the required dentil-edge gilding, and this applies to both Swansea and Nantgarw artefacts! The presence of the dentil edge gilding is not, therefore, a pre-requisite for assignment of the Swansea or Nantgarw porcelain artefact to a London decoration source atelier: the presence of dentil edge gilding definitively confers a London decoration upon the artefact but its absence does not conversely mean that the artefact has been locally decorated, as from the above analysis some three-quarters of the assigned London-decorated Nantgarw and Swansea porcelain pieces do not possess dentil edge gilding.

Thirdly, Dr. William John noted in his book (*Nantgarw Porcelain*, 1948) that the presence of a London decoration on Nantgarw porcelain invariably conferred upon the enamels a residual iridescence which could be observed visually at grazing angles of incidence: he attributed this to the leaching out of metal ions and volatiles into the glaze during the enamelling firing processes undertaken in the London ateliers through their adoption of reducing kiln conditions. It is certainly true that locally decorated pieces do not display this iridescence effect, which can also be used as a visual input into the argument for local *versus* London decoration and this issue has been amplified further in earlier works by this author (Edwards, *Nantgarw and Swansea Porcelains: A Scientific Reappraisal*, 2017; Edwards, *Swansea and Nantgarw Porcelains: An Analytical Perspective,* 2018). This iridescence effect has not been highlighted as a parallel indicator for Swansea porcelains decorated in London and it would be interesting therefore to embark upon a correlated study of this aspect, which might prove useful in the attribution of future armorial exemplars for Swansea porcelains that may emerge: could it perhaps arise from different kiln firing conditions used by the London enamellers for Swansea porcelain and could therefore be a function of the particular atelier involved?

Analysis of the compositions of the armorial porcelain bodies is also quite interesting: all seven exemplars of the Nantgarw armorial porcelain bodies are composed of the highly phosphatic soft paste porcelain, whereas for the Swansea armorials one is clearly of the steatitic magnesian body, the remainder being of the soft paste phosphatic duck-egg variety, with a further body composition afforded by the new armorial discovery of the Lliw punchbowl example. One, with the Lockwood impaling Manners-Sutton escutcheon, is composed of the phosphatic duck-egg variety accompanied by its border moulding of flowers and foliage, whereas the remaining nine crested exemplars are all composed of the phosphatic duck-egg porcelain which has always been regarded in higher esteem by clients. The punchbowl composition in a trident or hard paste porcelain would not be out of place in view of the heavier use that would be expected that would require its more robust construction but the difference between the higher quality of the crested and of the part-armorial escutcheoned Swansea china is noteworthy.

In summary, the application of forensic scientific principles to the holistic analysis of armorial Welsh porcelains has advanced the baseline for their identification and has not only been successful in the identification of many of those currently unknown or outstanding but has also indicated several new routes and pathways that are profitable to follow for the potentially successful attribution of new exemplars which may surface in the future. The field of armorial porcelains is indeed a fascinating one for specialist studies in ceramics history and the two Welsh porcelain manufactories of Swansea and Nantgarw are certainly no exception to this. Although several of the crests on the Swansea earthenwares and creamwares that have been explored and investigated as a background here have not achieved a match with known Swansea or Nantgarw armorial porcelain exemplars, the establishment of their familial origins will undoubtedly serve to form a novel database when new, as yet undiscovered, armorial examples arise and thence serve to effect a means of homing in on and elucidating their heraldic assignment.

As a result of this current study, therefore, 14 of the 18 known Nantgarw and Swansea armorial exemplars have now been accurately and unequivocally assigned using the literature armories and available associated historical documentation: during this process, it appeared that several armorials had been incorrectly or inadequately assigned hitherto in the literature and these attributions have now been corrected. Finally, the remaining four crests, all of which appear on Swansea porcelain armorials, have been tentatively but credibly assigned to potential local arms bearers and known supporters on a reasoned hypothetical basis and we must await the appearance of contemporary documentation in support of their attribution.

8.6 The Standing of Armorials Vis-à-Vis Named Services for Nantgarw and Swansea Porcelains

In an interesting conclusion to this research study, an evaluation of the incidence of armorial porcelains that have been found for the Nantgarw and Swansea China Works compared with the established number of named services that have appeared in the literature was undertaken. An earlier work by the author (Edwards, *Swansea and Nantgarw Porcelains: A Scientific Reappraisal*, 2017) explored the incidence of named services originating from the two Welsh porcelain manufactories and collated these for the first time and it is appropriate to further develop this theme to properly include the armorials that have formed the subject of this book. Table 8.1

lists the named services that have been identified by the author in the literature for Swansea and Nantgarw porcelains: in total these number 83, which includes the 18 armorials investigated here -- therefore, giving the armorials an overall percentage of 22%, which is composed as follows. Of the 52 named Nantgarw services the 7 armorials comprise 14% of the total, whereas of the fewer 31 Swansea named services the greater number of 11 armorials (35%) results in a significantly higher percentage overall, which can be attributed largely to the number of crested Swansea armorial exemplars that are extant compared with their Nantgarw analogues. The data harvested from the literature in Table 8.1. is of a naturally diverse reliability as several earlier ceramics historians, such as Dr. William John and Ernest Morton Nance, have sometimes mentioned that they have seen particular services when collecting material for their books and have relied upon information provided by the collectors at that time regarding their origin and provenance. Hence, several services mentioned do not have a description that is detailed enough to facilitate further research into their sourcing. Often, a rather sparse description of the "named service" is provided in earlier texts, such as ".... decorated with garden flowers in the manner of Thomas Pardoe", or "....containing groups characteristic of the painting of William Billingsley", or ".... a second service is possessed by the Duke of Newcastle", from which there is clearly insufficient information to identify positively today the artefacts concerned. Despite this setback, where possible, the "named service" has been included in the data assimilated for the analysis in Table 8.1. It must also be borne in mind that many of the originators of the service commission two hundred years ago have since largely had their effects divided thoruh inheritance or dispersed at auction sales and the integrity of the named service has thereby been lost: classic examples would be the *Duchess of Richmond* Nantgarw service and the *John Vivian, Sir John Nicholl* or *Bosahan* services in Swansea porcelain. Single items of these important services and several others might well now exist in museum and private collections and remain unrecognised for their important provenancing nomenclature. Other services have a generic importance, such as the *Brace*, *W.T. Wilde* and *Sir John and Lady Williams* Nantgarw services, which are now recognisable as such from their later appellations despite the details of their original commissions having been lost. In this context, it should be borne in mind, therefore, that the nomenclature of such a service is often unrelated to its original purchaser from the Swansea or Nantgarw China works or their agents in the early nineteenth Century: a good example of this is given by the *Lady Seaton* service in Swansea porcelain,which was originally purchased from the Swansea China Works by Arthur Jones of Bryn Newydd, Swansea, a distant relative, whence it passed to Sir Arthur Pendarves Vivian of Bosahan Manor, Cornwall, and thence to Lady Seaton after the dispersal of his effects in 1926, more than a hundred years after its original manufacture in Swansea. A Nantgarw analogy is provided by the *Mackintosh* service (Fig. 8.16), referred to in Table 8.1 as Mackintosh 1, which was purchased originally by Edward Priest Richards, the land agent of the second Marquess of Bute, John Crichton Stuart, who was at that time said to be the wealthiest man in Britain. Edward Priest Richards of Plas Newydd, Roath, Cardiff, married Harriet Georgina Tyler of Cottrell St Nicholas in 1856 and their daughter Harriet Diana

Fig. 8.16 Nantgarw porcelain, dessert plate from the *Mackintosh* service as ordered by Edward Priest Richards, Plas Newydd, Glamorgan, *ca.* 1818–1820, land agent for the second Marquess of Bute, London-decorated in the atelier of Robins & Randall, Islington, with a central avian feature probably by Thomas Martin Randall. *(Reproduced by courtesy of Bonhams Auctioneers and Valuers, London)*

Arabella Mary Richards, born in 1857, married Alfred Donald Mackintosh (1851–1938) in 1880, the 28th Captain and Chief of Clan Chattan and Clan Mackintosh, and known as The Mackintosh of Mackintosh of Moy, Inverness. The Clan Chattan, based at Torcastle, Inverness, is a unique confederation of 12 Scottish Highland Clans formed for their mutual protection including the Clans Mackintosh, Macpherson, Macbean, Farquharson and Maclean. The Nantgarw service purchased by Edward Priest Richards was given as a wedding present to his daughter Harriet (who some accounts call "Ella", as presumably a diminutive of her name Arabella?) upon her marriage to Alfred Mackintosh and thereafter this became known as the *Mackintosh* service. The service was dispersed in the late 1800s and several pieces were later re-purchased by the Mackintosh family in 1901. This is perhaps the most famous named service in Nantgarw porcelain, pieces of which are so highly sought after by collectors today that they command premium auction prices, and it was the subject of a special study by Dr. William John (*Nantgarw Porcelain*, 1948). Clearly, it is therefore not really of a Mackintosh source originally, as they were in fact its last family owners, unlike the *Ramsay of Straloch and Barra* Nantgarw services, and strictly, perhaps the *Mackintosh* service should now be referred to as the Priest Richards or Plas Newydd service in deference to its original purchaser and probable place of use for many years initially! The *Mackintosh* service, perhaps surprisingly, was not ordered locally by Edward Priest Richards but through John Mortlock 's agency in Oxford Street, London, and it was decorated at the Robins & Randall atelier of Barnsbury Street, Islington, with characteristic dentil-edge gilding; the avian subjects depicted on this service were initially ascribed to Francois Levaillant but Roger Edmundson (*Welsh Ceramics in Context I*, 2003, p. 204) has assigned them preferably to the hand of Thomas Martin Randall personally. A similar situation pertains to the three Nantgarw tea and coffee services presented to Edward Edmunds of Penrhos, the landlord of Tyla Gwyn, the site of the Nantgarw China Works and the Billingsley residence at Nantgarw; it is believed these were donated in lieu of rent for the site in its early days by William Billingsley, whose service names have now been recorded as the *Edwards, Twyning* and *Duncombe* services through the later inheritance and marriage of Edward Edmunds' daughters and granddaughters. Pieces from these *Edwards, Duncombe* and *Twyning* Nantgarw services are shown in Figs. 8.17, 8.18 and 8.19, respectively, all of which were painted by William Billingsley at Nantgarw and therefore have a full and special provenance

Fig. 8.17 Nantgarw porcelain, *ca.* 1817–1820, items from the *Edwards* tea and coffee service, comprising a teapot, stand, covered sucrier, milk jug, slop bowl, large plate and a tea cup, coffee cup and saucer trio with heart-shaped handles. Made for Edward Edmunds, Pen-y-Rhos, Nantgarw, and landlord of the Nantgarw China Works site at Tyla Gwyn, and decorated by William Billingsley with single pink roses on a bed of gilt seaweed. Presented to Frances Edmunds on the occasion of her marriage. (*Reproduced by courtesy of Bonhams Auctioneers and Valuers, London)*

Fig. 8.18 Nantgarw porcelain, *ca.* 1817–1820, tea cup and saucer with heart--shaped handle from the *Twyning* tea and coffee service: made for Edward Edmunds, Pen-y-Rhos, Nantgarw, and landlord of the Nantgarw China Works site at Tyla Gwyn. Decorated by William Billingsley with single pink roses and green foliage. Believed to have been presented to Edward Edmunds by William Billingsley in lieu of rent for the Nantgarw site. *(Reproduced by courtesy of Bonhams Auctioneers and Valuers, London)*

Fig. 8.19 Nantgarw porcelain, *ca.* 1817–1820, tea pot and stand from the *Duncombe* tea and coffee service: made for Edward Edmunds, Pen-y-Rhos, Nantgarw, and landlord of the Nantgarw China Works site at Tyla Gwyn. Decorated by William Billingsley with single pink roses alternating with gilt anthemions. Believed to have been presented to Edward Edmunds by William Billingsley in lieu of rent for the Nantgarw site. *(Reproduced by courtesy of Bonhams Auctioneers and Valuers, London)*

Fig. 8.20 Swansea porcelain, *ca.*1817–1820, dessert plate from the *Burdett-Coutts* service ordered by Sir Thomas Coutts, banker to King George III, to celebrate his marriage to the actress Harriet Mellon in 1818. Decorated in London by James Turner at John Sims' atelier, with a central basket of flowers and random butterflies and moths. It passed to his granddaughter Angela, Baroness Burdett – Coutts, and was sold at the auction of her effects after her death in 1922. *Private Collection*

in this respect. At Swansea, the nomenclature of perhaps the most famous named service, the *Burdett-Coutts* service, is also compromised in a similar fashion as originally it was commissioned in 1818 by Sir Thomas Coutts, banker to King George III, to celebrate his marriage to the actress Harriet Mellon. A large dinner-dessert service of almost 300 pieces originally, its ownership descended by inheritance to his daughter Sophia, and thence to his granddaughter Angela, who as Baroness Burdett-Coutts (she married a Sir Francis Burdett) gave her name to the service, which was sold at auction in London on her death in 1922. A beautiful London-decorated service, painted by James Turner in John Sims' atelier, again with its characteristic dentil gilt edging, a dessert plate from the *Burdett-Coutts* service is shown in Fig. 8.20. Technically, this service should more correctly be named the *Sir Thomas Coutts* service, but it is still referred to by collectors and connoisseurs as the *Burdett-Coutts* service.

Another important result emerges from inspection of Table 8.1. regarding the multiple number of services that were commissioned or owned by individuals in the list: 18 named clients of the Swansea and Nantgarw China Works (some 23% of the total surveyed here) owned more than one service according to the literature records and that excludes vague comments made in the earlier accounts of William Turner and Ernest Morton Nance to the effect that " X has several Swansea or Nantgarw services" (Morton Nance, *The Pottery and Porcelain of Swansea and Nantgarw*, 1942) and "The Marquess of Bute also has several services of Nantgarw porcelain for use in his country estates in Scotland (author: these being Mount Stuart on the Isle of Bute and Dumfries House in Ayrshire)" (Turner: *The Ceramics of Swansea and Nantgarw. A History of The Factories with Biographical Notices of the Artists and Others, Notes on the Merits of the Porcelain, the Marks Thereon Etc.*, 1897). The data from such vague comments have not been included in our survey as it is impossible to even attempt to identify either the number or the characteristics of the particular services referred to from the information provided. What is also clear is that only four of these owners of multiple services of Nantgarw and Swansea porcelains, namely Ramsay, Lewis, Lockwood and Clarke, actually have armorial porcelains that have been identified in our armorial research lists. So, the immediate question that now arises is, what has occurred with the other 14 entitled individuals who already owned multiple services of Welsh porcelain, and are their crested or armorial services on Welsh china still to be discovered?

In Table 8.2 a summary of the conclusions of this armorial research project are compiled in a tabular form and now gives the correct attribution of each armorial exemplar considered here. A column in this Table also gives the potential reason for the acquisition of the service, as gleaned from related genealogical searches: clearly, this cannot be definitive evidence in some cases but is nevertheless based on some celebratory or memorial events that occurred during the narrow periods of manufacture of the Nantgarw and Swansea China Works. There are parallels in special services raised at other manufactories – for instance, the famed *Pares* service in Derby porcelain, which has been recorded historically as a service ordered in 1812 for the funeral of a daughter of John Pares of Hopwell Hall, Ockbrook, Derbyshire (Twitchett, *Derby Porcelain: An Illustrated Guide, 1748–1848,* 2002). However, upon examination of the genealogy of the Pares family, all is not straightforward: John Pares (1749–1833), a wealthy hosiery manufacturer and banker, and Sheriff of Leicestershire, married Agnes Lightbody of Liverpool, born in 1759, who died on the 31st August 1812, on precisely the same date as her mother Elizabeth Lightbody (1735–1812) at the ages of 53 and 77, respectively! Agnes had several daughters, namely Agnes, Katherine, Anne, Hannah and Dorothea, all of whom lived until the late 1800s. So, despite the historical allusion that this was a funeral service for his daughter, could it be that it in fact it was commissioned by John Pares as a memorial to his wife and mother-in-law who both died on the same day – perhaps in a tragic fatal accident together? No further information seems to be forthcoming in pursuit of the genealogy and circumstances in this direction but it does highlight the need to check all possible sources to verify what at first seems to be otherwise incontrovertible evidence obtained even from manufactory records.

Nevertheless, the results of this research study on the armorial porcelains of Nantgarw and Swansea has established their rightful presence in the service

Table 8.2 Final Attribution of Welsh Armorial Porcelains

Factory	Description	Name	Figure	Reason for Commission[a]
Nantgarw	Escutcheon, motto, crest, supporters	Lord Thomas Thynne, Viscount Weymouth, Longleat House, Wiltshire	2.2	Marriage of Lord Thomas Thynne and Harriet Robbins, May 1820.
	Escutcheon, crest	John Wyndham-Quin of Dunraven Castle, Southerndown	4.2	Accession to title Earl of Dunraven and Dunraven estate in 1822.
	Crest, motto	Colonel John Ramsay of Straloch & Barra, Newmachar, Aberdeen	4.3	Succession to estate in 1817.
	Crest, motto	Colonel John Ramsay of Straloch & Barra, Newmachar, Aberdeen	4.4	?
	Crest	Baron Phipps of Normanby Mulgrave Castle, Whitby	4.5	Coming of age of Constantine Phipps in 1818.
	Crest	John Homfray of Penllyne Castle Cowbridge	4.7	Brother, Samuel, elected MP in 1819.
	Escutcheon	Jeffreys of Brecon Priory	4.9	?
Swansea	Escutcheon, crest, motto	Graham-Clarke of Newcastle upon-Tyne impaling Parkinson of Kinnersley Castle	5.1	Accession of Frocester Manor estate in 1818
	Crest, motto	Thomas Lloyd of Bronwydd Ceredgion	5.3	Marriage of Thomas Lloyd and Ann Davies in 1819
	Crest	Orme of Northampton	5.7	?
	Crest	John Edwards of Rheola House, Glyn Neath, Vale of Neath	5.5	?
	Crest	Sir Watcyn Williams-Wynn of Wynnstay, Ruabon	5.10	?
	Crest, motto	Wyndham Lewis MP, Greenmeadow, Llantrisant	5.12	?
	Crest	John Graham-Clarke of Newcastle-Upon-Tyne	5.13	?
	Escutcheon	Thomas Lockwood of Dews Hall Lambourne, impaling Manners-Sutton	5.20	Thomas Lockwood aged 50 and granddaughter's birth in 1818.
	Crest	Thomas Lockwood	5.16	Birth of grand daughter
	Crest	William Booth-Grey of Duffryn St Nicholas, Vale of Glamorgan	5.22	?
	Crest	William Vaughan, Lanelay Hall, Llantrisant	5.23	?
	Escutcheon, crest, motto	Llwchwr (Lliw) Corporation[b], Loughor, punchbowl	3.10	Gift from T &J Bevington in January, 1825; painted by William Pollard at Swansea

[a]This column proposes a potential reason from genealogical studies for the commissioning of the armorial artefact within the time frame of the factory production

[b]The Llwchwr punchbowl has been included here as an "addendum" since it has been verified to be porcelain and has been documented as originating in the Swansea China Works but until analytical chemical data are forthcoming it cannot be assumed to be of a Swansea porcelain body and could possibly be an interloper purchased elsewhere and decorated at Swansea. Nevertheless, it could prove to be another example of armorial Welsh porcelain, which then would bring the current database of exemplars to 19

nomenclature for these two manufactories, although in all cases the armorial services have been dispersed and sometimes only single exemplars have re-surfaced in our listing. Some readers may well be surprised that the two Welsh porcelain manufactories can claim as large a number of named services as are demonstrated in Table 8.1, namely 83: this measure compares favourably numerically with other major eighteenth Century contemporary china producers such as Derby and Coalport (which have survived into the twenty-first century) and itself is a testimony to the interest in their purchase and acquisition that was manifest by a demanding clientele over the rather brief period of their porcelain production operations, which for Nantgarw was only 3 years and for Swansea only 5 years. It is a matter of conjecture to estimate how many more "named services", and by implication, armorial crested services, would have been identified if the factory workbooks for the Nantgarw and Swansea China Works had been preserved and were still available for consultation and if John Mortlock's agency sales records in London had survived. Nevertheless, the overall conclusion of the present study is that now most, if not all, of the armorial services currently known for the Swansea and Nantgarw manufactories have been identified or at least have had reasonable and credible suggestions made for their attribution.

Quod Erat Demonstrandum!
(Transl.: Which was to have been shown!)

An ancient expression which was generally made at the conclusion of a logical scientific research investigation.Or alternatively:

Finis Coronas Opus
(Transl.: The End Crowns the Work / The Goal Gives Value to the Labour that produced It)
Ovid (Publius Ovidius Naso, 43 BCE – 17/18 CE)

Surely a relevant quotation at the conclusion of a successful heraldic research study in which crowns and coronets play such a significant role!

References

Armorial Worcester Porcelain of the First Period: Specimens from the Marshall Collection in the Ashmolean Museum and Other Sources, Published by Percy Lund, Humphries & Co, and Illustrated by H. (Norman Eastaugh, London, 1964)

W. Bemrose, *Bow, Chelsea and Derby Porcelain: Being Further Information Relating to the Factories Obtained from Original Documents Not Hitherto Published* (Bemrose & Sons, London, 1898)

Sir J. Bernard Burke, *Burke's General Armory—The General Armory of England, Scotland, Ireland and Wales: Comprising a Registry of Armorial Bearings from the Earliest to the Present Time* (Harrison & Sons, Pall Mall, London, 1884)

Bonhams Auctioneers and Valuers (Fergus Gambon, Director of British Ceramics), New Bond Street, London

P. Bradshaw, *Derby Porcelain Figures 1770–1848* (Faber & Faber, London, 1990)

R. Denyer, M. Denyer and H.G.M. Edwards, *The Pendock-Barry Derby Porcelain Service: A Reappraisal*, to be published (2022)

R.S. Edmundson, Billingsley, Randall and Rose, in *Welsh Ceramics in Context, Part I*, ed. by J. Gray, (Swansea, Royal Institution of South Wales, 2003), pp. 193–214

H.G.M. Edwards, *Nantgarw and Swansea Porcelains: A Scientific Reappraisal* (Springer – Nature, Dordrecht, 2017)

H.G.M. Edwards, *Swansea and Nantgarw Porcelains: An Analytical Perspective* (Springer Nature, Dordrecht, 2018)

H.G.M. Edwards, *Porcelain to Silica Bricks: The Extreme Ceramics of William Weston Young, 1776–1847* (Springer-Nature, Dordrecht, 2019)

H.G.M. Edwards, *Porcelain Analysis and its Role in the Forensic Attribution of Ceramic Specimens* (Springer -Nature, Dordrecht, 2021)

W. Elis Jenkins, *Swansea Porcelain* (D. Brown & Sons, Cowbridge, UK, 1970)

C.N. Elvin, *A Handbook of Mottoes Borne by the Nobility, Gentry, Cities, Public Companies Etc.*, Bell & Daldy, Fleet Street, London,1860. Reprinted with Supplement and an Index by R.Pinches, Heraldry Today, Beauchamp Place, (London and Genealogical Publishers, Baltimore, USA, 1971)

J. Fairbairn, *Crests of the Families of Great Britain and Ireland and Their Kindred in Other Lands,* Volumes I and II, (T.C. and E.C. Jack, Edinburgh and London, 1905 and New Orchard Publishing, London, 1986)

O. Fairclough, The use of set pattern decoration at the Swansea China Works. *Welsh Ceramics in Context,* Volume II, ed. J. Gray (The Royal Institution of South Wales, Swansea, pp. 192–208, 2005)

F. Gambon, Porslen Abertawe a Nantgarw (Swansea and Nantgarw Porcelain), (Oriel Plas Glyn-y-Weddw, Llanbedrog, Gwynedd, 2016)

Gartre'n Ol*: Coming Home - An Exhibition of the Finest Nantgarw Porcelain, Nantgarw China Works Museum, Tyla Gwyn, Nantgarw,* (2019)

S. Hanscombe, Armorial porcelain decorated in the London workshop of James Giles. Transactions of the English Ceramics Circle **20**(1), 119–135 (2008)

J. Haslem, *The Old Derby China Factory: The Workmen and their Productions. Containing Details of their Chief Artist Workmen, the Various Marks Used, Fac-Similes Copied from the Old Derby Pattern Books, the Original Price List of More than 400 Figures and Groups, Etc., Etc.* (George Bell & Sons Covent Garden, London, 1876: reprinted by E.P. Publishing, Wakefield, 1973)

M. Hillis, The development of welsh porcelain bodies, in *Welsh Ceramics in Context, Part I*, ed. by J. Gray, (Swansea, Royal Institution of South Wales, 2005), pp. 170–192

L. Jewitt, The Ceramic Art of Great Britain from the Prehistoric Times Down to the Present Day, Vols. I and II, (Virtue & Co. Ltd., Paternoster Row, London, 1878)

W.D. John, *Nantgarw Porcelain* (The Ceramic Book Company, Newport, 1948)

W.D. John, *Swansea Porcelain* (The Ceramic Book Company, Newport, 1958)

W.D. John, K. Coombes, G.J. Coombes, *Nantgarw Porcelain Album* (The Ceramic Book Company, Newport, Gwent, 1975)

A.E. Jones, S.L. Joseph, *Swansea Porcelain: Shapes and Decoration* (D. Brown, Cowbridge, UK, 1988)

E. Longford, The duchess of Richmond's ball, in *The Oxford Book of Military Anecdotes*, Anecdote No. 194, ed. by M. Hastings, (Oxford University Press, Oxford and New York, 1986), pp. 230–234

R. Massey, An unrecorded Derby figure. Transactions of the English Ceramic Circle **20**(1), 91–94 (2008)

D. Miller, *The Duchess of Richmond's Ball, 15th June, 1815* (Spellmount Publications and The History Press, Stroud, Gloucestershire, 2004)

E. Morton Nance, *The Pottery and Porcelain of Swansea and Nantgarw* (B.T. Batsford & Son, London, 1942)

A. Renton, Thomas Pardoe and William Weston Young, in *Welsh Ceramics in Context, Part I*, ed. by J. Gray, (Royal Institution of South Wales, Swansea, 2003), pp. 120–146

A. Renton, National Museum of Wales, Amgueddfa Cymru, Cardiff, personal communication, (2021)

J. Sandon, *The Ewers-Tyne Collection of Worcester Porcelain at Cheekwood* (Antique Collectors Club, Woodbridge, Suffolk, 2007)

W. Turner, *The Ceramics of Swansea and Nantgarw. A History of the Factories with Biographical Notices of the Artists and Others, Notes on the Merits of the Porcelain, the Marks Thereon Etc* (Bemrose & Sons Ltd., The Old Bailey, London, 1897)

J. Twitchett, *Derby Porcelain: An Illustrated Guide, 1748–1848* (The Antique Collectors Club, Woodbridge, Suffolk, 2002)

W. W. Young, *The Diaries of William Weston Young, 1776–1847*(1802-1843), 30 volumes, West Glamorgan Archive Service, Swansea, SA1 3SN. https://arcgiveshub.jisc.sc.uk/data/gb216-d/dxch/ddxch/i/hub

Appendices

Appendix I: China and Goss China

What Is China?

China is a generic term that has been applied to fired ceramics often quite indiscriminately by historians over many centuries and covers different forms of fired ceramics ranging across the spectrum from porcelains to earthenwares: typically, creamwares, faience, celadons, majolica, jasperware, salt-glazed pottery, pearlware, stonewares and porcelain have at times all been described as "china". Often, adjectives have been used in to distinguish between selected "china" artefacts, such as ironstone china, semi-opaque china and bone china, and this does produce something approaching a partial differentiation of the ceramic material when used in its proper context. The narrower field of porcelains, which forms the theme of this book, has also been subjected to the use of adjectives in an attempt to differentiate between the different types whilst avoiding the more intricate details of their compositional and chemical differences; these have varied significantly through the ages and are only now being rationalised and quantified (Edwards, *18th and 19th Century Porcelain Analysis: A Forensic Provenancing Assessment,* 2020; Edwards, *Porcelain Analysis and Its Role in the Forensic Attribution of Ceramic Specimens,* 2021). For example, the original Chinese porcelain manufactured in one of the Imperial Dynastic kilns, such as that at Jingdezhen, has been described historically as "true" porcelain whereas all other porcelains were then categorised as "artificial" porcelains (Brogniart, *Traite des Artes Ceramiques, ou des Poteries, Considerees dans Leur Histoire, Leur Pratique et Leur Theorie*, 2 Vols.,1854: Edwards, *Porcelain Analysis and Its Role in the Forensic Attribution of Ceramic Specimens,* 2021) because the latter contained synthetic components such as lime, glass frit, calcined bone ash, potash, soda ash and other additives as raw materials in their formulation. Although this is certainly a valid basis for the differentiation genorically between

H. G. M. Edwards, *Welsh Armorial Porcelain*,
https://doi.org/10.1007/978-3-030-97439-8

Chinese and European porcelains, for example, it does not facilitate the further distinction between the other types of porcelain, such as phosphatic, glassy, siliceous and steatitic or magnesian porcelains. Also, it could be argued that since all porcelain is synthetic and, unlike glass, has no natural equivalent then all porcelains including Chinese porcelain should be categorised as artificial. The fact that all porcelains must be translucent to some degree, whereas earthenwares are clearly not so, partly addresses the problem by providing a visual differentiation between porcelain and earthenware artefacts, without providing a distinctive separation into the type of porcelain comprising an artefact. It has been argued that all china that is deemed to be porcelain should contain china clay as a raw material component, but this also falls foul of the analytical chemical evidence compositionally, when it is realised that most early earthenwares and porcelains contained secondary ball clays but no china clay, until its presence in the manufacture of Chinese porcelains was announced to the world in the second decade of the eighteenth Century by the Jesuit priest to the Peking mission, Pere Xavier D'Entrecolles, in his letters of 1717 and 1722. Thereafter, sources of high quality fine white china clays were obtained from diverse locations and local outcrops, such as those in the Cherokee Lands (South Carolina), Ebengebirge (Saxony), Saint-Yrieux-la-Perche (France) and Gewgraze (Cornwall). In fact, the earliest English and European porcelains did not contain any china clay at all, unlike the Chinese porcelains then being imported into Europe, which compositionally comprised simply just china clay (kaolin) and *petuntse* (a micaceous, feldspathic rock found in China and related in composition to but not identical with china stone, which was found elsewhere geographically). The logical statement has hence been made that:

All porcelain is china but not all china is porcelain.

Whilst this summarises the situation perfectly it unfortunately does little to assist us in our objective differentiation between the two extremes of porcelain and earthenware and the types of ceramic that are found in between these categories.

The origin of the word "porcelain" is believed to have been derived from the Italian *porcellana,* which was used to describe the lustrous, hard and texturally smooth white shell of the cowrie shellfish, although an alternative source has been suggested from Marco Polo's observations of the fine *qingbai* porcelain that was being made in China towards the end of the thirteenth Century, in which he referred to *porcellino,* which in Italian is used to describe a young piglet whose skin it was thought to resemble (Pelliot, *Notes on Marco Polo*, 1959). In the sixteenth Century it appears that the term "porcelain" was used not just to describe fired ceramics but more broadly included natural biological materials such as mother of pearl (nacre) and seashells. The inventory of King Philip II of Spain's ceramics collection carried out after his death in 1598, which during his lifetime held some 3000 pieces of Chinese porcelain comprising the largest collection in Europe along with some European Medici porcelain and other faience, included items comprising agates and stonewares (which clearly although not porcelain at least chemically are all highly siliceous in composition). In several Dutch paintings of the early seventeenth Century, particularly from the Antwerp and Rotterdam Schools, blue and white decorated Chinese porcelain artefacts are included in groups along with seashells and faience, thereby stressing the perceived commonality of the natural biomaterials and the synthetic ceramic products. Two classic examples of these are, firstly, *The Chamber of Curiosities*, by Frans Franken the Younger, Antwerp, 1636, which depicts a *kraak* porcelain wine cup, sea shells, mother of pearl and ceramic tiles (Kunsthistorisches Museum, Vienna) and a second example is entitled *Pronkstilleven* (Sumptuous Still Life), by Willem Kalf, Rotterdam, 1678 (National Gallery, Copenhagen, Denmark), depicting a Ming porcelain bowl, a nautilus shell cup, a glass goblet and fruit.

It was indicated earlier that some historians hitherto considered porcelains and earthenwares collectively as china, which can be confusing for modern readers. For example, in the context of the theme of this book, Swansea china has been taken by some ceramics historians to include collectively the output of the Swansea China Works (which was exclusively porcelain) and also that of the neighbouring Cambrian and Glamorgan Potteries in Swansea (which included earthenwares, creamwares, salt-glazed pottery and pearlwares, but never porcelain). This situation does not arise at Nantgarw where the Nantgarw China Works was established on a new site where there had been no previous ceramics manufacturing history, although certainly later in its history the kilns at Nantgarw produced earthenware artefacts and clay pipes in abundance under the stewardship of William Henry Pardoe and Percival Pardoe. It is even recorded that William Henry Pardoe attempted to revive porcelain manufacture unsuccessfully at Nantgarw in the late 1850s, some 30–40 years after the formal closure of the Nantgarw China Works, alongside his ongoing successful earthenware manufacturing business there. Hence, this is another good reason for examining the crested armorials on "china" as discussed here, especially those that were produced on earthenwares at the Cambrian Pottery in Swansea, to ascertain if any exemplars are comparable with their porcelain analogues and perhaps then to make use of this information in cross-referencing and comparison with the porcelain research that was being undertaken for this book.

For a fuller description of the different types of porcelain and their chemical composition the reader is referred to an earlier text by the author (Edwards, *Porcelain Analysis and Its Role in the Forensic Attribution of Ceramic Specimens,* 2021) where a protocol has been developed to distinguish between the various types of porcelains on the basis of their analytical chemical composition. The essential requirement for a material to be called porcelain is that it is a ceramic composed of several raw materials that have been fired at a high temperature and is translucent to some degree, unlike brick, earthenware, stoneware or terracotta. For this current text, the Welsh porcelains produced at Swansea and Nantgarw are both classified as soft-paste porcelains; the raw materials comprising them have been twice-fired in a kiln at a high temperature, and they can be defined as glassy, phosphatic and soap-rock /magnesian porcelain (for Swansea) and phosphatic porcelain (for Nantgarw) in type. The phosphate content, which has been supplied as a raw material component through the use of calcined bone ash (hydroxyapatite; calcium phosphate, $Ca_3(PO_4)_2$), for both the Swansea (duck-egg) and Nantgarw porcelains is regarded analytically as being high, that of Nantgarw being one of the highest recorded for

British soft paste porcelain manufacture in which the calcined bone ash comprised around 42%–45% of the raw materials in composition. Bone china, by definition, also must have at least a minimum of 40% of calcined bone ash in its composition but it is fired in a technically very different one-stage process to that adopted for the soft paste porcelains described above and by-passes the preliminary initial firing of some components of the soft paste porcelain to make a frit, which is then finely ground and mixed further with other components for a second firing process in soft paste porcelain manufacture. The determining factor for the classification of a high-fired ceramic artefact as porcelain is its translucency, measured on a scale of 0–100%, which must be differentiated from a glass which is also fired at a high temperature but generally has a transmission to visible light approaching 100%, i.e it is termed transparent. So, porcelain is translucent, wherein the transmission to visible light is <100%, whereas glass is usually transparent wherein its transmission to visible light is =100%: however, certain categories of glass, such as milk-glass, do have a transparency of <100% and then they can only be differentiated from porcelains by a chemical compositional variation. It is interesting that milk-glass was first made in Venice in the sixteenth Century using opacifiers such as tin oxides in the form of suspended particulates to scatter light and was known as "opaque glass", "opalescent" or "translucent glass" and was marketed as an alternative for porcelain.

Goss Crested China (1858–1939)

The growth of seaside resorts, such as Blackpool, Scarborough, Weston-super-Mare, Clacton, Yarmouth and Southend, and family holidays taken there and at many other resorts by factory workers from the growing cities in an industrialised Britain in the mid- to late-nineteenth Century, was a social phenomenon which rapidly escalated into a pattern whereby the heavy industries, such as the mills and steelworks, would often close down for a set 2 weeks in the summer to allow the workforce and their families to participate in a mass migration to the seaside, aided by the growth of the emerging passenger rail networks and road transportation by large horse-drawn or motorised vehicles such as charabancs (French: char-a-banc, meaning a carriage with wooden benches). Holidaymakers would tend to buy small souvenir gifts and mementoes of their holiday in the form of small, white, glazed porcelain models which often carried the heraldic coat-of-arms or crest appropriate to the places they had visited on their holiday. The classic purveyor of these small heraldic souvenirs was W.H. Goss – a firm founded by William Goss (1833–1906) in 1858, first at the Cook Road site then later at the Falcon Pottery, Stoke-on-Trent, which was adjacent to the Newcastle to Stoke Canal and to the neighbouring Minton's China Works (Pine, *The Concise Encyclopaedia and Price Guide to Goss China*, 2000; Pine, *Goss and Souvenir Heraldic China*, 2005).

William Henry Goss' son, Adolphus, was instrumental in creating the idea for these small souvenir heraldic china items bearing the name of the seaside and holiday resorts for which the firm soon became synonymous with British holiday mementoes in the late Victorian era, especially from the 1870s: over 2500 different miniature china models were made by W.H. Goss, bearing over 10,000 crests and coats-of-arms, some having a particular association with notabilities of the selected resort, such as Captain James Cook and Whitby. It was estimated that 90% of Victorian homes possessed at least one or more items of Goss china in the late nineteenth Century and many had dressers or cabinets literally full of these mementoes which had been purchased personally or exchanged as gifts by friends and family members. Contrary to some historic opinion, Goss did not lead the way in the manufacture of souvenir porcelain items, as the Lowestoft porcelain manufactory had already marketed similar porcelain mementoes a hundred years before, but these were essentially not heraldic in type (Pine & Pine, *William Henry Goss: The Story of the Staffordshire Family of Potters who Invented Heraldic Porcelain,* 1987). Usually, the Lowestoft porcelain "trifles", as they were termed, were useful wares, such as jugs and beakers, which bore a simple inscription, such as a cream jug bearing the painted words "*A Trifle from Lowestoft*", often bearing a year date. These porcelain trifles have a strong collectors' base now for early English porcelains; they were often dated and sometimes bore monogrammed initials. The Lowestoft factory started up in 1757 and closed in 1799 (Godden, *An Illustrated Guide to Lowestoft Porcelain,* 1969). The title of the Pines' book (L. Pine and N. Pine, *William Henry Goss: the Story of the Staffordshire Family of Potters who Invented Heraldic Porcelain,* 1987) is perhaps rather misleading in this respect in that it implies that the W.H. Goss family "invented heraldic porcelain", which they clearly did not, and the true origin of armorial porcelains can be traced back to Duke Francesco de Medici of Tuscany in the 1580s, almost three hundred years before their Goss china analogues, with the manufacture of his armorial diplomatic porcelain gifts, and perhaps even earlier to King Manuel I of Portugal in 1517.

A typical example of a W.H. Goss china miniature is shown in Fig. A1.1, which shows a bone china model of a leather bottle which was found on the battlefield at Newbury in 1644 during the English Civil War and here bearing the arms of Aldershot: namely, an escutcheon with three pyramidal mounds of cannonballs sable on a fess gules with an oak tree near Boscobel House, symbolising the Royal Oak where the future King Charles II hid after the Battle of Worcester whilst evading the Parliamentarian forces of Oliver Cromwell that were hunting him and his father.

The other major point of interest relating to Goss china that is of particular relevance here is that the factory mark employed on the underside was the crest of a goshawk on a torse (Fig. A1.2), which was presented as a proper heraldic symbol and has been claimed to arise by several ceramics historians from a pun on the family name and is, therefore, rather more akin to a rebus than a proper heraldic crest.

Comparison of the W.H. Goss trade mark crest and torse based on a goshawk with wings extended with other similar crests found in Fairbairn's armory (Fairbairn, *Crests of the Families of Great Britain and Ireland and Their Kindred in Other Lands,* Volumes I and II, 1905) reveals that there is actually nothing identical to be

Fig. A1.1 Armorial Goss china which shows a china model of a leather bottle found on the battle-field at Newbury in 1644 and bearing the arms of Aldershot: an escutcheon with three pyramidal mounds of cannonballs sable on a fess gules with an oak tree, the Royal Oak at Boscobel House, symbolising the refuge of the future King Charles II hiding from Parliamentary Cromwellian forces after the Battle of Worcester in 1651

Fig. A1.2 Stencil mark of armorial Goss china with the rebus of a goshawk on a torse

seen there: however, a very close similarity is seen in the crest shown in Fig. A1.3 on the immediate left of the group and numbered "4", taken from Fairbairn's armory, Plate 136/4. Also, in Fig. A1.3 are shown examples of slightly different crests, which all display birds of prey in different heraldic attitudes, numbered 1, 3, 7 and 12, and all featuring in Fairbairn's armory, Plate 224/1,3,7 and 12, respectively. These other four crests illustrated all show the raptor bird of prey (which could be regarded as an eagle, or perhaps a falcon) in different poses, some with head erect and with tongue exposed (beaked), whilst others are sitting on a nest (nos. 1 and 12) or with their captured prey, depicting a rabbit or a grouse held between their talons (nos. 3 and 7). In addition, the crest shown as no. 12 in Plate 224 is engorged heraldically, meaning that it has a coronet placed around its neck. These pictorial

Fig. A1.3 Raptor birds of prey depicted on crests in Fairbairn (Fairbairn, *Crests of the Families of Great Britain and Ireland and Their Kindred in Other Lands,* Volumes I and II, 1905). From left, Plate 136/4, followed by Plate 224/1, Plate 224/3. Plate 224/7 and Plate 224/12

differences between otherwise several similar crests serve to differentiate between the appropriate arms-bearing individual or families and illustrate the need to examine and to research carefully the exact composition of the crest that is displayed on the armorial porcelain artefact to determine the precise attribution of the arms and their assignment to a particular arms bearer and the family concerned (Fairbairn, *Crests of the Families of Great Britain and Ireland and Their Kindred in Other Lands,* Volumes I and II, 1905). This is critically important in armorial research and its use will be accentuated several times in this book as part of the decipherment process of the heraldic symbols which comprise the coat-of-arms of an armigerous family member.

Appendix II: The Named Decorators and Gilders of Welsh Armorial Porcelains.

Survey of the Named and Identified Decorators and Gilders and the Difficulty Experienced in Their Attribution

Of the 18 identified known armorial services or artefacts in Welsh porcelain, numbering 7 from Nantgarw and 11 from Swansea, which have been discussed here only four armorial artefacts can be unequivocally assigned to three known decorators or gilders working at the manufactories, namely Henry Morris at Swansea (Clarke of co. Hereford/Graham-Clarke of Newcastle impaling Parkinson of Kinnersley Castle, Herefordshire, service, Figs. 5.1 and 5.2) and William Billingsley (Homfray of Penllyne Castle service, Figs. 4.6 and 4.7) and Thomas Pardoe (the second Ramsay of Straloch and Barra service, Fig. 4.3, and Jeffreys of Brecon Priory service, Figs. 4.8 and 7.5) at Nantgarw. It is appropriate here to survey the careers of each of these to input information into the holistic assessment of the armorials that have been assigned to them at each manufactory. Unlike the Cambrian Pottery exemplars for which there is often subsidiary information forthcoming from the associated inscriptions on the base of several armorial items, the armorial porcelain analogues from the sister Swansea and Nantgarw porcelain manufactories are completely devoid of such ancillary information. For example, the Richardson-Francis punchbowl is dated 1845 and provides details of the date of commissioning, the person who initiated the order, the purpose of the commissioning and the actual painter of the heraldic arms and devices. At best, for the porcelain armorial exemplars the appropriate factory marks suitably stencilled in enamels or impressed can be found and, whereas these do provide some chronological context in respect of the porcelain body and potential date of its manufacture, they do not inform the researcher about the identity of the enameller or gilder of the arms.

In addition, it has already been stated that the pattern and order books for both the Swansea China Works and the Nantgarw China Works are no longer extant and the only two sources of contemporary information about production schedules and porcelain composition of the products are the work books and diaries of Lewis Weston Dillwyn at Swansea (Dillwyn, *Notes of Recipes for Porcelain Bodies and Glazes 1815–1817,* Library of the Victoria & Albert Museum, South Kensington, London, deposited by John Campbell Esq. in 1920; reproduced in Eccles & Rackham, *Analyses of English Porcelain in the Victoria and Albert Museum Collection,* 1922) and William Weston Young at Nantgarw (William Weston Young, *The Diaries of William Weston Young (1776–1847), 1802–1843,* 30 Volumes, West Glamorgan Archives Service, Swansea, SA13SN; discussed in Edwards, *Porcelain to Silica Bricks: the Extreme Ceramics of William Weston Young, 1776–1847,* 2019). The former provides a wealth of information concerning trial experiments carried out by Dillwyn between 1815 and 1817 on the variation of his Swansea porcelain body composition but no information is given about specific commissions or service patterns; in contrast, the Young papers give much information about his travels in

South Wales in pursuit of his business as land agent and surveyor but none is provided about the compositional formulation of Nantgarw porcelain. This is in complete contrast with other contemporary factories such as the Derby China Works, where the pattern books and associated notes relating to the porcelain commissions undertaken by the factory, particularly under the proprietorship of William Duesbury, are still maintained, archived and available for consultation by ceramics researchers in the Royal Crown Derby Museum (Twitchett, *Derby Porcelain: An Illustrated Guide, 1748–1848,* 2002). As a result, the literature relating to the patterns of the services produced at Swansea and at Nantgarw is extremely sparse: the only comprehensive coverage that has been attempted recently, and then only for Swansea porcelain, is that of Jones and Joseph (*Swansea Porcelain: Shapes and Decoration,* 1988) in which 60 recorded pattern numbers have been assimilated and identified thus far from the many existing, unnumbered patterned services. The highest pattern number recorded on Swansea porcelain is 706, so if we assume reasonably that the pattern numbers were accorded to manufactory work on a sequential basis the identification of known pattern numbers is rather small and probably represents perhaps less than 10% of the manufactured patterns. The recognition of the identity of the enameller at Swansea or Nantgarw, therefore, is totally dependent upon the stylistic appreciation of the painting and its composition and upon the graphological interpretation of the cursive handwritten script marks that may accompany appropriate pieces. Sir Leslie Joseph has accomplished a comprehensive graphological analysis of these handwritten script marks on Swansea porcelain and has thereby successfully identified the painter or gilder of the pieces concerned, including Henry Morris and William Billingsley, two of the named decorators of Welsh armorial porcelains here, and he has indicated the idiosyncrasies noted in their handwritten script marks which will aid in their future identification and attribution. In addition, Edwards (Edwards, *Swansea and Nantgarw Porcelains: A Scientific Reappraisal,* 2017) has constructed an alphabetical list of handwritten script marks used in the upper and lower cases by William Billingsley dissected from a service of Swansea porcelain he painted with named scenes described in his hand on the reverse and, as a result, we now have a database of 80% + of the script handwriting that William Billingsley adopted to name his work on Swansea porcelain, including capitals and lower case letters.

This situation is found to be significantly different at other factories, where the enameller may have signed his work or where the records indicate that a certain artist was recorded as being tasked with painting a particular pattern or a service commission. At the Derby China Works, for example, comprehensive pattern books exist for their tea and dinner-dessert services which considerably assist in the identification of an artist and gilder when these have been recorded against the pattern numbers. For example, the *Prince of Wales* service, a dessert plate from which is illustrated in Fig. 1.14, was commissioned by Prince George, eldest son of King George III, from the Derby China Works in 1786 and the Derby work books confirm that William Billingsley would have decorated this service with a single rose and foliage placed centrally in a circlet of gilt dots with a dawn pink ground edging and that the gilder involved was William Longden, who would have placed his gilder's number "8" under the foot-rim, as indeed he did for the plate shown here in Fig. 1.14. It is interesting that in the Derby lists, William Billingsley is also given a special gilder's number of "7" to indicate his gilding prowess, so he must have been an accomplished gilder as well as an esteemed flower painter to have been awarded this accolade. This information provides a very useful cross-check of a factory painter's style and accomplishments matched with his assigned work tasks and these can thereby afford an identification procedure for his more anonymous work on unnamed or unnumbered services. As with all lists, however, several different versions can exist in historical records and ceramic historians have noted inconsistencies between the information and descriptors in the lists and actual exemplars that were produced by the factory: this is particularly the case for the prolific figures lists produced by the Derby China Works in the eighteenth Century, where there are clear mismatches and omissions occurring between the different citations made by several ceramic historians. John Haslem, who worked as a decorator at the Derby China Works and who obtained several items of porcelain of historical importance at its closure in 1848, including the famous *Prentice Plate* produced by William Billingsley personally as a model for rose painting to be used by apprentices in the enamelling workshop there (Edwards, *Porcelain Analysis and Its Role in the Forensic Attribution of Ceramic Specimens*, 2021), quotes a list of figures which does not match that of William Bemrose from a list used in the litigation proceedings between Duesbury and Kean (W. Bemrose, *Bow, Chelsea and Derby Porcelain: Being Further Information Relating to the Factories Obtained from Original Documents Not Hitherto Published,* 1898). John Twitchett (*Derby Porcelain: An Illustrated Guide, 1748–1848*, 2002) has made a detailed analysis of these two lists cited by John Haslem and by William Bemrose and highlighted the discrepancies between them (Haslem, *The Old Derby China Factory: The Workmen and their Productions. Containing Details of their Chief Artist Workmen, the Various Marks Used, Fac-Similes Copied from the Old Derby Pattern Books, the Original Price List of More than 400 Figures and Groups, Etc., Etc.*; 1876). A classic example of such an omission is the beautiful Derby China Works William Duesbury II period biscuit porcelain figure, made *ca.* 1795, and illustrated in Fig. 8.7, showing an elegant young woman dressed as *Flora* in a classical dance pose wearing a long dress and typical shoes of the period with buckles and low heels and carrying a garland of flowers. Height 12 inches (30 cm). Marked on the base with the Duesbury Derby mark of incised crown, crossed batons and six dots, a number "390" and the incised triangle mark of the "repairer" Joseph Hill, who constructed the figure. The figure was modelled by Jean-Jacques Spangler after a character of Angelika Kaufmann and entitled "*A Paris Opera Girl in the Role of Flora*". This figure curiously does not appear in either of Haslem's or Bemrose's lists and this is certainly not a unique occurrence as another Derby biscuit figure discovered recently by Roger Massey (2008) provides another example of a similar omission. Dr Peter Bradshaw (*English Porcelain Figures of the 18th Century*, 1981) has illustrated the figure of *Flora* shown here in Fig. 8.7 in his book and commented that it is indeed an extremely rare if not unique product of the Derby factory, having never encountered another one during his comprehensive research into English porcelain figures.

The means that we have, therefore, outside of any surviving documentary evidence in family records for the identification of the Swansea and Nantgarw armorials, their artists and enamellers, only the stylistic matching of their associated floral painting and perhaps gilding patterns that were favoured for their work to serve as an aid to the attribution of the artists. Up until now, therefore, only the following assignments can be made confidently for the artists involved in the decoration of the known Welsh armorial porcelain exemplars:

- Henry Morris: Clarke of co. Hereford/Graham-Clarke of Newcastle impaling Parkinson of Kinnersley Castle, Swansea porcelain, Figs. 5.1 and 5.2.
- Thomas Pardoe: Ramsay of Straloch and Barra, service number 2, Nantgarw porcelain, Fig. 4.3.
- Thomas Pardoe: Nicholas Jeffreys of Brecon Priory, Nantgarw porcelain, Figs. 4.8 and 7.5.

The assignment of William Billingsley to the Homfray Nantgarw service illustrated in Figs. 4.6 and 4.7 is perhaps much more conjectural evidentially but this has now become an accepted feature of this armorial service, arising from a statement made to that effect by Ernest Morton Nance, who conducted extensive research in the early part of the twentieth Century into Swansea and Nantgarw china. Forensically, it seems that there is no evidence for this assertion except that there would have been no one else other than William Billingsley available locally at the Nantgarw China works at that time to decorate such a prestigious service commission to the standard required and that must remain a powerful argument in favour of the correctness of this attribution. A former employee, Richard Millward, of the Nantgarw China Works in the Billingsley/Walker era commented to William Turner (*The Ceramics of Swansea and Nantgarw*, 1897) on the composition of the workforce there between 1817 and 1820 and it is clear that Billingsley was undoubtedly the only enameller of repute based in house locally, although several young accomplished apprentices and workers were identified, including Lavinia Billingsley and someone called "Kitty", who signed her painting on a specimen Nantgarw spill vase shown at the recent *Gartre'n Ol Exhibition* at the Nantgarw China Works Museum in 2019. We shall have occasion to discuss Lavinia Billingsley at Nantgarw later and show a putative example of her work (Fig. A2.3). A certain William Pegg (not the William "Quaker" Pegg who was also at Derby and whose high-quality botanical work especially is immediately recognised), who was an accomplished enameller and who knew William Billingsley from his Derby China Works days, apparently left Derby and joined Billingsley at Nantgarw after the start-up in 1817, but his work has never been identified at Nantgarw. This would certainly imply that there was another first-class enameller employed at the Nantgarw China Works alongside William Billingsley: the problem is that William Pegg's work at Derby has not yet been definitively identified for comparison with his local enamelling at Nantgarw. Also, the heraldic crest decoration on the *Homfray* armorial service is exclusively gilded and fine gilding was a speciality of Billingsley, as we have seen from his time at the Derby China Works: it should be pointed out also that gilding porcelain was a very expensive and meticulous specialist task which involved the application of

24-carat gold leaf which had been triturated with an organic resin or honey and was then burned off in the glost firing kiln, followed by burnishing. In some cases recorded at other china factories, noble families who commissioned large armorial services did so with an express instruction for the exclusion or minimisation of the gilding to economise on production costs, such as the armorial crested service ordered by the Duke of Devonshire from the Derby China Works in 1800 and mentioned in Twitchett (Twitchett, *Derby Porcelain: An Illustrated Guide, 1748–1848*, 2002). Hence, in summary, it is extremely likely, even without hard evidential support documentation being available now that William Billingsley did execute the painting of the gilded crest of the speared otter on the Homfray service at Nantgarw. In a parallel scenario, an invoice exists which records William Billingsley's gilding of the emblems on a set of six Masonic beakers in Nantgarw porcelain in December, 1819, perhaps the last personal commission he undertook chronologically at the Nantgarw China Works before he departed to join John Rose at Coalport early the next year; clearly, therefore, Billingsley did gild Nantgarw porcelain and this must confirm the assignment of his gilding of the crest on the Homfray of Penllyne Castle Nantgarw tea and coffee service.

What is challenging, however, is that there is no information at all about the decoration of the other Nantgarw armorials that were surveyed earlier in the text, namely the *Viscount Weymouth* service, the *Baron Phipps* service and the *Wyndham* service. We shall now consider each of these individually as follows:

- The *Viscount Weymouth* service: illustrated here in Fig. 2.1 and in John *et al.* (*The Nantgarw Porcelain Album*, 1975). This Nantgarw armorial plate is unique amongst the Welsh armorial services in having all the heraldic components present which enables the identification and assignment of the armigerous family to be made uniquely and unequivocally. It has not been discussed as to whether this is London or locally decorated, but the consensus of opinion historically seems to be that was commissioned in London through John Mortlock, the Nantgarw agent, in Oxford Street. Perhaps, a rather unusual feature with the *Viscount Weymouth* service is that the dinner plate shown here does not possess the dentil – edge gilding that has been deemed characteristic of many superb London-decorated Nantgarw services, such as the *Lady Seaton* service (Fig. 8.8), but not in the *Sir John and Lady Williams* service (shown in Fig. A2.1), both of which are London-decorated (John, *Nantgarw Porcelain,* 1948). This aspect will be amplified and discussed further in Appendix III. Unfortunately, the records of Mortlock's china agency were lost in the Blitz of London in the 1940s, these having been in archival storage when the business finally closed in the 1930s. In retrospect, the definitive statement made by Dr John about dentil gilding should now perhaps be examined further and perhaps qualified as there are several examples cited of London-decorated Nantgarw porcelain in the *Gartre'n Ol* catalogue of the finest Nantgarw porcelain which do not have this dentil- edged gilding (*Gartre'n Ol,* 2019)—see further in Appendix III for a discussion and numerical analytical assessment of this particular point which has been carried out with a survey of literature examples.

Fig. A2.1 Nantgarw porcelain dessert plate from the *Sir John and Lady Williams* service, *ca.* 1817–1820, impressed NANT-GARW C.W., decorated in a London atelier with garden flowers including the bearded tulip that is characteristic of de Junic (or de Juinnie) and the accompanying profuse gilding in the French Empire style. *Private Collection*

- The *Baron Phipps of Normanby* service: illustrated in Fig. 4.4. A simply decorated armorial service which displays Nantgarw porcelain translucency at its finest, this service has a simple gilt crest and torse, which again if locally decorated would have to be assigned to the work of William Billingsley, but again it is not evident as to whether this was a London originated commission or not. It is quite clear that the requirements of the London-based Nantgarw clientele were for heavily gilded china (as exemplified with examples cited here, such as the *Sir John and Lady Williams* service) and this particular example certainly does not meet that criterion in being tastefully presented with the only decorative ornamentation on an otherwise plain porcelain artefact being the carefully drawn gilded crest of a demi-lion rampant holding a palm frond in both paws. Only one Nantgarw service was mentioned in the diligent research of Morton Nance (Morton Nance, *The Pottery and Porcelain of Swansea and Nantgarw*, 1942) as having no decoration at all to demonstrate the superior translucency of Nantgarw porcelain and that was commissioned by John Crichton-Stuart, The Marquess of

Bute, but no trace of this service now remains. The author proposes, therefore, that this *Baron Phipps* of Normanby service could also be locally decorated (or more properly gilded), probably by William Billingsley. This specimen plate has been analysed non-destructively in a previous study and has been confirmed to possess the *Nantgarw No.1 Glaze* used by Billingsley and Walker for their china between 1817 and 1820.

- The *Wyndham* service: illustrated in Fig. 4.1. The only known example surviving that the author has found in the literature is in the John Andrews Collection, Plas Oriel Glyn-y-Weddw, Llanbedrog, Abersoch, Gwynedd. Fergus Gambon (*Porslen Abertawe y Nantgarw*, 2016) comments that armorial porcelain on Nantgarw is exceedingly rare but does not specify either a London or a local decoration for this piece. Henry Wyndham – Quin, Earl of Dunraven, lived locally to Nantgarw at Dunraven Castle, Southerndown, only some 15 miles distant from the manufactory, so it is quite reasonable to propose that this would have been a locally ordered and decorated service, in which case it would probably have been painted by either William Billingsley or perhaps later by Thomas Pardoe. Here, the application of non-destructive analytical laser molecular spectroscopy to interrogate the glaze on this unique example would be advantageous for discriminating between early or later Nantgarw porcelains as it has been shown previously that it is possible to distinguish unequivocally between the Billingsley/Walker *Nantgarw No. 1* glaze and the Pardoe/Young *Nantgarw No. 2* glaze on finished decorated perfect specimens without the need to take samples (Edwards, *Porcelain Analysis and Its Role in the Forensic Attribution of Ceramic Specimens,* 2021).This experiment would also input chronological information to the commissioning of this service as the Billingsley/Walker glaze can be firmly dated to 1817–1820, whereas the Pardoe/Young glaze was used between 1820 and 1823 by William Weston Young and Thomas Pardoe to decorate remnant stocks of unglazed Nantgarw porcelain from the William Billingsley Nantgarw production era. In terms of armorial chronological historiation, Henry Windham Wyndham-Quin acceded to the Dunraven estate in 1815 and to the Earldom of Dunraven in 1822, so either event could have occasioned the commissioning locally of a Nantgarw service in celebration either from William Billingsley or from William Weston Young and Thomas Pardoe.

A similar analysis for the Swansea armorial porcelains described earlier reveals that the preponderance of crested china over the other partial armorial artefacts often makes it difficult to arrive at unique solutions for their attribution. One crested Swansea armorial of note, in addition to the specimen of Clarke of co. Hereford/Graham-Clarke of Newcastle impaling Parkinson of Kinnersley Castle, is the *Thomas Lloyd of Bronwydd* service, which has a unique polychrome crest and motto, and this has been definitively identified unequivocally hitherto. However, from the point of view of assignment of a painter this particular armorial service suffers in that it comprises the *Mandarin* chinoiserie pattern (No. 164) which is essentially a filled-in transfer pattern: the application of coloured pigments to the initial transfer pattern such as this was usually undertaken by younger, less experienced hands in the factory decorating workshop but here, in contrast, the detailed drawing and enamelling of the crest in the verge (Figs. 5.3 and 5.4) would have necessitated the input of an accomplished decorator and enameller, who thus far remains unidentified. In comparison, the Nantgarw crested plate shown in Fig. 4.8 (with the enlargement of the simply drawn crest shown in Fig. 7.5) is easily assignable to Thomas Pardoe on account of his characteristic painting of the birds and roses which accompany the gilded crest decoration at the verge. The only negative point here is that the crest on the Nantgarw plate in contrast to the rest of the decoration seems simply drawn, but again this may have been at the request of the commissioning agent or titled landowner, in this instance, who has been attributed to Nicholas Jeffreys of Brecon Priory.

It is appreciated that the decoration of a ceramic artefact could involve more than one artist or painter: a relevant example is the *Richardson-Francis* punchbowl from the Cambrian Pottery, Swansea, discussed in Sect. 3.3.2.1. and shown in Fig. 3.8, where the crest painting is visibly superior to that of the rest of the decoration and has been inscribed thereon helpfully to Stephen Dingley, who was an esteemed painter at the Cambrian Pottery, Swansea, factory.

Concise History of Swansea and Nantgarw Armorial Decorators

Useful information for the holistic appreciation of armorial porcelain decoration is often provided by some historical and chronological timelines of the activities of the decorators. Here, we shall review the essential facts which serve to establish the appropriate chronologies of identified decorators for their activities at the Swansea and Nantgarw China Works.

Henry Morris

Born 1799, died 1880, in Swansea. He is best remembered for his *viva voce* statement given to Colonel George Grant Francis on the 14th August 1850, which was finally published in *The Cambrian* newspaper on the 30th January 1896. In this he provided his recollections of the Swansea China Works from his starting an apprenticeship there at the Cambrian Pottery under Lewis Weston Dillwyn at the age of 14 in 1813 and his transfer to the Swansea China Works upon its foundation in 1814 until its final closure sale in 1826 (Jones & Joseph, *Swansea Porcelain: Shapes and Decoration,* 1988). This statement is ostensibly factual in content but also contains personal comments and reveals much about the interactions of personalities at the Swansea China Works; it was first taken as being correct in every detail by earlier ceramic historians such as William Turner but has since been questioned over several inconsistencies in its chronology and misconceptions. A recent text by Edwards (*Porcelain Analysis and Its Role in the Forensic Attribution of Ceramic Specimens,* 2021) reproduces the Morris statement in its entirety and points out the unacceptability of some of the glaring inconsistencies of fact and chronology. William Turner, the earliest chronicler of Swansea and Nantgarw china (Turner, *The*

Ceramics of Swansea and Nantgarw, 1897) accepted all of Morris' statement, which was reinforced by his own personal interviews with Henry Morris that would have been carried out in the 1870s in the preparation of his book and he built his own history of the Swansea China Works around these accordingly. Henry Morris was trained by William Billingsley in the decorating workshop upon the arrival in Swansea of Billingsley and Walker from their failed start-up in Nantgarw late in 1814. Turner reports that Morris was impressed by Billingsley's kindness towards him and recalled with affection Billingsley's regular lectures to the apprentices at Swansea on the *modus operandi* of their art. Morris admitted to imitating closely the painting style of William Billingsley and that as a result his work was often mistaken for that of the "Master" which he regarded as the highest compliment that could have been paid to him. Most modern connoisseurs of Swansea porcelain would not concur with this statement, as Henry Morris' painting, although excellent in accomplishment, is easily recognisable for itself but it is arguably certainly not stylistically in the same class as that of William Billingsley. Edwards (*Porcelain Analysis and Its Role in the Forensic Attribution of Ceramic Specimens,* 2021) has also pointed out several other inconsistencies in Morris' statement to Colonel Grant Francis, which demonstrates the danger forensically in accepting verbal truths and reminiscences made by eye-witnesses some 30 or 40 years after the event. In one case, Morris claimed to have painted at Swansea the prestigious *Burdett-Coutts* service of Swansea porceain which was commissioned by Thomas Coutts, banker to King George III, on the occasion of his marriage to the American actress Harriet Mellon in 1818, and what is more, that he was "assisted" in this enterprise by William Billingsley! There are several incorrect statements of fact here: the *Burdett-Coutts* service is on record as being commissioned in London through John Mortlock's agency and painted there in the atelier of Robins & Randall. The date of commissioning is entirely in conflict with Morris' statement also, as Billingsley had already departed from his employment with Dillwyn at Swansea in 1816, and by 1818, the Swansea China Works had been taken over by Timothy and John Bevington. This self-aggrandisement by an already able and accomplished enameller such as Henry Morris must therefore be taken *cum grano salis* and one cannot conceive that William Billingsley's artistic perfection in his decoration at Swansea, which is so much appreciated by collectors nowadays, would ever be deemed to be subservient to that of his then apprentice, Henry Morris! The danger of accepting this seemingly innocent statement of Henry Morris as an absolute fact is that for many years it was then naturally assumed by ceramic historians that the *Burdett-Coutts* service was decorated by Henry Morris locally at Swansea, when now it is realised that it was not.

Lewis Weston Dillwyn relinquished the proprietorship of the Swansea China Works and Cambrian Pottery in 1817 to Timothy and John Bevington who were never interested in the continued manufacture of porcelain but rather wished to acquire the site and kilns for an expansion of their production of earthenwares. Between 1817 and the final auction sale in 1826, the Bevingtons therefore arranged for the decoration of the remnant stocks of Swansea porcelain from the Dillwyn era in their workshops and Henry Morris was one of the artists involved in this enterprise. Morris claimed that he "painted the very last portions of remnant Swansea porcelain". Turner (*The Ceramics of Swansea and Nantgarw*, 1897) records that Morris told him that Lord Swansea admired his work so much that he bought a Swansea porcelain plate from him bearing his signature for the then princely sum of £10 (equivalent to £1200 today). Morris then left Swansea in 1827 and worked in Staffordshire decorating an assemblage of porcelains—it being thought now that his painting decreased in quality during this period as his output increased. He returned to Swansea finally in 1841 and based himself at his home in Pleasant Street, where he erected a muffle furnace and decorated porcelain bought in from Copelands and other Staffordshire manufactories. He diversified his artistic talents thereafter and painted flowers in oils until he became blind in 1875. Elis Jenkins (*Swansea Porcelain*, 1975) firmly believed that "*Morris may have been less painstaking when churning out quantities of mediocre stuff for the local market*". The excellence of his painting on the finest Swansea duck-egg porcelain, however, can be appreciated by his execution of the *Lysaght* service, which, like the *Burdett-Coutts* service also depicts a central wicker basket of garden flowers, so perhaps his recollection of claiming to paint the *Burdett-Coutts* service can be ascribed to a faulty memory of an event that occurred some 30 or 40 years hitherto! Nevertheless, Henry Morris is an accomplished flower painter on Swansea porcelain and a fine example of his work on Dillwyn's duck-egg porcelain is shown in Fig. A2.2: a Swansea dessert plate with Nantgarw type ribbon and foliage moulding and decorated with six vignettes of garden flowers, containing Morris' signature auriculas, and a central bouquet with C-scroll gilding.

Fig. A2.2 Swansea duck-egg porcelain dessert plate decorated locally by Henry Morris, *ca.* 1817–1826, with Nantgarw type ribbons and foliage moulding and having garden flowers in six vignettes at the verge and a central bouquet with prominent C-scroll gilding. *Private Collection*

William Billingsley

Born in Derby 1758, died at Coalport, 1828. His father was a china decorator who moved from Chelsea to Derby and established a decorating business with a muffle furnace as a sideline in the family home in Bridge Street. At an early age, William Billingsley and his friend Zacariah Boreman undertook the decoration of porcelain at his home, where the upper attic rooms had been converted into a studio with good natural lighting. He was apprenticed in 1774 at the age of 16 to Edward Withers in the Derby China Works, where he introduced an entirely new naturalistic style of flower painting to that currently being adopted by Chinese, Continental European and English porcelain artists. On the 4th November 1780 he married Sarah Wrigley and they had three children, a boy who died in infancy, Sarah and Lavinia, who grew up to join their father in the decoration and gilding of porcelain (Edwards, *Swansea and Nantgarw Porcelains: A Scientific Reappraisal*, 2017; Edwards, *Nantgarw and Swansea Porcelains: An Analytical Perspective*, 2018; John, *William Billingsley*, 1968). William Billingsley soon was held in high esteem for his painting, especially for his signature pink roses and rosebuds with their slightly faded and washed-out appearance, and a characteristic of his work was the depiction of the underside of the rose and of a browning of the edges of the leaves and foliage. John Haslem, who arrived as an apprentice at the Derby China Works after Billingsley had departed for Pinxton wrote (Haslem, *The Old Derby China Factory: The Workmen and their Productions. Containing Details of their Chief Artist Workmen, the Various Marks Used, Fac-Similes Copied from the Old Derby Pattern Books, the Original Price List of More than 400 Figures and Groups, Etc., Etc.*, 1876):

> *The soft, delicate and graceful manner in which he painted flowers on china has seldom been equalled and can scarcely ever be surpassed.*

His first major commission at Derby was the decoration of the *Prince of Wales* dessert service for the future King George IV in 1786, which is still appreciated by connoisseurs for its simplicity and elegance, see Fig. 1.14 for an illustration of a dessert plate from this service. In 1790 Billingsley took over as chief decorator in the Derby China Works workshop in succession to his mentor Edward Withers—a shell dish painted with roses by Edward Withers, *ca.* 1780, is shown in Fig. A2.3, which demonstrates his stylistic influence upon the young Billingsley but also reveals the great steps forward Billingsley had made in his execution of the *Prince of Wales* service naturalistic rose painting just a few years later.

William Billingsley cherished the idea of creating the world's finest and most translucent porcelain body as a canvas for his exquisite painting and he now set about to achieve this, firstly by joining John Coke at his neighbouring Pinxton China Works in 1795 where he was plagued with high kiln losses and experiencing a not very satisfactory porcelain body. He then departed to Mansfield, where he endeavoured to establish a decorating business for porcelain bought in from elsewhere. George Hancock, who joined Billingsley in Mansfield, recalled that Billingsley was a hard taskmaster who exhibited great impatience if the flower painting was not done to his satisfaction, commenting that "*If a flower was not portrayed to his liking*

Fig. A2.3 Derby porcelain shell dish, puce mark of crown over crossed swords and cursive *D*, painted by Edward Withers, head of the decorating workshop at Derby and mentor to William Billingsley who was apprenticed to him in 1774 by William Duesbury. Comprising an attractive *oeil-de-perdrix* ground with vignettes of pink roses – compare the rather "stilted" and formalised style of rose painting which was typical of the period with the innovative naturalistic enamelling of William Billingsley coming a few years later as shown here in Figs. 1.11, 1.14 and 1.16. *Private Collection*

he would say – is there not light, shade and roundness in that?" In 1802, or possibly early 1803, Billingsley moved to Brampton-in-Torksey, where accompanied by his daughters Sarah and Lavinia, he engaged a young engineer from a local farm, Samuel Walker, to help manufacture porcelain at Hall House Farm. His wife Sarah meanwhile remained at the family home in Bridge Street, Derby. Whilst undertaking experiments on his own porcelain body composition he still acquired porcelain from other manufactories for decoration, notably French hard paste porcelain from the de la Courtille factory, Paris. The de la Courtille manufactory was founded by Jean-Baptiste Locre and Laurent Russinger in 1772 and made exclusively hard paste porcelain (Dawson, *French Porcelain: A Catalogue of the British Museum Collection*, pp. 339–340, 1994; R. de Plinval de Guillebon, *Paris Porcelain, 1770–1850,* 1972).

Several ceramics historians seem to adopt divergent theories as to why Billingsley moved to join Martin Barr at the Royal Worcester China Works in 1808: one suggestion is that he found himself in serious financial difficulties and changed his name to Beeley to escape the attentions of bailiffs and the law. Others suggest that in this name change he was attempting to hide some murky secret and for some reason decided to decamp to Worcester with his family. The diminution of his name to Beeley from Billingsley could be totally unrelated to any attempt to hide his identity, as this is quite a common occurrence with longer English surnames: for example, Featherstonehaugh is abbreviated to Fanshawe, Cholmondeley to Chumley, and Auchinleck to Affleck, with no hint of any secret that needs to be hidden by this change of name. Other ceramic historians have suggested that possibly Billingsley had engaged in a bigamous marriage in Bristol where he allegedly set up a decorating business whilst separated from his wife Sarah, who had remained in Derby, but it is now believed this is a bogus assertion which arises from an historical confusion with Thomas Pardoe who resided at the same address in Bristol given for Billingsley when he left Swansea for Bristol in 1809! It is a relevant fact that William Billingsley was mis-spelled in the Derby China Works pattern books in 1790 as Billensley, then he was correctly named as Billingsley at Pinxton in his correspondence with John Coke, and similarly at Mansfield (where he signed his work on porcelain as Billingsley), at Brampton-in-Torksey and finally at Worcester, where correspondence with Martin Barr and with Messrs Flight refer to him with the use of his true name and not Beeley or any other diminutive form.

The Billingsley family, William, Sarah and Lavinia, with Samuel Walker, it is true left Brampton-in-Torksey in rather extreme financial circumstances in early 1808 and travelled seeking work: walking first to Worcester, then on to Bristol, crossing the Bristol Channel to Swansea by boat in an exceedingly bad storm and then walking back to Worcester where their engagement in the Royal Worcester China Works was eventually successful, a total distance of more than 400 miles (650 Km !!) on foot. At Worcester, Billingsley and Walker were engaged by Martin Barr to perfect a new porcelain body using a novel experimental high temperature reverberatory kiln and it is here that they perfected their version of an experimental soft paste highly phosphatic porcelain body that was to give eventually Nantgarw porcelain its finest accolade of the most translucent and finest soft paste porcelain in the world. After their departure from Worcester in 1814, Billingsley and Walker arrived at Nantgarw, some say via Bristol again, thence to Cardiff and therefrom up the Taff Vale to Nantgarw. There has been a suggestion that Billingsley diverted to Swansea to visit the Cambrian Pottery, where he renewed his acquaintance with Thomas Pardoe, whom he had mentored earlier in the workshop at Derby, and seeking work there: the author feels this is a rather incongruous tale as surely the appearance of an experienced manufacturer and decorator of porcelain arriving in Dillwyn's ceramic sphere at that particular time would have occasioned an immediate offer being made by Dillwyn for his services to manufacture porcelain at Swansea – which had been Dillwyn's ambition for some years previously according to Henry Morris in his statement to Colonel Grant Francis (see above), in which the unsatisfactory engagement of two porcelain experts from Coalport was effected by Dillwyn in 1812. Billingsley rented the farmhouse of Tyla Gwyn in Nantgarw, adjacent to the newly constructed Glamorgan Canal which had been built in 1794, from Edward Edmunds and set about establishing the porcelain manufactory using a grant from Martin Barr for services rendered by himself and Samuel Walker at Worcester and with the additional personal financial support from William Weston Young. A biscuit porcelain kiln was constructed and a subsidiary glost kiln along with a range of potting and proving sheds and outbuildings for the storage of raw materials. They there produced specimen examples of the first Nantgarw china from the newly built kiln which they used to try and obtain support from the British government through their *Memorial* submitted in September 1814 and signed by all three: Walker, Young and now Beeley! Despite the strong support of a local landowner, Sir John Nicholl of Merthyrmawr, who was an influential member of the Committee of Trade and Plantations which considered the application, this bid was unsuccessful and resulted in the closure of the site at Nantgarw but fortuitously then resulted in the association of Billingsley and Walker with Lewis Weston Dillwyn, involving an introduction through William Weston Young, at Swansea towards the end of 1814, where their expertise helped to successfully deliver the first Swansea porcelains in 1815. Dillwyn's adoption of the beautifully translucent duck-egg porcelain body later in 1815 provided a suitable base for some of Billingsley's finest ceramic painting on Swansea porcelain but sadly the decision to replace this beautiful soft paste phosphatic china by the perceived inferior but more robust magnesian steatitic "trident body" by Dillwyn was not well received by Billingsley, who departed Swansea with Lavinia to try to start up again in Nantgarw, probably around the middle to late 1816. Walker stayed on at Swansea with Dillwyn until Dillwyn leased the Swansea China Works site to the Bevingtons later in 1817; meanwhile Samuel Walker's wife, Sarah, who was Billingsley's elder daughter, had died in Swansea in January 1817. Then, later that year Walker joined Billingsley and Lavinia at Nantgarw at the conclusion of Dillwyn's experiments and the Nantgarw Phase II operation commenced with new financial support from William Weston Young, the third partner in the enterprise. Young, a chartered land surveyor based in Nottage, Porthcawl, had useful contacts with landed gentry and landowners in South Wales which helped to establish support locally for the Nantgarw China Works. The eccentric Dr William Price of Llantrisant (1800–1893), an extrovert medical practitioner who later held surgery and consultations in a small room in the Pardoe family residence at Tyla Gwyn, Nantgarw, stated to William Turner (*The Ceramics of Swansea and Nantgarw*, 1897, p. 100) that Lavinia Billingsley had been a junior decorator employed at the Nantgarw China works:

> *To her has been ascribed rather lightly decorated examples of simple design which apparently did not fit in well with the known mannerisms of the more authenticated artists.*

The identity of these "more authenticated artists" then working at Nantgarw as asserted by Dr Price is itself rather mysterious as outside of William Billingsley and possibly William Pegg, one would be hard put to name any others! Lavinia's output at Nantgarw would have been very small anyway as she died in September 1817 in Nantgarw from seemingly a heavy cold – but see the interpretation of Edwards (*Swansea and Nantgarw Porcelains: A Scientific Reappraisal*, 2017), who has advanced a further observation on the serious implications of her malady as resulting from a prolonged exposure to mercury vapour from the described symptoms related to her employment in the gilding workshop at Worcester. This double blow affected William Billingsley very deeply and in September 1817 he wrote a very emotional and grieving letter to his wife Sarah, who was still in Derby. A simple locally decorated Nantgarw plate showing sprays of delphiniums and roses, which has been attributed tentatively to Lavinia Billingsley, is shown in Fig. A2.4.

Fig. A2.4 Nantgarw dessert plate, locally decorated and attributed by connoisseurs to Lavinia Billinsgley, comprising three groups of blue delphiniums and apricot – pink coloured roses simply painted and with a plain gilt edging. The plate exhibits several small carbon spot blemishes with pitting and has been subjected to some "sagging" which arises from the inadequate temperature control of the kiln in excess of 1420 °C, both of which would have rendered it unsuitable for sending to John Mortlock for sale in London. This would have therefore been sent out for local consumption after decoration by one of the younger hands in the workshop but was still judged to be decoratively suitable for expensive gilding and is historically important for that reason. *Private Collection*

The successful creation of Nantgarw porcelain commercially was realised beyond bounds when John Mortlock of Oxford Street, London, negotiated the sole agency for its sale in the capital and he took more than 90% of Billingsley's production in the white for decoration there in the esteemed ateliers of John Sims and Robins & Randall. Demand quickly outstripped supply for Nantgarw porcelain and Mortlock wanted even more Nantgarw porcelain to satisfy his clients, which Billingsley and Walker just could not maintain because of their horrendously high kiln firing losses approaching 90%. As Elis Jenkins (*Swansea Porcelain*, 1975) so assiduously commented, the Nantgarw operation was artistically sublimely successful but economically a disaster, with nine out of ten pieces made being unsaleable – this would not be viable for any business proposition. Eventually, Billingsley and Walker had to close down the Nantgarw China Works operation and they decamped to join John Rose in the Coalport China Works in April 1820, where Billingsley died a pauper and buried in an unmarked grave in 1828.

Thomas Pardoe

Thomas Pardoe was born July 3rd 1770, died July 22nd 1823 at Nantgarw and is buried in Eglwysilan Church. Like William Billingsley, Thomas Pardoe was born in St Alkmund's, Derby, to Thomas and Alice Pardoe, and he was apprenticed at the age of 15 in 1785 to William Duesbury in the decorating workshop at the Derby China Works. There, he would have met William Billingsley, who was a rising star and ascendant in the decorating workshop under the headship of Edward Withers. He must have come into close contact with Billingsley at this early stage since Pardoe's notebook from that period shows six drawings of floral groups composed by Edward Withers and William Billingsley given to him and signed and dated by them in 1777 and 1778 (Morton Nance, *The Pottery and Porcelain of Swansea and Nantgarw*, 1942) – a most treasured possession of Thomas Pardoe's. His apprenticeship at Derby ended in 1791, by which time William Billingsley was in charge at the decorating workshop at the Derby China Works, and Pardoe was then employed at the Royal Worcester China Works. It has been difficult to identify Pardoe's earliest work at Derby as he would only have been accorded the simplest tasks as an apprentice, but his later painting shows the influence both of Edward Withers and William Billingsley.

At some unspecified date he joined the earthenware manufactory of William and John Turner of Lane End, Staffordshire (Jewitt, *The Ceramic Art of Great Britain from the Prehistoric Times Down to the Present Day*, pp. 574–5, 1878) where he gained further experience in the decoration of china and by 1795 he had joined the Cambrian Pottery of George Haynes at Swansea. In October 1795, he married Mary Landeg, eldest daughter of Roger Landeg, an ironmonger, colliery owner and banker in the Glamorgan Old Bank (with his brother Colonel John Landeg of Brynwilan, Swansea). In July 1798, they lived in Goat Street, Swansea. On 10th May their first daughter, Fanny, was born, but in 1801 Mary Pardoe died during the childbirth of their son, Lewis, on the 27th April, who also died at that time. On December 8th, 1802, Thomas Pardoe married Sarah Ann Allum of Swansea and they then lived in a house owned by William Dillwyn close to the Cambrian Pottery. Around this time, William Weston Young arrived in Swansea to work at the Cambrian Pottery and he lodged with the Pardoes; Lewis Weston Dillwyn had also arrived in Swansea at this time and he appreciated very much the botanical draughtsmanship of Young, whom he diverted onto his pet project on the British *Confervae*, which resulted in the publication of his botanical book in 1807 and the award of an FRS in 1808 (Edwards, *Porcelain to Silica Bricks: the Extreme Ceramics of William Weston Young, 1776–1847*, 2019; Dillwyn, *British Conferva or Coloured Figures and Description of British Plants Referred by Botanists to the Genus Conferva*, 1809). At Swansea, William Henry Pardoe was born on December 8th 1803, Thomas Jr. on September 13th 1805, Emma on January 21st 1807 Henry Allum at the end of 1807 and Charles on March 30th 1812. In 1809 Thomas Pardoe then departed Swansea for Bristol to establish his own china decorating business (Owen, *Two Centuries of Ceramic Art in Bristol*, 1873) at his home "under the Bank", now St Augustine's Back. He is recorded as trading in china, a painter, a glass stainer and a gilder based at 28 Bath

Street, Bristol between 1812 and 1816. Pardoe used the kilns at the Temple Pottery in Bristol to fire the glazes and enamels on his decorated china. He then seems to disappear from the central Bristol archives after 1816, probably because he is then based outside Bristol at Hanham. On April 5th 1814, Alice Burford Rawlings Pardoe was born in Bristol and on December 12th 1819, Mary Sarah Frances Sully Pardoe was born in Bristol. An interesting statement has been made in Pountney (Pountney, *Old Bristol Potteries*, pp. 115–116, 1920) that Thomas Pardoe painted on Swansea porcelain purchased in the white when he was based at Bristol but this is now thought to be highly improbable as no exemplars have ever been found. It could be that this has arisen from Pountney's loose usage of the generic word "china", which technically covers earthenwares, creamwares and porcelain (in Welsh, the word "china" translates as "llestri", which describes "crockery"—an apt descriptor generically for all these ceramics !) and it could be that Pardoe purchased pottery from the Cambrian Pottery, Swansea, for decoration in Bristol and this was noted somewhere as Swansea "china", therefore being mis-interpreted later perhaps as Swansea porcelain? Harbour records from Swansea docks in 1819 show that some 140,000 pieces of pottery and porcelain were shipped out from Swansea that year in small coasters, some of only 30 tons burthen, many of which were destined for ports in Cornwall, such as Falmouth, and Bristol where they were then shipped overseas to the USA. It is a reasonable supposition that a small part of this cargo might have been Cambrian Pottery earthenwares or creamwares destined for decoration by Thomas Pardoe in Bristol.

Early in 1821, Thomas Pardoe responded to an invitation from William Weston Young, who had just assumed the proprietorship of the financially compromised manufactory before the auction sale in October 1820, to come to the Nantgarw China Works to decorate the remnant stock of Nantgarw porcelain residual from the Billingsley/Walker era for future sale at auction. He was assisted by his son, William Henry, then 17 years old, and he moved into Tyla Gwyn, Billingsley's former residence there. Mrs Sarah Pardoe remained in Bristol with the other children. William Weston Young became disillusioned after the final auction sale in October 1822 and he departed Nantgarw later that same year to concentrate on his land and estate surveying business, leaving Pardoe in control at Nantgarw until he died in July 1823.

William Weston Young tried to revive china production at Nantgarw but was severely hampered by his lack of knowledge of the raw materials and their formulation, especially regarding the presence of the major and critical component of calcined bone ash, which had always been kept a strict secret by William Billingsley. Young managed to create a new hard paste Nantgarw porcelain based upon feldspar but this was not well received by potential purchasers, who were anticipating a soft-paste Nantgarw porcelain: however, Young and Pardoe did need to formulate a new glaze for the biscuit porcelain stock remaining at Nantgarw – this proved to be a thinner, creamier and texturally softer glaze than the harder white glaze that was used by Billingsley and Walker and signifies to modern connoisseurs of Nantgarw porcelain the items that were decorated locally by Pardoe and possibly Young between 1821 and 1823. The Pardoe decorating business in Bristol was carried on by Thomas Pardoe Jr. thereafter.

In 1823, after his father's death, William Henry Pardoe, aged 20, stayed on in the family home at Nantgarw, called Pottery House, and then moved to Neath where he started up his own china decorating business. In 1833, William Henry Pardoe with his wife Mary and 14 children returned to Nantgarw and took out a new lease on the site, establishing The Old Nantgarw China Works, manufacturing salt glazed pottery, brown glazed earthenwares, stoneware bottles and clay pipes. The factory business did very well indeed and in 1858, William Henry Pardoe advertised that he had re-instituted the manufacture of a hard paste Nantgarw porcelain there but it transpires that nothing really seems to have come of this venture: indeed, there are no examples extant if he did succeed in its manufacture. William Henry Pardoe died in 1867 and his wife Mary carried on the management of the two businesses at Nantgarw and at Bristol for many years. Census returns attest that both Pardoe businesses at Nantgarw and at Bristol were successful and were expanding their work force. The arrival of automated cigarette rolling machines by American Tobacco in the 1880s prefaced the death of the clay pipe industry and Percival Pardoe carried on with the manufacture of earthenwares at Tyla Gwyn, Nantgarw, until he died in 1920. The Pardoe family stayed in their home in Tyla Gwyn until the 1970s, after which the Nantgarw China Works Trust established their Museum there on the site, where archaeological projects and excavations have revealed evidence of the kilns and outbuildings used from the time of William Billingsley and Samuel Walker and by Pardoe. An especially important discovery was the waste pit of china shards found by Isaac Williams (Williams, *The Nantgarw Pottery and its Products: An Examination of the Site*, 1932) some 30 feet from the potting sheds in his archaeological excavation of the site in 1931, which has yielded quantities of porcelain and ceramic shards from the earliest days of Billingsley and Walker's porcelain production at Nantgarw.

Appendix III: Local *versus* London Decoration of Swansea and Nantgarw Porcelains

Much discussion has centred upon the two types of decoration encountered on Swansea and Nantgarw porcelains, namely decoration at the factory by local artists and decoration in the London ateliers through commissions received by the London agents. For the case of armorial porcelain attribution, the identification of the source of the decoration could be an indicator of a potential assignment particularly for entitled families who have seats near the factories or alternatively who are based closer to London. Dr John in his seminal works on Nantgarw porcelain (John, *Nantgarw Porcelain*, 1948; John *et al.*, *The Nantgarw Porcelain Album*, 1975) first commented on dentil-edged gilding (from the Latin, *dens*, a tooth) being a characteristic feature of London decoration: although in architecture *dentil* refers to small rectangular projections in cornices, in porcelain *dentil* describes the small hemispherical gold pattern which is placed at the rim of a porcelain artefact resembling

Table A3.1 Distribution of the Numbers of Locally Decorated and London Decorated Porcelain Artefacts from the Nantgarw and Swansea China Works in Selected Literature

Factory	Citation & Year	London decorated	Locally decorated	Dentil edging	Armorial pieces
Nantgarw	John, Nantgarw Porcelain 1948	74	95	20	2[a]
	John *et al.*, Nantgarw Porcelain Album 1975	213	54	62	1
	Gambon, Porslen Abertawe Nantgarw 2016	64	6	20	1
	Gartre'n Ol 2019	45	25	10	1
Swansea	John, Swansea Porcelain 1958	91	699	23	5[b]
	Jones & Joseph 1988	[c]	528	7	3[d]
	Gambon, Porslen Abertawe Nantgarw 2016	31	58	16	6[e]

[a]Two distinct and individual armorials are listed: these comprise the Ramsay of Straloch crest and an unidentified escutcheon

[b]Only four distinct and individual armorials are listed: a dolphin naiant 10; Clarke of co. Hereford 11; demi-leopard 1; boar's head couped 1

[c]Jones & Joseph illustrate set patterns and shapes produced at the Swansea factory without addressing specifically London *versus* local decoration, hence a ratio expressed here would not be cogent to the discussion

[d]Two distinct armorial bearings are listed: these are Clarke of co. Hereford 1 and an unidentified escutcheon 1

[e]Only seven distinct and individual armorials are listed: a dolphin naiant 1; Clarke of co. Hereford 1; demi-leopard 1; boar's head couped 3; escallop shell 1; bird on sprouting tree trunk 1; rampant lion 2

rounded hemispherical teeth. It was decided that it may be a useful analytical exercise to try and evaluate the occurrence of dentil edge gilding on Nantgarw and Swansea porcelain artefacts that have been London decorated and locally decorated. Also, it would be useful to investigate the occurrence of dentil edging on Welsh porcelains decorated in London to see if it was a general occurrence, a feature of all ateliers, or perhaps of individual decorator(s) or enamelling workshops.

In Table A3.1 are listed some seven of the specialist books on Nantgarw and Swansea porcelains which illustrate relatively large numbers of porcelain artefacts with identified local and London decorations and some conclusions and more general points can now be made from their survey in this context:

- The distribution of different decorative origins and sourcing opportunities that were available for Nantgarw and Swansea porcelains is now immediately apparent from a comparison of the relative columns listing the numbers of London decorated and locally decorated pieces cited in each of the books listed in Table A3.1. For Nantgarw porcelain, the occurrence of London decorated items vastly exceeds that of locally decorated pieces as would have been expected from John Mortlock's negotiation of the majority of unfinished Nantgarw china pieces directly from William Billingsley (John *et al.*, *The Nantgarw Porcelain Album*, 1975; Gambon, *Porslen Abertawe y Nantgarw*, 2016) for decoration in his favoured London ateliers, such as those of Robins & Randall and John Sims. However, in some cases, such as the earlier work of Dr William John (*Nantgarw Porcelain*, 1948) and in the *Gartre'n Ol Catalogue* (2019), it appears that this London: local ratio has been significantly distorted numerically by the inclusion of local examples over a more limited sampling range and the favoured inclusion of locally decorated pieces that are now much collected by connoisseurs of Welsh porcelain. Dr John's classic book (*Nantgarw Porcelain*, 1948) cites and illustrates many locally decorated examples within it, especially favouring the local work of Thomas Pardoe, William Weston Young and William Billingsley on Nantgarw porcelain, which he wished to bring to the attention of his readers. For example, he maintains that there would have been relatively little Nantgarw porcelain remaining for local consumption and commissions for decoration directly from the Nantgarw China Works during the Billingsley/Walker era as John Mortlock effectively took more than 90% of the perfect porcelain available from the kilns for his London showrooms and for decoration in the ateliers of Robins & Randall and of John Sims – and still he wanted more because of the unprecedented demand for this wondrous china in the capital ! So keen was John Mortlock to acquire this very saleable commodity for his showrooms in London that he personally visited the Nantgarw China Works site in the autumn of 1817 and negotiated to take **all** of the porcelain being produced there by Billingsley. Hence, the number of pieces that would be available for decoration locally would necessarily have been very few in number and, of course, it could be reasoned that William Billingsley would be concentrating his efforts necessarily on the manufacturing and production side of the business where he had a ready sales market rather than on his personal decoration of the remaining china locally. A summary of Mortlock's retailing business in the capital is given in Appendix VI, where it is seen that in fact there were two Mortlocks retailing porcelain in London at the same time from 1803, under the names William Mortlock of Regent Street and John Mortlock of Oxford Street – two brothers in competition who had inherited the original John Mortlock business in 1801 then separated some 2 years later!

- Richard Millward, a former young employee of the Nantgarw China Works, told the early ceramic historian William Turner (*The Ceramics of Swansea and Nantgarw: A History of the Factories*, 1897) many years afterwards that Billingsley was so eager to protect the secrecy of his manufacturing operations at Nantgarw that he undertook much of the raw materials preparation and mixing processes for the raw materials personally: for example, he supervised the fine milling of the bones and the pulverisation of the calcined bone ash by his local miller, David Jones, and he undertook himself the mixing of the finely ground and calcined bone ash with other raw materials behind closed doors in the cellar of his home at Tyla Gwyn—the calcined bone ash being his "secret ingredient". He would surely then not have had the time to undertake much enamelling of the local pieces that remained at Nantgarw and examples of his work there are now very scarce and eagerly sought after by collectors. Financially, the ready purchase of as much fired Nantgarw porcelain that was available by John Mortlock for his London outlet would be advantageous for Billingsley, whose business viability was compromised by the distortion of most of the pieces (>/= 90 %) in the kiln charge which would have rendered these items unsaleable by Mortlock in his London agency.

- William Weston Young, although a fully-fledged partner in the Nantgarw China Works along with William Billingsley and Samuel Walker, according to his very comprehensive diaries (Young, *The Diaries of William Weston Young (1776–1847), 1802–1843,* 30 Volumes, West Glamorgan Archives Service, Swansea, SA13SN; Edwards, *Porcelain to Silica Bricks: The Extreme Ceramics of William Weston Young, 1776–1847,* 2019) spent very little time at Nantgarw during Phase II of its operation from 1817 as he travelled extensively with his ongoing land surveying business, which necessitated his visiting the estates of local landowners and aristocracy, where it is certain that he must have extolled the virtues of his new porcelain venture and attempted to gain client support for the Billingsley/Walker/Young enterprise. A careful perusal of his Diaries reveals that between 1817 and 1820 Young only spent overnight at Nantgarw on a very few occasions whilst travelling to or from Merthyr, Cardiff and Brecon with his land surveying duties, so he would not physically have had the opportunity to exercise his own painting skills on much of the Nantgarw porcelain that remained on site at this time, although it has already been noted that he was himself an accomplished ceramics decorator from his time spent at the Cambrian Pottery in the first decade of the nineteenth Century. It would have been a rather different scenario when Young took over the tenure of the Nantgarw China Works in 1820, but then his interest was primarily in the re-instigation of the manufacture of the china body itself and in the glazing and enamelling of the residual biscuit porcelain stocks from the Billingsley/Walker production era. He employed Thomas Pardoe from Bristol to decorate the remnant china stocks from the Billingsley and Walker era, which was his full-time occupation there. There is no doubt that William Weston Young did decorate Nantgarw china but this did not probably occur chronologically during the Billingsley /Walker period: John (*Nantgarw Porcelain*, 1948, Illustrations 35A and B) gives two examples of Young's painting of local scenes on Nantgarw porcelain and in the *Gartre'n Ol Catalogue* (2019, pages 113, 115 and 117) three more examples are given from a service that Young painted for the Rev. Williams. Hence, numerically, Dr John's citation and illustration of 95 pieces of locally decorated Nantgarw china compared with only 74 London decorated analogues in the same text effectively lists a greater proportion of the known exemplars of this type at that time and this just cannot be a true representation of the amount of locally decorated china that was in circulation relative to the number of London decorated specimens – remembering that John Mortlock had already taken some 90% of the total output production of perfect Nantgarw porcelain.
- Similarly, the *Gartre'n Ol Catalogue* (2019) lists some 70 carefully selected pieces of the finest Nantgarw porcelain made in celebration of the bicentenary of china manufacture at Nantgarw, in which no fewer than 25 pieces are locally decorated – and these mainly comprise some classic works from the hand of Pardoe, Young and Billingsley.

- A more realistic overall picture is probably given by Fergus Gambon (*Porslen Abertawe y Nantgarw*, 2016) for the Andrews Collection at Plas Glyn-y-Weddw, Llanbedrog, Abersoch, Gwynedd, which cites only 6 locally decorated examples of Nantgarw porcelain compared with 64 analogues decorated in London, a 10% ratio which must surely more typically reflect the production and decorating schedules which were then operating at the Nantgarw site between 1817 and 1822? It is also the case that although many collectors some years ago especially favoured the exotic, heavily gilded and beautifully enamelled London-decorated Nantgarw china, and this is still popular and eagerly sought after at auction today, a subtle shift has been noted more recently towards the appreciation of the locally decorated variants – on the grounds that this china represents the **real** Nantgarw porcelain in that it was made, glazed, decorated by local artists and sold from the Nantgarw China Works site, 7 miles north of Cardiff, between 1817 and 1822.
- The only agent acting for the acquisition of Nantgarw porcelain in London was John Mortlock, with showrooms in 47, Oxford Street, who received Nantgarw porcelain in the white, glazed and unglazed, from William Billingsley at the Nantgarw site. Mortlocks was an established retailer of porcelains founded in 1747 which later acquired the Royal Warrant from Queen Victoria as "Artists and Designers to HM the Queen" until they closed down in 1933: they kept meticulous records in an archive which was sadly completely destroyed in the Blitz bombing of London during 1941–1944, when the records of the details of the commissions that they had received from clients were lost. From 1817, John Mortlock sent the Nantgarw porcelain that he received from William Billingsley for decoration to two London ateliers; Richard Robins & Thomas Martin Randall of Barnsbury Street, Spa Fields, Islington and John Sims of Five Fields' Row, Pimlico, who each employed several esteemed artists for its decoration. These London ateliers were both significant operations, and in the case of Robins & Randall over 40 artists were employed in their atelier decorating porcelain for Mortlock and other retailers. Both ateliers operated a strict code of practice forbidding the individual artists' placement of signatures on their painted works of art and, therefore, the input of connoisseurs towards the recognition of the styles of the artists working in these ateliers from their painting is now invaluable. Thus far, the names of Moses Webster, James Plant, James Turner, Charles Muss and Thomas Randall himself have been identified on Nantgarw porcelains painted in these two ateliers. In addition, to this established list we can now add the name "de Junic" or "de Juinnie", who was for a long time a mysterious and rather ephemeral artist on locally decorated Swansea porcelain, and his existence was called into doubt by some historians. Following on from the initial investigative research of Jimmy Jones and Sir Leslie Joseph carried out at the Royal Sevres Manufactory (Jones & Joseph, *Swansea Porcelain: Shapes and Decoration*,1988), it is now confirmed that de Junic left Sevres before 1810 and was employed at Swansea until about 1815/16 when he then disappeared from the scene. Later research by Edwards (*Porcelain Analysis and Its Role in the Forensic Attribution of Ceramic Specimens,* 2021) places de Junic in London, painting Nantgarw por-

celain in the atelier of either John Sims or Robins & Randall and incorporating his signature bearded tulip in his floral groups, as exemplified in the *Sir John and Lady Williams* Nantgarw dessert service illustrated in Fig. A2.1, which is also accompanied by de Junic's effusive and baroque French Empire style of gilding which was then so much in vogue with the London Georgian clientele.

- The situation for Swansea porcelain is far more nebulous in this respect as they had five known established retail outlets in London, the largest being those of Bradley and Co., 47, Pall Mall and Apsley Pellatt of St Paul's Churchyard. Whereas the restriction on the appendage of the individual artist's names to their painting was still rigorously forbidden in these ateliers, the inclusion of a stamp or script mark indicating the retailer's name was sometimes affixed to the highest quality porcelain, the decoration of which they were extremely proud, such as that of "Bradley & Co., 47, Pall Mall, London" on the beautifully painted Swansea plate shown in Fig. A3.1 This plate, which is now in the John Andrews Collection at Plas Glyn-y Weddw, Llanbedrog Abersoch, Gwynedd, was painted by an unknown artist in a London atelier for Bradley & Co. depicting *The Red-Coot Footed Tringa* bird taken from the drawings of George Edwards (G. Edwards, *A Natural History of Uncommon Birds: And of Some Other Rare and Undescribed Animals, Quadrupeds, Fishes, Reptiles, Insects, &c., Exhibited in Two Hundred and Ten Copper-Plates, from Designs Copied Immediately from Nature and curiously Coloured After Life,* 4 Volumes, 1743–1751; G. Edwards, *Gleanings of Natural History, Exhibiting Figures of Quadrupeds, Birds, Insects, Plants &c.,* Volumes I and II, 1758–1760). The Andrews Collection at Plas Glyn-y-Weddw has three more dessert plates from this service and a square dish *en suite*, which have all been decorated to the highest standard by the same artist, but only the first-mentioned bears the retailer's mark; these other plates and the square dish from this same service are shown in Figs. A3.2, A3.3, A3.4 and A3.5, from which their superb quality can be appreciated.
- There are instances noted in historical records of outside decorators purchasing Welsh porcelains in the white and enamelling them for sale afterwards: the Swansea artists Henry Morris, William Pollard and George Beddow have all been identified on Nantgarw marked porcelain, which they presumably purchased at the three auction sales on 8th November 1820, 9th–11th May 1821 and the 28th October 1822. Also, some rather curious examples surface occasionally, such as that illustrated in John et al. (*The Nantgarw Porcelain Album*, 1975, Illustration 24) of a London-decorated Nantgarw plate with the underglaze cobalt blue border that was typical of Thomas Pardoe's decoration but bearing a central atypical floral decoration and the script signature of "*J. Powell, Wimpole Street, London*", who decorated the plate which was seemingly only partially decorated at Nantgarw during the Young/Pardoe period, finished in London and then sold on through Powell's atelier in the capital. John Powell, a small independent porcelain decorator, had premises in 91, Wimpole Street, between May 1817 and 1820 and it is interesting to speculate on his sourcing of this Nantgarw plate (Fairclough 1997). The dates of Powell's tenancy at Wimpole Street coincide with the first Nantgarw auction sale and it seems at first that he purchased some

Fig. A3.1 Swansea dessert plate, with Nantgarw type moulding, London-decorated with dentil edge gilding, *ca.* 1817–1822, superbly painted and retailed by Bradley and Co, No. 47 Pall Mall, London, as inscribed in red enamel script on the base after ornithological specimens taken from George Edwards' *A Natural History of Uncommon Birds: And of Some Other Rare and Undescribed Animals, Quadrupeds, Fishes, Reptiles, Insects, &c., Exhibited in Two Hundred and Ten Copper-Plates, from Designs Copied Immediately from Nature and curiously Coloured After Life,*, 4 Volumes, published between 1743 and 1751: this plate was also inscribed *"The Red Coot-Footed Tringa"*. One of four dessert plates and a square dessert dish from the same service now in the Andrews Collection, Plas Glyn-y-Weddw, Llanbedrog, Gwynedd. *Reproduced with permission from Gwyn Jones, Director of the John Andrews Charitable Trust Collection, Oriel Plas Glyn-y-Weddw, Llanbedrog, Abersoch, Gwynedd*

china there – yet, Thomas Pardoe did not start work at Nantgarw until January 1821, which was outside the residency of Powell at Wimpole Street: Powell would not have given his address there after 1820 so the logical conclusion must be that the blue border was not painted by Pardoe but earlier at Nantgarw by another local decorator?

- It was thought useful in this research to investigate the occurrence of *dentil* gilding on London-decorated Nantgarw and Swansea porcelains to see how prevalent this was and, additionally, if perhaps this could be attributed to a particular atelier or artists operating in the period between 1817 and 1822. Dr John (*Nantgarw Porcelain*, 1948) expressed the view that dentil edge gilding was undoubtedly a characteristic of London decorated Nantgarw porcelain and we need to examine the correctness of this statement and the meaning of characteristic in this context; is it to be taken as a fact that all London decorated Nantgarw porcelain has dentil edged gilding, or perhaps that those artefacts that have dentil-edge gilding are all London decorated Nantgarw porcelain –

Fig. A3.2 Another plate from the same service shown in Fig. A3.1. This plate shows two birds: inscribed *"The Little Brown Humming Bird"* and *"The Long-tail'd Red Humming Bird"*. *Reproduced with permission from Gwyn Jones, Director of the John Andrews Charitable Trust Collection, Oriel Plas Glyn-y-Weddw, Llanbedrog, Abersoch, Gwynedd*

Fig. A3.3 Another plate from the same service shown in Fig. A3.1. This plate has the inscription *"The Brown and Spotted Indian Cuckow"*. *Reproduced with permission from Gwyn Jones, Director of the John Andrews Charitable Trust Collection, Oriel Plas Glyn-y-Weddw, Llanbedrog, Abersoch, Gwynedd*

Fig. A3.4 Another plate from the same service shown in Fig. A3.1. This plate has the inscription *"The Red Bird from Surinam". Reproduced with permission from Gwyn Jones, Director of the John Andrews Charitable Trust Collection, Oriel Plas Glyn-y-Weddw, Llanbedrog, Abersoch, Gwynedd*

Fig. A3.5 A square dessert dish from the same service as the plate shown in Appendix III Figure inscribed *"The King of the Greater Birds of Paradise". Reproduced with permission from Gwyn Jones, Director of the John Andrews Charitable Trust Collection, Oriel Plas Glyn-y-Weddw, Llanbedrog, Abersoch, Gwynedd*

these are indeed two very different conclusions? Table A3.1 includes a survey of seven books in the literature which contain significant pictorial representation of London-decorated and locally decorated porcelain pieces from the Nantgarw and Swansea factories, comprising four books that describe Nantgarw porcelain and three that describe Swansea porcelain. In this literature selection, for this particular purpose the book of Jones & Joseph (Jones & Joseph, *Swansea Porcelain: Shapes and Decoration*, 1988) is included in the survey as it describes many artefacts of Swansea porcelain but the data cannot be included in the numerical analysis because the purpose of their book is to consider primarily the shapes and set patterns of Swansea porcelain produced at the factory and does not specifically address the incidence of local *versus* London decoration on these pieces. Of the remaining six texts cited in Table A3.1 the ratio of specifically dentil-edged decorated porcelains to the total number of London-decorated items is in the range of 22–29% for Nantgarw and 22–26% for Swansea porcelains, giving an overall mean ratio of 25 +/−3%. This means that only approximately one in every four London-decorated Welsh porcelain specimens has dentil edge gilding, which is supposedly a characteristic of the London atelier decorating studios; this statement is therefore potentially misleading as most of the London decoration from the literature examples, like its local analogues, is devoid of dentil edge gilding! It is still true, however, that dentil-edge gilding does not appear on locally decorated examples. This "characteristic" of London decoration, therefore, that has been adopted into the annals of Nantgarw and Swansea porcelains is a classic example of a unidirectional statement forensically which should now be reinstated more correctly and strictly as: *all dentil edge gilding on Nantgarw and Swansea porcelains is attributable to a London decoration but not all London decoration bears dentil edge gilding.* This can also be supplemented further by the corollary that the local decoration at Swansea and Nantgarw does not bear dentil-edge gilding. As an example of a superbly decorated Swansea plate from a presumed London atelier, Fig. A3.6 shows a birds' nest containing eggs surrounded by strands of a rambling pink rose: the absence of the dentil edge gilding and its replacement with an attractive geometric gilding pattern is to be noted here.

- The next stage of our investigation is a study of the London ateliers and their associated identified artists and enamellers to ascertain if this dentil edge gilding that was adopted there is a particular attribute of an individual artist or their workshop. A perusal of the exemplars cited in the literature is again revealing in that for all the exemplars that have been unequivocally identified as being painted by a known artist we have the clear situation that there is not one particular atelier that is associated with this type of applied dentil gilding: furthermore, individual artists' work when examined carefully shows that in some cases they used dentil edge gilding and in other cases not – this applies for all the artists mentioned above in both the ateliers of Robins and Randall and John Sims for Nantgarw porcelain, and where appropriate, in other places for the Swansea porcelains too. The general conclusion that one can draw from this analysis, therefore, is that the dentil edge gilding, as perhaps applied also to the other specific

Fig. A3.6 Superbly painted Swansea plate from a London atelier, showing a central bird's nest with eggs surrounded by a rambling pink rose: note the absence of dentil edge gilding and its substitution by an attractive geometric gilded border. *Reproduced with permission from Gwyn Jones, Director of the John Andrews Charitable Trust Collection, Oriel Plas Glyn-y-Weddw, Llanbedrog, Abersoch, Gwynedd*

gilding patterns chosen for a particular commission, would possibly be selected by the client concerned who may even have left that selection to the retailer on commission or to the particular atelier to better complement the type of artistic enamelling that was being used. For example, the Nantgarw coffee cup and saucer illustrated in Fig. A3.7 has the characteristic heart shaped handle and flat unglazed base lacking a foot-rim – it was decorated in the atelier of Robins & Randall in Islington, London, by Moses Webster, probably around 1819–1820, with his beautiful recognisable rose painting and including a dentil-edge gilding. Other examples of Webster's painting on Nantgarw china, however, from the same atelier do not have this dentil edge gilding. Likewise, James Plant, working in John Sims' atelier in Pimlico, similarly has dentil edge gilding on his work illustrated in John *et al.* (*The Nantgarw Porcelain Album,* 1975): in Illustration No.1 he has dentil-edged gilding, whereas later in Illustrations Nos. 13 and 25 he does not employ dentil edge gilding for equally superbly decorated plates. For his work on *The Three Graces*, which is the most expensive single Nantgarw piece of porcelain to sell at auction and which was shown at the *Gartre'n Ol* Exhibition of the highest quality Nantgarw porcelain at Tyla Gwyn in 2019 and is now shown here in Fig. A3.8, James Plant again employed the dentil edge gilding pattern in John Sims' workshop.

- In a particularly interesting and relevant example of the occurrence of dentil-edge gilding being undertaken by a specific London decorating atelier we can consider one Swansea duck-egg porcelain plate and one Nantgarw porcelain plate which have both been decorated in the workshop of Bradley & Co., Pall

Fig. A3.7 Nantgarw porcelain coffee cup and saucer, *ca.* 1819–1820, with a heart-shaped handle and characteristic plain unglazed base without a footrim. From a well-known service decorated by Moses Webster in Robins & Randall's atelier in Islington, London, with sprays of roses simulating the famous Derby "three-rose" pattern. Note the dentil gold edging that is indicative of a London-decorated Nantgarw piece. *Private Collection*

Mall, London. Firstly, there is an example of the *Biddulph* service, made for Lord Biddulph of Ledbury Park, Malvern, Worcestershire, which was decorated with English landscapes: this particular specimen is labelled "*Stanmer Park, Sussex. The Seat of the Earl of Chichester*" in red script along with Bradley's script mark (Fig. A3.9). A known decorator for this service at Bradley's was Philip Ballard (Edmondson, *Welsh Ceramics in Context*, I, p. 203, 2003) and another example of his work is found with a landscape scene, unnamed, but on a Nantgarw plate impressed Nantgarw C.W., which at first sight seems to be *en suite* with its Swansea analogue and again is marked with the Bradley and Co. stencil mark (Fig. A3.10). The Nantgarw plate has the same gilding pattern as its Swansea analogue on the cavetto but has dentil edge gilding at the edge; however, the Swansea plate has an additional repeat gilding sequence at the edge accompanied by a plain gilt-edged line in place of the dentil edging. This gives us an interesting scenario: two plates which are apparently very similar in design and sourced from the same enameller working at the same decorating atelier, and with some distinct gilding border similarities, one being Swansea and the other being Nantgarw porcelain, differ in that the Nantgarw plate has dentil-edged gilding and the Swansea version does not! A second point of interest that emerges from this study is that the Nantgarw plate has been decorated at Bradley's atelier, yet we know already that John Mortlock had acquired the sole agency for Nantgarw porcelain in London and he exclusively used the ateliers of Robins & Randall or John Sims for his decorating commissions. There are three possible explanations for this apparent breach of normal business practice: firstly, Mortlock may have had a time delay for completion from Robins & Randall's and John Sims' ate-

Fig. A3.8 Nantgarw porcelain, *ca.* 1817–1820, *The Three Graces* dinner plate, described as "one of the finest pieces of Nantgarw porcelain in existence" (Gambon, *Porslen Abertawe y Nantgarw*, 2016) and as a result commanded the highest price paid for a single piece of Nantgarw porcelain when it was auctioned some years ago. The figures are believed to represent the three daughters of Thomas Coutts, banker to King George III, and his first wife, Susannah Starkie, namely Susannah, Frances and Sophia, although Fergus Gambon maintains that this is somewhat conjectural (Gambon, *Porslen Abertawe y Nantgarw*, 2016). The third daughter of Sophia, Angela, became Baroness Burdett-Coutts, the Victorian philanthropist. The artist was James Plant and the work was accomplished in John Sims' atelier in Pimlico, London. This plate now resides in the Andrews Collection, Plas Glyn-y-Weddw, Llanbedrog, Gwynedd. *Reproduced with permission from Gwyn Jones, Director of the John Andrews Charitable Trust Collection, Oriel Plas Glyn-y-Weddw, Llanbedrog, Abersoch, Gwynedd*

liers because they had too many enamelling commissions already on their books, secondly, Mortlock's may have sold Nantgarw china in the white to Bradley's – which is a highly unlikely scenario as this would be seen to be a divestment of his acquisition of the sole agency for Nantgarw china and its decoration in the capital – and finally, Bradley's may have acquired Nantgarw porcelain in the white directly from William Weston Young and Thomas Pardoe at Nantgarw China Works in the post Billingsley/Walker era,or perhaps even at one of the sales of Nantgarw china "in the white " held in 1820 and 1821?

- Occasionally a specimen example of finely decorated Nantgarw or Swansea porcelain surfaces which apparently seems to be standard but in consideration of the novel research and conclusions cited above now requires a re-examination and attribution. Such a case is provided by the splendidly painted Nantgarw cruciform dish illustrated in John *et al.* (*The Nantgarw Porcelain Album*, 1975, Illustration no. 71) which is there described as being painted by William

Fig. A3.9 Swansea duck-egg porcelain plate from the *Biddulph* service, decorated by Philip Ballard at the atelier of Bradley & Co., 47, Pall Mall, London and bearing the legend "*Stanmer Park, Sussex. Seat of the Earl of Chichester*" on the reverse in red stencil script. Lord Biddulph commissioned this service of over 100 pieces bearing named landscapes for his seat at Ledbury Park, Malvern, and it was decorated by local artist Philip Ballard at Bradley's London atelier. *Reproduced by courtesy of Bonhams Auctioneers and Valuers, London*

Billingsley with his characteristic style of floral decoration, comprising four groups of garden flowers, one of which is a full-blown rose accompanied by rosebuds, which has been clearly assigned to a locally decorated exemplar of Billingsley's exceptionally rare work at the Nantgarw China Works. However, the presence of the *dentil edge gilding* on this attractive dish must now ring some alarm bells, because as we have surveyed above, dentil-edged gilding is not normally found on locally painted examples! This exemplar must then, according to our hypothesis derived in the above discussion, be concluded to be a London-decorated piece and it cannot therefore have been painted by William Billingsley at Nantgarw; we should, therefore, assign another London-based decorator for its artistic attribution. The obvious candidate for this painting is Moses Webster at the Robbins and Randall workshop in Islington, whose depiction of roses is indeed very similar to that of Billingsley, but his working base was in London at this time and as seen above, his painting can be accompanied by dentil edged gilding, or not, as the case may be. A comparison of Billingsley's rose painting at Nantgarw is provided by the example cited by John *et al.* just a few pages later (*The Nantgarw Porcelain Album*, 1975, Illustration No.75) which does accord with all we now know about Billingsley's rose painting and has additionally a plain gilded border as we might have expected from a locally decorated Nantgarw

Fig. A3.10 Nantgarw porcelain plate, impressed Nantgarw C.W., which is apparently *en suite* with the Swansea plate shown in Fig. A3.9: although the landscape is presented centrally and attributed to the same artist, *viz.* Philip Ballard, and the gilt border is identical in the cavetto of both plates, the matching gilt edge border is missing from the Nantgarw plate and has been replaced with dentil-edge gilding. *Reproduced by courtesy of Bonhams Auctioneers and Valuers, London*

piece. To be precise, it should be recognised that it is within the bounds of possibility that a client may have specified dentil-edged gilding to be applied to a locally decorated piece, but this would seem to be a very rare if not exceptional occurrence indeed from the analytical data described above from the Swansea and Nantgarw factories.

- A beautiful Swansea duck-egg porcelain cabinet cup and saucer which is now in the ceramics collection of the V&A Museum, South Kensington, London, provides another example of Lindon decorated porcelain which bears the name of the retailer: in this case, John Powell, 91 Wimpole Street, London, in red enamelled script on its base (Fig. A3.11). Despite this, it is catalogued as being painted by William Billingsley, which clearly implies a definitive local decoration but with a London retailer's mark! The decoration comprises single roses minus their foliage in vignettes on a caerulean blue ground but an interesting feature is that although the gilding at the edge of the cup and the saucer is simple line-gilding, that surrounding the roses in each of the vignettes is dentil-edged !! This is most curious and probably negates its attribution to Billingsley and confers upon this item a hybrid character in having both dentil and plain edge gilding on the same piece. The author also finds it curious that each rose is depicted very unnaturally without any stems or foliage, unlike other representations of William Billingsley which are shown in this book, for example, the single roses on the *Prince of Wales* service (Fig. 1.16). It could possibly be argued that John Powell commissioned this cabinet cup and saucer to be painted by Billingsley at Swansea and sent to London for his client, but this also seems unlikely, and we must conclude that it was decorated at Powell's London atelier in imitation of Billingsley's style.

Two particularly fine examples of London-decorated Swansea duck-egg porcelain, *ca.* 1815–1817, are the *Gosford Castle* and *Marquess of Exeter* services which must have been commissioned at a similar time chronologically and painted exquisitely in the same atelier with botanical and floral arrangements and possessing identical gilded borders comprising an array of Adam-style urns linked with curlicues. An example of a superb dinner plate from the Earl of Gosford's *Gosford Castle* service is shown in Fig. A3.12. Significantly, neither service has dentil-edge gilding. The only distinction between the two services is that the *Gosford Castle* service is completely unmarked and the *Marquess of Exeter* service is fully marked underneath with SWANSEA in red stencil enamel. Gosford Castle, Markethill, Armagh, Ireland, was the seat of the Earls of Gosford and Archibald Acheson, the 2nd Earl of Gosford (1776–1849) was for some

Fig. A3.11 Swansea porcelain chocolate cup and saucer, duck-egg porcelain, *ca.* 1815–1817, attributed to William Billingsley at Swansea but bearing the script mark *Powell, 91 Wimpole Street,* at its base. French Empire style with cup mounted on three gilded feet. In the Victoria and Albert Museum Collection, accession number C.48 & A -1950. The usual reference to cups and saucers of this type are "cabinet cups and saucers" but this example has been catalogued as a chocolate cup and saucer. *Reproduced with courtesy of the V&A Museum, South Kensington, London*

time the Lieutenant – Governor of Canada and Governor -General of British North America, attaining the Earldom by inheritance in 1807. His crest is a cock gules on a trumpet or and his motto is *"Vigilantibus"*, meaning *"To be watchful"*. Brownlow Cecil, Lord Burghley and the 2nd Marquess of Exeter (1795–1867), which is the premier Marquessate of the British peerage, was a noted porcelain collector and some fine examples of porcelain now reside in the Burghley House Collection. His seat was Burghley House, Stamford, Lincolnshire. His arms are a barry of ten argent and azure containing six escutcheons sable in a 3-2-1 array each charged with a lion rampant argent. Crest: on a chapeau gules turned up ermine a garb (wheatsheaf, a symbol of plenty) or supported by two lions, the dexter argent and the sinister azure. Motto; "*Cor unum via una*", meaning *"One heart, one way"*. Despite the magnificence of these two aristocratic services, they are not armorial in that they do not bear any heraldic devices.

Gilding on Armorials

An important message conveyed in Table A3.1 for the specialised armorial theme of this book is the last column, which cites the relatively small number of the porcelain artefacts appearing in the literature which have armorial bearings compared with the overall survey of other pieces: out of nearly 2000 pieces of Swansea and Nantgarw

Fig. A3.12 Dinner plate from the *Gosford Castle* service, Swansea duck-egg porcelain, *ca.* 1815–1817, commissioned by the 2nd Earl of Gosford, Archibald Acheson; London decorated in an unknown atelier by an unknown painter. *Reproduced by courtesy of Bonhams Auctioneers and Valuers, London*

porcelain reported in the literature that have been surveyed for the data in Table A3.1 only 19 have an armorial bearing and several of these are duplicated (only 12 of the armorials are distinct in this data set), thus the armorials represent only 1% of the total number of porcelain artefacts surveyed. The extreme rarity of these armorial porcelains for the Swansea and Nantgarw manufactories is thus confirmed by these data.

But what does this exercise in gilding analysis tell us about armorial porcelain attribution: can it be informative in our investigation of its source of origin? Firstly, not one of the 35 or so exemplars known of Nantgarw and Swansea armorial porcelain artefacts bears dentil gilding: this, hitherto, would perhaps have naturally led one to assume that all have been locally decorated either at Swansea or at Nantgarw, because of the potential ambiguity generated by the term "characteristic" that was applied to the presence of dentil edge gilding and London decoration. However, as the result of the above analysis has demonstrated, it can now be stated that all of these armorials could well have been decorated either at London or locally decorated. In some cases, such as the Swansea armorial service of Clarke of co. Hereford/Graham-Clarke of Newcastle impaling Parkinson of Kinnersley Castle (Fig. 5.1), the painter has been identified as Henry Morris, who was strictly a local Swansea enameller and in addition, the trident porcelain body, whose manufacture dated from after 1817, was not popular outside of the Swansea China Works locale. This can then be unequivocally assigned to a local Swansea decoration and commission. A similar situation applies to the Thomas Pardoe painting accompanying the escutcheon of Jeffreys of Brecon in Fig. 4.8, which cannot be other than a local decoration. Similarly, the Thomas Lloyd of Bronwydd Swansea service (Figs. 5.3 and 5.4) comprises a filled-in transfer *Mandarin* pattern with a beautifully enamelled crest at the verge, surely a combination locally of relatively unskilled and skilled painters, which marks this also as a potentially locally decorated service. The situation regarding several other armorial pieces is not as clear, however, and several points can now be made in this context:

- The Viscount Weymouth armorial service in Nantgarw porcelain (Fig. 2.1) is finely decorated in beautiful detail but its only occurrence in Dr John's book (John *et al., The Nantgarw Porcelain Album,* 1975) avoids mentioning its attribution to either a local or to a London atelier decoration.
- The Lockwood impaling Manners-Sutton Swansea armorial cannot be definitively adjudged to be either locally or London-decorated: although the armorial escutcheon is superbly enamelled there is no clue about this sourcing provided from a consideration of the gilding, which is in the form of a plain edge-gilding. Thomas Lockwood's seat was in Dawes Hall, Essex, so it would be more reasonable to assume that he negotiated this commission with one of the London Swansea china agents but as he also owned the site on which the Glamorgan Pottery was established in Swansea, he still would have maintained business links with the town and could have placed an order for his armorial Swansea porcelain service either personally or perhaps through his local agent there.
- The crested china is even more difficult to source and place definitively between a local workshop and a London-decorated workshop: evidence from the presence of a dentil edge gilding does not help here as in all cases the gilding is either plain

edged or seen to be part of a more complex geometric pattern. Hence, the crested porcelain could have been commissioned locally or in London. Here, we have theorised in the text that local families would reasonably have sourced their china directly from either the Nantgarw China Works or Swansea China Works sites, bearing in mind the difficulties experienced in transportation and communication that would have been experienced in the early nineteenth Century. Hence, the Wyndham-Quin, the Homfray and Jeffreys of Brecon armorial services would most likely have originated at Nantgarw whereas the Phipps armorial service would perhaps most likely to have been commissioned in London. The two Ramsay of Straloch armorial services pose an interesting hybrid in this context as the earlier service we could hypothesise would most likely have been commissioned in London as their family seat was in Scotland (?): however, there is no doubt that the second Ramsay of Straloch service was locally decorated by Thomas Pardoe and therefore would have had to have been commissioned locally from the Nantgarw China Works!

- For the Swansea armorial services, it is clear that the Clarke of co. Hereford/ Graham-Clarke of Newcastle impaling Parkinson of Kinnersley Castle must have been locally commissioned as it has the enamelling of Henry Morris associated with the escutcheon; the case of the other escutcheon on Swansea porcelain, namely that of Lockwood impaling Manners-Sutton could be either local or London-decorated and all the crested Swansea armorial china is likewise source indeterminate in this respect, However, it is to be considered that perhaps the Orme of Northampton service (which displays the dolphin *naiant* crest) could possibly be better considered as a London decorated service from the viewpoint of ease of travel from the country seat to London? Other Swansea crests, such as the Paschal lamb assigned to Wyndham Lewis of Llantrisant, Thomas Lloyd of Bronwydd and the Turbervills of Ewenny Priory were sufficiently local to realistically suggest that they were commissioned directly from the Swansea China Works. The case of Sir Watcyn Williams-Wynn of Wynnstay in Denbighshire is problematic in this respect as despite the distance involved from Swansea, Henry Morris has confirmed that he was seen at Swansea personally purchasing his china, having made an arduous journey of 200 miles from North Wales by coach, which would have taken him several days to accomplish! Other local contacts we have assigned here to the Swansea crests were all within reasonable geographic travelling distances of the Swansea China Works, such as Merthyr, the Vale of Neath and the Vale of Glamorgan.

Appendix IV: The Ironmasters of Merthyr Tydfil and The Birth of Steam Locomotion

Although it may at first seem entirely unrelated and peripheral to porcelain manufacturing and particularly the activities of the Nantgarw and Swansea China Works, the two themes of porcelain manufacture and heavy industry are linked through four supporters of especially the Nantgarw China Works whom we have identified in Chapter 8 as potentially strong contenders for the commissioning of armorial Welsh porcelains. It is apposite, therefore, to consider the theme here of the local iron-works and how they effected the transportation of materials in the early nineteenth Century. The four nominated persons of interest were the industrialists Richard Crawshay of Cyfarthfa Castle, Sir Josiah Guest of Dowlais, Samuel Homfray of Penydarren and Richard Hill of the Plymouth Ironworks. All of these heavy industrial sites were located in the urban area of Merthyr Tydfil, a place that William Weston Young had identified in his Diaries for frequent visits and which he must have considered as a potential source of support locally for the china being manufactured at Nantgarw (Young, *The Diaries of William Weston Young, 1776–1847(1802–1843)*, 30 volumes, West Glamorgan Archive Service, Swansea, SA1 3SN).

Merthyr Tydfil to Nantgarw is a distance of only 14 miles and thence to Cardiff docks a further 7 miles. In the mid- to late-eighteenth Century the establishment of iron works and forges at Merthyr Tydfil was generated because of the local abundance of essential raw materials such as high- quality Welsh anthracite steam coal, which burned at a high temperature and with low ash residues and was in plentiful supply in the Rhondda Valleys nearby, along with rich beds of iron ore and limestone outcrops. The availability of a deep-water port at Cardiff required the construction of a feeder canal, the Glamorgan Canal, which was completed in 1794, linking Merthyr Tydfil and Cardiff docks. This was essential for the importation of iron ore as the industries expanded, which with the local supplies of abundant limestone and coal, was necessary for feeding the blast furnaces of the ironworks. The War of American Independence and the later Napoleonic Wars between 1776 and 1815 required large supplies of iron cannonballs, ships' anchor chains and rails for horse-drawn wagons hauling supplies in heavy industry and in the collieries. Heraldically, the arms of the Crawshays of Cyfarthfa Castle appropriately feature a pyramid of cannonballs in their escutcheon. Expansion of the ironworks to satisfy the needs of mechanisation in the Industrial Revolution kept apace and by the early 1800s Merthyr was the largest producer of pig iron, cast iron, wrought iron and later steel in Britain, if not in the world, and the four ironmasters named above were then among the richest men in Britain.

The manufacture of iron in Merthyr Tydfil started in 1759 with the founding of the Dowlais Iron Co. and the engagement of John Guest from Broseley, Staffordshire, as its manager in 1760. Anthony Bacon (1717–1786) of London in 1765 leased the mineral rights on 4000 acres of land around Merthyr Tydfil for £100 per annum over 99 years, stretching from Cyfarthfa down the Taff Valley, which held the best Welsh steam coal, iron ore and limestone and possessed an abundant supply of water available locally for power generation. The first blast furnace was erected immediately adjacent to these mineral outcrops. In 1777 Anthony Bacon acquired further mineral rights from the Earl of Plymouth and he established the Plymouth Ironworks and Hirwaun Ironworks, which Francis Homfray came to manage in 1782 before establishing his own ironworks at Penydarren. Richard Crawshay took over at Cyfarthfa and Richard Hill at the Plymouth Ironworks. Anthony Bacon died in 1786, at that time it is said that he was the wealthiest man in Britain, and his son Anthony Bacon II succeeded him in the Aberaman estate – he was the "Bacon of Aberaman" alluded to in William Weston Young's Diaries as a keen supporter of the Nantgarw China Works. Anthony Bacon II married Lady Charlotte Mary, daughter of Edward Harley, the Earl of Oxford. After the death of Anthony Bacon II in 1827, Crawshay Bailey (another name cited by William Weston Young in his Diaries), of the Nantyglo and Beaufort Ironworks, took over the Aberaman House estate. Richard Crawshay died in 1810 and his grandson, William Crawshay II, took over the Cyfarthfa Ironworks, whilst Francis, Jeremiah and Samuel Homfray still ran the Penydarren Ironworks; the Homfrays were noted for their litigation suits in the law courts, particularly with the Crawshays.

In 1801, after the expiry of James Watt's patent for steam propulsion, a Cornish engineer, Richard Trevithick (1771–1833), built a steam locomotive named the "Puffing Devil" and on Christmas Eve it ran along Fore Street in Camborne, over Camborne Hill and on to the village of Camborne Beacon carrying a wagon with six passengers. This was the first demonstration of transportation using steam power. His major triumph occurred on the 21st February, 1804, when he answered a call from Samuel Homfray and Richard Crawshay who had wagered each other 1000 guineas (which is equivalent today to about £120,000) that it would be possible to construct a steam engine to move freight and passengers from the Penydarren Ironworks to a wharf at Abercynon in the Rhondda Valley, a distance of almost 10 miles (16 Km), and back again. The wager was held by Richard Hill of the Plymouth Ironworks. Josiah Guest does not seem to feature in this enterprise at all, perhaps because he objected to the wagering element, being a Freemason. Trevithick's "Puffing Devil", weighing 8 tons, completed the task outwards in 4 h 5 min on rails with a gauge of 4 ft (1.2 m) at an average speed of 2.5 miles per hour (4 Km per hour). Trevithick noted that:

> *Yesterday, we proceeded on our journey with the engine and we carried 10 tons of iron in 5 wagons with 70 men riding on them the whole of the journey ... the engine while working reached nearly 5 miles an hour, there was no water put into the boiler from the time we started until our journey's end..... the coal consumed was 2 hundredweights*
>
> (author: 100 kilos).

On the return journey a bolt sheared, the boiler overheated and the firebox in the locomotive had to be damped down until the repair was made, whereupon the engine, passengers and empty wagons returned to Penydarren the next day. It is a matter of conjecture as to whether or not the wager was ever redeemed from Richard Hill because of this breakdown. It was also found that the rails were too brittle to sustain the heavily loaded locomotive and freight train. The famous "Puffing Billy", the world's oldest surviving steam locomotive, was constructed 10 years later in 1813–1814 by William Hedley, Jonathan Forster and Timothy Hackworth to haul coal on tracks at Wylam Colliery, Newcastle upon-Tyne. However, George Stephenson (1781–1848) has been accorded the historical credit for constructing the first locomotive, carrying passengers and wagons, to run on a public railway between Stockton and Darlington on the 27th September 1825; *Locomotion No.1*, weighing

7.3 tons and with a drive wheel configuration of 0-4-0, hauled a train weight of 80 tons comprising 11 wagons of coal and 20 wagons of passengers (an estimated 550 people) with an average speed of 8 mph and a top speed of 12 mph for the inaugural opening of the railway (see Fig. A4.1, a watercolour by John Dobbin of the train crossing the Skerne Bridge in Darlington on its inaugural run). The procession was led by a man on horseback with a red flag. The train was prepared at Aycliffe and a coach named *Experiment* was attached to the locomotive and tender: at 12.30 it started from Stockton and arrived at Darlington with an average speed of 8 mph, after making two unscheduled stops to jettison a wagon that had lost a wheel and to repair the locomotive, to be welcomed there by a crowd of 10,000 people. Although George Stephenson has acquired the accolade of "Father of the Railway", there is some sentiment that this title should be transferred to Richard Trevithick, for the first ever rail journey involving steam locomotion carrying freight and passengers on a railway track. In the scope of our current text it is interesting to note that the coat-of-arms of Darlington features an escutcheon with *Locomotion No. 1* and its tender in a chief argent position (Fig. A4.2). As an indicator of how dangerous the railways were perceived to be in their early days, in addition to the man on horseback with a red flag riding in front of the train, the adopted motto of the Stockton to Darlington Railway was *"Periculum privatum utilitas publicia"*, which translates as *"At private risk for public services"*.

So, we may ask what relevance does this digression have to the situation regarding sponsorship and support for the Welsh porcelain manufactories? Aside from the appearance of the names of these major industrialists, who were based in Merthyr Tydfil only a few miles from the Nantgarw China Works, and their recorded connections with William Weston Young, research into their family interests reveals that several were interested in porcelain and its synthesis (Edwards, *Porcelain to Silica Bricks: The Extreme Ceramics of William Weston Young, 1776–1847,* 2019). The doyen of early English porcelain collectors was Lady Charlotte Guest, wife of Sir Josiah Guest of the Dowlais Ironworks, who after his death in 1851 assumed control of the management of the works until his heir, Sir Ivor Guest Bt., could take over. In her Diaries (*The Lady Charlotte Schreiber Journals: Confidences of a Collector of Ceramics and Antiques Through Britain, France, Holland, Belgium, Spain, Portugal, Turkey, Austria and Germany from the Year 1869–1885*, 1911), Lady Charlotte comments on several visits that she made with Sir Josiah to European manufactories, such as Sevres and Meissen, to discover and understand more of the intricacies of fine porcelain synthesis and its production (Jacob-Hanson, *Charlotte Schreiber – The Unforgotten Grand Dame of the "Chasse"*, pp. 50–61, 2012; Owens, *Lady Charlotte Guest: The Exceptional Life of a Female Industrialist*, 2020). The porcelain collection, along with glass and enamels, amassed by Lady Charlotte Guest, or Lady Charlotte Schreiber as she became known after her second marriage to Charles Schreiber in 1852, was presented to the South Kensington Museum (later to become the Victoria & Albert Museum) in 1884, where it was catalogued by Bernard Rackham (Rackham, *The Catalogue of The Schreiber Collection of English Porcelain, Pottery, Glass and Enamels collected by Charles Schreiber MP and the Lady Charlotte Elizabeth Schreiber and Presented to The South Kensington Museum*

Fig. A4.1 Watercolour by John Dobbin of *Locomotion No.1* crossing the Skerne Bridge on the inaugural opening of the Stockton to Darlington Railway on the 27th September 1825, the first locomotive, passengers and wagons to run on a public railway, at an average speed of 8 mph despite making two unscheduled stops to jettison a wagon that lost a wheel and to repair the locomotive. The procession was led by a man on horseback with a red flag. Many thousands of people turned out to watch this historic event

Fig. A4.2 The coat-of-arms of Darlington features an escutcheon with *Locomotion No. 1* and its tender in a chief argent position

in 1884, Volume I: Porcelain, Volume II: Earthenwares, Volume III: Enamels and Glass, 1915). This collection is extremely important historically as it provided the basis for the first chemical analyses of English porcelain by Sir Arthur Church (Church, *English Porcelain: A Handbook to the China Made in England During the 18th Century as Illustrated by Specimens Chiefly in the National Collection*, 1885) and later more comprehensive chemical analyses of English and Welsh porcelains by Herbert Eccles and Bernard Rackham in 1922 (Eccles & Rackham, *Analysed Specimens of English Porcelain in the Victoria & Albert Museum Collection*, 1922).

The invention of the silica brick by William Weston Young using the highly siliceous *Dinas rock* from an outcrop in the Vale of Neath on land owned by the Marquess of Bute (Edwards, *Porcelain to Silica Bricks: The Extreme Ceramics of William Weston Young, 1776–1847*, 2019), another avid Nantgarw China Works supporter, depended critically upon his trial experiments undertaken using the Billingsley /Walker high temperature biscuit porcelain kiln at the Nantgarw China Works during his tenure there from 1820 and its seems that he even made some hard paste porcelain using this same kiln (Colomban et al. 2020).

As an additional exercise to see what might emerge from a consideration of the crests adopted by these industrialists, the information in Table A4.1 has been collected from a consultation of Fairbairn (Fairbairn, *Crests of the Families of Great Britain and Ireland and Their Kindred in Other Lands*, Volumes I and II, T.C. and E.C. Jack, Edinburgh and London, 1905), which reveals some interesting results. In the context of the research into the potential crest bearers of supporters of Welsh porcelain synthesis, Appendix V considers in more detail the identities and heraldic arms of the "Ten True Men of Glamorgan" as nominated by William Weston Young (Young, *The Diaries of William Weston Young, 1776–1847(1802–1843)*, 30 volumes, West Glamorgan Archive Service, Swansea, SA1 3SN; discussed in Edwards, *Porcelain to Silica Bricks: The Extreme Ceramics of William Weston Young, 1776–1847*, 2019) and referred to by Ernest Morton Nance (*The Pottery and Porcelain of Swansea and Nantgarw*, 1942). It is pointed out in Appendix V that William Weston Young's "Ten True Men of Glamorgan" actually number 12 and that the £100 that he collected as a donation from each of these for the Nantgarw start-up in 1817 totalled £1000, so there were two redundancies in this list. In Appendix V it is concluded that as Thomas Mansel-Talbot of Margam Park and Thomas Wyndham of Dunraven Castle had died prior to 1817 then they would not have been able to make a contribution to the Nantgarw China Works start-up fund but that William Weston Young would perhaps have kept them on his list of donors anyway as a gesture of their benevolent intentions.

The research undertaken into Welsh porcelain armorials here has established a firm database for the attribution of new exemplars that present themselves for future discovery. It is interesting that the investigation of the local landowners, industrialists and recorded benefactors and supporters of the Swansea and Nantgarw China Works who had the right to bear heraldic arms has realised a sound basis for the assignment of all the hitherto unidentified crests and affords a means of identifying other as yet unknown crests, of which there could be several extant. It is appreciated that the assignments of the crests described above is perhaps not as hard and

Table A4.1 Crests of the Ironmasters of Merthyr Tydfil

Family name	Seat	No. of families[a]	Description of crest[a]	Fairbairn reference[a]
Crawshay	Gwaelodygarth / Cyfarthfa	2	A greyhound, sable	Plate 28/7
Homfray	Penllyne Castle/ Cowbridge	2	An otter, ppr., wounded in shoulder	Plate 9/5
Hill	Llandaff Court	31	Trunk of tree (in fess) or, falcon, ppr., beaked and belled, gold.	Plate78/8
			A boar's head, couped sable in mouth an acorn or, leaved vert	Plate 125/7
Guest	Dowlais House	3	A swan's head and neck erased, ppr., gorged with collar and charged with a cross moline sable between two ostrich feathers or	Plate 66/10
			A swan, wings expanded ppr.	Plate 122/13

[a]Fairbairn: J. Fairbairn, *Crests of the Families of Great Britain and Ireland and Their Kindred in Other Lands,* Volumes I and II, T.C. and E.C. Jack, Edinburgh and London, 1905 and New Orchard Publishing, London, 1986. The number signifies the arms-bearing families of that name in the armory

evidential as they could be in the absence of supporting documentation from the manufactories and from the individual country estate archives, but a reasoned argument for their attribution has nevertheless been proposed here. The other major feature which has been uncovered during this work is the role of the powerful ironmasters of Merthyr Tydfil in the sponsorship of the two manufactories, as might have been expected from their standing in the local communities. This correlation between people and ceramic art is an important aspect of the work and further development in this direction would not be amiss: the classic association in this respect is provided in the person of Lady Charlotte Guest of the Dowlais Ironworks, who started to collect porcelains in the first half of the nineteenth Century, it is realised not contemporary with the production runs of the Swansea and Nantgarw China Works, but her appreciation of fine ceramic art which continued as Lady Charlotte Schreiber formed the basis of her donation of the Schreiber Collection to South Kensington Museum in 1884. Later, this provided Sir Arthur Church and Herbert Eccles the opportunity to chemically analyse the specimens she had acquired, which included the first ever analyses of Nantgarw and Swansea porcelains (Church, *English Porcelain: A Handbook to the China Made in England During the 18th Century as Illustrated by Specimens Chiefly in the National Collection*, 1885; Eccles & Rackham, *Analysed Specimens of English Porcelain in the Victoria & Albert Museum Collection,* 1922; Edwards, *Porcelain to Silica Bricks: The Extreme Ceramics of William Weston Young, 1776–1847*, 2019).

Appendix V: The "Ten True Men of Glamorgan"

It is very appropriate here that we should consider in more detail the identities of "*The Ten True Men of Glamorgan*" who were nominated as such by William Weston Young in his Diaries (W. W. Young, *The Diaries of William Weston Young, 1776–1847(1802–1843)*, 30 volumes, West Glamorgan Archive Service, Swansea, SA1 3SN; and further discussed in Edwards, *Porcelain to Silica Bricks: the Extreme Ceramics of William Weston Young, 1776–1847*, 2019) as being so vital in their support of the Nantgarw China Works in their Phase II of start-up operations in 1817. Firstly, we should look at the reason for twelve names being on this list when Young specifically states that each nominee gave £100 towards the venture, realising a total funding capacity of £1000, which along with his own donation of £1100 meant that the sum of £2100 was available for the start-up of Phase II at the Nantgarw China Works by William Billingsley and Samuel Walker in 1817. This sum would equate to a capital of approximately £250,000 today. Many of these names are recognisable landowners who would have had major estates in South Wales, but the only Merthyr Tydfil industrial ironmaster listed here by William Weston Young is William Crawshay of Cyfarthfa Castle.

Careful examination of the names on William Weston Young's list of twelve reveals that two potential subscribers had actually died before the Nantgarw China Works Phase II operations commenced in 1817; presumably, therefore, their pledge of support given to Young would not have been legally binding upon their heirs and would not have been forthcoming for the venture. Thomas Wyndham of Dunraven Castle (1763–1814) was Vice-Lieutenant of Glamorgan and whose two sons died in their youth; his daughter and heiress, Caroline, married Windham Henry Quin, son of the Earl of Dunraven and Mountearl of Adare, County Limerick, Ireland. on the 27th December, 1810. Thomas Wyndham died on the 8th November 1814, being referred to as "a model of benevolence" in his *Obituary with Anecdotes of Remarkable Persons* cited in the *Gentleman's Magazine,* which also declared that "the title was now extinct in the county of Glamorgan". That changed when Henry Quin took the name of Wyndham-Quin on the 7th April 1815 upon inheriting the Dunraven estate and he also succeeded to the title of 2nd Earl of Dunraven and Mountearl on the 24th August 1824. Caroline and Henry found the estate to be heavily encumbered financially through Thomas Wyndham's local benevolence and generosity and on account of his expenditure on improvements to the fabric of Dunraven Castle (*Gentlemen's Magazine*, 1814). In view of their financial insecurity at that time it is highly unlikely, therefore, that they would have supported the pledge on the late Thomas Wyndham's behalf and donated to the Nantgarw China Works project set up by William Weston Young.

Similarly, Thomas Mansel-Talbot of Margam Park and Penrice Castle (1747–1813), the son of John Talbot of Lacock Abbey and Mary Mansel, daughter of the 1st Baron Mansel, inherited both estates of Margam Park and Penrice Castle. He married Lady Lucy Fox-Strangeways, daughter of Henry, the 2nd Earl of Ilchester, in 1794. Their son, Christopher Rice Mansel- Talbot (1803–1890), inherited both these estates at the age of 10 after the death of his father in 1813. William Weston Young was commissioned by the family to design and build a mausoleum for Thomas Mansel Talbot in Margam Abbey Church using local stone, which he completed in 1819/20 after making frequent site visits to Margam Park. Again, it would be rash to assume that Thomas Mansel-Talbot's heir would accede to the request for a donation to the Nantgarw China Works as his successor, Christopher, would have been barely 14 years old at the time of the set-up of Phase II at the Nantgarw China Works. Hence, we can reasonably delete the names of Thomas Wyndham and Thomas Mansel-Talbot as donors to William Weston Young's project and this then leaves us with the ten donations of £100 each as calculated above, which agrees with the sum of £1000 collected as mentioned in Young's Diaries. The second point worth noting is that Thomas Wyndham and Thomas Mansel-Talbot had died some years before the start-up at Nantgarw for Phase II, actually in 1814 and 1813, respectively, so Young's compilation of his list of financial supporters for the Nantgarw China Works must have been ongoing for several years before it was actually needed in 1817 and this must credibly have been prior to 1813. This chronology is of some interest as the very first concept of a start-up at Phase I in Nantgarw was only accomplished in late 1813/ early 1814, which resulted in the unsuccessful *Memorial* submission of Billingsley, Walker and Young to the British government for financial support in September 1814. This means that although the financial support for the first venture at Nantgarw came personally and exclusively from Young, Billingsley and Walker, it appears that Young must have already been planning for a further top-up financial commitment for the Nantgarw China Works from his list of potential subscribers as early as 1813/14, when both Thomas Mansel-Talbot and Thomas Wyndham were still alive. This cannot have happened before 1813 as Billingsley and Walker were still engaged at Worcester until Martin Barr of Flight, Barr & Barr, died in October of that year and therefore they would not have been contemplating a departure from Worcester before this time. So, William Weston Young could not realistically have been contemplating support for an enterprise that did not even exist before 1814, yet by that time one of his named supporters, namely Thomas Mansel-Talbot, had already died! It remains conjectural as to why Young's list was not "activated" later in 1814, following the unsuccessful attempt made at achieving government support in September of that year, although events did move remarkably rapidly with the intervention of Lewis Weston Dillwyn according to Young's Diaries; it was only a matter of some 3 weeks before Dillwyn was meeting with Billingsley and Walker and by November 1814, they were engaged in the project to make porcelain at Swansea. It seems, therefore, that by September 1814 Young's list for potential Nantgarw China Works support was not complete and that he added to it over the next 3 years, so that by 1817 he was then able to call in the promises and thereby acquire the financial commitment to facilitate the Phase II start-up at Nantgarw. It is still rather intriguing, therefore, to try and explain why Thomas Mansel-Talbot of Margam Park, who died on the 4th May 1813 at Penrice Castle, would have been named by William Weston Young as a financial supporter for the Nantgarw China Works, which technically did not even exist in concept let alone reality until many months afterwards!

In Table A5.1, the "Ten True Men of Glamorgan", but actually numbering twelve as listed by William Weston Young as noted above, are given and their crests have been taken from Fairbairn (*Crests of the Families of Great Britain and Ireland and Their Kindred in Other Lands,* Volumes I and II, 1905). Clearly not all the crests available for potential use have been assimilated here: whereas some are unique, for example that of Wyndham, others such as that for William Booth-Grey have been selectively taken from the 37 crests available, comprising 15 for Booth and 22 for Grey. It is revealing to examine the crests for the families listed in Table A5.1: it is seen that several of these crests match closely those which have remained unidentified in the literature thus far on armorial Welsh porcelains, which lends credibility to the hypothesis that the financial support pledged by these men actually perhaps extended even further to the commissioning of porcelain artefacts from the Nantgarw China Works, if not especially armorial then perhaps in the form of more anonymous porcelain items, such as tea sets, dessert services and perhaps more specialist items such as cabaret sets and cabinet cups, which in the absence of factory records would be difficulty to ascribe to individual clients.

It will be apparent that most of the unidentified crests pertain to the Swansea China Works rather than Nantgarw, so the relevance of the Nantgarw China Works support specifically may be questionable in this regard. However, the fact that ten wealthy gentlemen of considerable estate were prepared to support financially a fledgling porcelain manufactory at Nantgarw and would, as we have hypothesised above, probably seen fit to acquire porcelain items which would have perhaps included services and cabinet pieces, is indicative that they may well have supported the ceramic activities at Swansea as well, which would have been equally "local" to many of them. Hence, we might have expected to see evidence of their heraldic crests on items of Swansea porcelain too, and this seems to be the case: in fact, no fewer than six of these accredited Nantgarw supporters have versions of crests that appear on unidentified Swansea porcelains, out of some nine potential crests that are of note in this connection. Furthermore, of the Nantgarw supporters cited by William Weston Young and listed in Table A5.1, four already have accredited named services in Nantgarw and Swansea porcelains so it is entirely feasible that several of the remainder could also be formally included in this category when unidentified armorial and anonymous porcelain pieces can be attributed to their commissioning. Although some may consider this theme to be rather speculative, it should perhaps at least still merit some consideration as new potential evidence emerges of their objective participation in the ceramic activities of the Nantgarw and Swansea China Works. Sadly, Lewis Weston Dillwyn's notebooks do not reflect a similar account of supporters for his Swansea China Works: earlier, we proposed two reasons for this, one being that Dillwyn himself was already from a wealthy landowning estate and background, so possibly he did not need to acquire investment funds for his business venture in this way. Certainly, at the start-up of his operations with William Billingsley and Samuel Walker in late 1814 and early 1815 he was able to embark straightaway upon the construction of new kilns and ancillary buildings, as confirmed by the ex-employee of the Swansea China Works, Henry Morris, in his interview with Colonel Grant Francis on the 14th August, 1850, which was later cited by William Turner and discussed by Edwards (Turner, *The Ceramics of Swansea and*

Table A5.1 The "Ten True Men of Glamorgan" and Their Crests[a]

Name	Alternative[b]	Seat	Description of crest(s)	Fairbairn reference[a]
William Booth-Grey	Earl of Stamford	Dyffryn St Nicholas	Booth: Boar's head, erased, erect	16/11
			Demi-leopard rampant	12/14
			Lion rampant	67/5
			Unicorn salient ermin.	37/7
			Grey: Scaling ladder ar.	98/15
			Out of ducal coronet or a demi-swan ppr.	100 /7
			Out of ducal coronet azure a demi-peacock in pride ppr.	92/11
			On a chapeau gules turned up ermin. a wyvern or	104/5
John Crichton-Stuart	Marquess of Bute Lord Dumfries	Cardiff Castle	Stuart: Demi-lion rampant gules	67/10
			Crichton: Wyvern wings elevated and addorsed, fire issuant from mouth all ppr.	51/2
Sir John Nicholls		Merthyrmawr	Dove close ppr.	66/12
			Wolf's head erased sable	14/6
			On a mural coronet a Cornish chough, wings addorsed ppr.	84/3
Robert Jones		Fonmon Castle	Boar's head, erect and erased or	21/7
			Boar's head, erect and erased in fess or	16/11
			Demi-lion rampant	67/10
			Lion rampant azure holding shield or	125/2
Wyndham Lewis		Greenmeadow, Llantrisant	Lion rampant sejant or	60/4
			On a chapeau gules turned up ermin. a heraldic tiger ppr.	7/5
Dr Whitlock Nicholl		Cowbridge, Adamsdown	A squirrel	16/9
			Eagle rising sustaining a cross fitched or	65/3
			Dove close ppr.	66/12
			Demi-lion rampant between paws a human heart	55/13
Thomas Mansel-Talbot[c]	Earl of Shrewsbury	Margam Park	Talbot: Demi-eagle displayed sable	22/11
			A talbot passant sable	120/8
			Lion statant or tail extended	118/10

(continued)

Table A5.1 continued

Name	Alternative[b]	Seat	Description of crest(s)	Fairbairn reference[a]
			Mansel: Eagle rising ppr.	67/4
			Arm embowed and vested in hand a pair of scales equally poised	39/1
			Shrewsbury: On a chapeau gules turned up ermin. a lion statant or, tail extended	107/1
Griffith Llewellyn		Baglan Hall	Demi-lion rampant or	67/10
			Holy lamb	48/13
Rev. Wm. Perkins Lisle		Llandaff & St Fagans	Perkins: Pineapple stalked and leaved vert	–
			Out of ducal coronet a unicorn's head	45/14
			Lion passant and fleur de lys gules	4/14
			Demi-eagle displayed sable	22/11
			Lisle: Stag or collared and chained attired or	68/2
			Lion's head couped	126/1
			Stag trippant attired or	68/2
Lord Windsor	Earl of Plymouth	St Fagans	Sheaf of seven arrows enfiled by ducal coronet all ppr.	37/5
			Stag's head gardant couped at neck	111/13
			Buck's head affrontee ar. couped at neck attired or	–
R. Franklin		–	Conger eel's head erect or erased per fess gules finned or between branches vert	112/8
			Dolphin naiant ppr. finned gules pierced through sides by two spears tied together at top	48/9
Thomas Wyndham[c]	Earl of Dunraven	Dunraven Castle	Wyndham: Lion's head erased within fetterlock and chain or	90/14
	Viscount Mountearl and Baron Adare		Quin: Wolf's head couped or	91/9

[a] Fairbairn: J. Fairbairn, *Crests of the Families of Great Britain and Ireland and Their Kindred in Other Lands,* Volumes I and II, T.C. and E.C. Jack, Edinburgh and London, 1905 and New Orchard Publishing, London, 1986. The number/number refers to the Plate /Crest illustrated in Volume II

[b]The Alternative Name refers to other noble family connections (e.g., William Booth-Grey was the second son of Earl Grey of Stamford, so they might be expected to share coats-of-arms as appropriate) or to other titles held (e,g., John Crichton-Stuart was Marquess of Bute and also Lord Dumfries). Thomas Mansel-Talbot of Margam Park was a member of the cadet branch of the Earl of Shrewsbury and several heraldic achievements are shared

[c] Thomas Mansel-Talbot and Thomas Wyndham are included in this list as they appeared in the original list of William Weston Young but, as discussed in Appendix IV, they could not have participated in the required financial support of the Nantgarw China Works in 1817. The heraldic details of Thomas Wyndham here refer to his successor, Henry Wyndham-Quin, who has a recorded armorial Nantgarw service

Fig. A5.1 Crest of Lewis Weston Dillwyn, as described in Fairbairn Plate 121/5. From Fairbairn, *Crests of the Families of Great Britain and Ireland and Their Kindred in Other Lands,* Volumes I and II, T.C. and E.C. Jack, Edinburgh and London, 1905 and New Orchard Publishing, London, 1986

Nantgarw, 1897; Edwards, *Porcelain Analysis and Its Role in the Forensic Attribution of Ceramic Specimens,* Appendix IV, 2021). Secondly, unlike Young, Billingsley and Walker, Dillwyn was himself an entitled landowner with the right to bear arms; his crest is given in Fairbairn (Fairbairn, *Crests of the Families of Great Britain and Ireland and Their Kindred in Other Lands,* Volumes I and II, 1905, Plate 121/5) and is described as a stag's head couped ppr. with three trefoils slipped vert, with the motto *"Craignez honte"*, meaning "*Dread shame*", and is shown here in Fig. A5.1. This would imply that he moved in different circles of Georgian society and would have counted similar wealthy landowners among his dinner guests: it could be argued that should he have needed some additional financial support to set up the Swansea China Works then he could have approached his circle of friends to secure this quite easily without resorting to the landowning clients who employed William Weston Young in his land agent and surveyor capacity.

Appendix VI: Mortlocks of London

It is timely to consider the topic of Mortlocks of London, undoubtedly the premier china retailer in Georgian London and a key factor in the sales success of the Nantgarw porcelain manufactory business between 1817 and 1820. In an advertisement, Mortlocks promote themselves as "*A Pottery for Presents: The Largest Selection of Ornamental China and Glass in London*" (see Fig. A6.1). We have already encountered the enthusiasm of Mrs Hester Thrale Piozzi Salusbury for fine porcelain and ceramics in the first decade of the nineteenth Century and in a letter to Sir James Fellowes she writes:

> *When you feel your purse too heavy take it to MORTLOCK'S in Oxford Street and carry Lady Fellowes a beautiful specimen of South Wales china and tell him I am panting for my ice-pails and large dishes to use this day tonight.*

As with many early statements in the ceramics literature, some over-simplification has been generated about Mortlock's which has led to some unfortunate and incorrect assumptions having been made. An earlier advertisement (Fig. A6.2) from 1873

Fig. A6.1 Advertisement for Mortlock's China and Glass showrooms in Oxford Street, London, dated June 1882. *Public domain*

Fig. A6.2 Advertisement for Mortlock's showrooms at Oxford Street and Orchard Street, London, 1873

shows the figure of a Chinaman with a flowing queue (pigtail), in which the Mortlock outlets at 204 Oxford Street and 31 Orchard Street, Portman Square, are mentioned.

Although the china business of Mortlock's was established by John Mortlock (1698–1758) in 1746 in 47, Oxford Street, London, it will come as a surprise to many readers that there were two Mortlock china retailing businesses in the capital in the early nineteenth Century. John Mortlock was a very successful businessman and he established the family line at Meldreth Court, Swavesey, Cambridgeshire, in 1772, commuting by train to the Mortlock China Co. in Oxford Street, London. In 1785, John Mortlock's grandson William (1747–1801) ran the business until 1801 when he handed over to his sons in partnership, William (1778–1833) and John (1776–1837). In 1803, the partnership was dissolved, leaving John Mortlock to carry on the business at 250, Oxford Street whilst William set up a new china business nearby at 18, Regent Street. This second Mortlock business also seemed to thrive and in 1860 they provided a Coalport porcelain armorial service for Queen Victoria for use in Buckingham Palace which was simply decorated with a blue Garter band with eth motto "*Honi Soit Qui Mal y Pense*": a plate from this service is shown in Fig. A6.3, with William Mortlock's stamp from his 18, Regent Street showrooms printed on the base (Fig. A6.4). A second Royal armorial service was ordered from William Mortlock for use at Windsor Castle also in Coalport porcelain of similar design but bearing the full Garter heraldic badge and motto (Fig. A6.5) and carrying the appropriate backstamp (Fig. A6.6).

Fig. A6.3 Armorial service in Coalport porcelain, *ca.* 1860, depicting an Oder of the Garter band in blue with the motto "*Honi Soit Qui Mal y Pense*" located centrally, which was used in Buckingham Palace. *Reproduced with Courtesy of Kingschina Antiques Ltd., Mayfair, London*

Fig. A6.4 Reverse of the armorial plate shown in Fig. A6.3, showing the stamp of William Mortlock, 18 Regent Street, London, and the place of use, Buckingham Palace. *Reproduced with Courtesy of Kingschina Antiques Ltd., Mayfair, London*

Fig. A6.5 Armorial service in Coalport porcelain, *ca.* 1860, similar to that shown in Fig. A6.3 but here depicting a full Order of the Garter badge bearing the cross of St George with the motto "*Honi Soit Qui Mal y Pense*" in a blue garter located centrally in the silver eight-pointed Garter Star., which was used in Windsor Castle. *Reproduced with Courtesy of Kingschina Antiques Ltd., Mayfair, London*

Fig. A6.6 Reverse of the armorial plate shown in Fig. A6.5, showing the stamp of William Mortlock, 18 Regent Street, London, and the place of use, Windsor castle. *Reproduced with Courtesy of Kingschina Antiques Ltd., Mayfair, London*

John Mortlock, of the Oxford Street business, was the man who made the arrangement with William Billingsley to take most of his Nantgarw output in the white for decorating in London according to the wishes of his clientele. He was a deeply religious man and never married: after his death he left over £200,000 to charity and in February 1834 he gave a London house to be used as a home for "Female Servants". In 1835 his nephew, also John (1808–1888), and son of William Mortlock Jr took over the Oxford Street business, so effectively bringing both businesses together again under the Mortlock name although they still ran as separate entities. In 1896, the Regent Street business, then under the management of Frederick Mortlock (1840–1915) closed, but John George Mortlock carried on at Oxford Street until that ceased trading in 1933. In 1879, there were two London china and glass retailers listed under the name Mortlock, namely, W. Mortlock & Sons of 18, Regent Street, Waterloo Place (next to Elkington's) and Mortlock J. & Co. of 204 Oxford Street and 31, Orchard Street, Portman Square (Fig. A6.2) (*Dickens' Dictionary of London: An Unconventional Handbook*, ed. C. Dickens Jr., Charles Dickens Publishers, London, 1879).

John Mortlock of Oxford Street acquired a self-granted "Royal Warrant" in the second decade of the nineteenth Century, signifying in his advertisements that he supplied "Coalbrookdale (Coalport) porcelain to Queen Charlotte and the Royal Family". This was followed by the equally grandiose claim that he manufactured porcelain with the phrase "Colebrookdale china manufacturers" in 1825! Mortlock's were the first to perfect the process of printing photographs onto porcelain and this was of interest to Queen Alexandra in 1908, who asked them to reproduce photographs that she had herself taken of Royal residences onto Coalport china. A genuine Royal Warrant for Mortlock's was issued by Queen Victoria and a second followed from Queen Alexandra in 1908.

John Mortlock was delighted to acquire the sole agency for Nantgarw porcelain in the capital from William Billingsley in 1817 and is said to have commented that:

> ***Billingsley's Nantgarw designs borrowed from the Brighton Pavilion with mouldings and beadings animal motifs and acanthus leaves jostling promiscuously for attention. And the customers loved it!***

An interesting statement made by Robert Griffiths, who assembled the *Mortlock Encyclopedia* a genealogical history and tour de force researched between 2000 and 2019, refers to the fact that Mortlocks used the workshops of Robins and Randall at Islington, John Sims at Pimlico, Thomas Pardoe at Bristol, Thomas Baxter at Clerkenwell (independent up to 1814, then based at the Swansea China Works) and John Powell at 91, Wimpole Street. The inclusion of Pardoe at Bristol is revealing and apart from the first two, Robins & Randall and Sims, the others have not been mentioned hitherto as decorators of Nantgarw porcelain – which certainly would not have applied for Thomas Baxter as he had joined Dillwyn at Swansea before the Nantgarw China Works Phase II had commenced in 1817. Another comment that Mortlock's had the sole agency of Swansea porcelain also in the capital is clearly incorrect and perhaps the inference that can be drawn is that they decorated Swansea porcelain along with others such as Bradley's of Pall Mall and Apsley Pellatt of St Paul's Churchyard. A Swansea chocolate cup in fine duck-egg porcelain (Fig. A3.11) and attributed to William Billingsley bears the legend on its base "*Powell, 91, Wimpole Street*" so does this mean that Powell did not just have a decorating business but actually sold porcelain in London perhaps decorated locally at Swansea! The attribution of William Billingsley to the rose decoration is perhaps a moot point in this regard and it probably was imitated by an artist in Powell's workshop, which is perhaps a more likely scenario: in the author's opinion the roses do not quite come up to Billingsley's standard signature and look rather artificial without any accompanying stems or foliage, which he very much favoured. Whether or not Thomas Pardoe decorated any Nantgarw or Swansea china whilst at Bristol is also debatable but we have postulated that he could well have decorated Cambrian Pottery creamwares there and we have noted already that Mortlock's had adopted the agency for the Cambrian Pottery from 1808. Another rather gruesome detail is a report under the heading "*Death Sentence for Stealing Mortlock's China*" that appeared in local papers dated 1820 in which two ex-employees of John Mortlock's business were arrested for stealing china worth about £80 over a 2-year period. Following their trial for this serious offence at The Old Bailey both "villains" were sentenced to death by hanging. Clearly, the theft of china was a serious crime and perpetrators were punished accordingly.

It is clear that the two Mortlock china businesses and showrooms were in competition with each other, although this has not been stated specifically hitherto, and perhaps affords an explanation of John Mortlock's visit to Nantgarw in 1817 as noted by William Weston Young to arrange his contract with Billingsley to take most of the porcelain production (>90%). By this means he would deny other china businesses, including that of his namesake, Mortlock's of Regent Street, access to the Nantgarw china that was so much in demand by wealthy clientele in the capital.

Glossary

Armorial china High-fired ceramics decorated with the heraldic achievements of an arms-bearing family which could constitute the full coat-of-arms comprising the escutcheon, motto, supporters, helm, lambrequins, crest and torse or any number of these or more usually just consisting of a crest and torse. In some cases, the crest and torse, with or without a motto, is encompassed in other decoration and may be accompanied by additional gilding and other geometric motifs.

Biscuit porcelain The unglazed product of the high-temperature firing process in the manufacture of porcelain using a "biscuit kiln" which operated typically at a temperature in the region of 1100–1400 °C. The resultant porcelain is of a pure creamy-white or ivory colour and possessed a texture which, if blemish-free and perfectly shaped, was very highly prized by ceramic artists and modellers particularly for the construction of ornamental figurines in simulation of marble statuary as first perfected by the Meissen manufactory in the mid-eighteenth Century. In the mid- to late -eighteenth Century, the finest biscuit porcelain artefacts were commonly placed on dining tables for admiration by the guests and to stimulate conversation during dinner. Generally, the biscuit porcelain, after cooling from the kiln firing process, was painted, glazed and gilded – during which process any small defects or blemishes in the body or the glaze could often be masked by strategically placed enamelling: hence, painted and enamelled figurines are found much more commonly than their biscuit analogues, which if found to be imperfect after their firing, were usually destroyed or sent to be decorated and glazed. The figures were designed by a "modeller" or sculptor and assembled in the workshop by a "repairer".

Blemish A blemish is a fault in the fired porcelain, whether this is in biscuit or glazed form, arising from several problems which could arise during the firing process such as inadequate kiln temperature control leading to firing cracks, the introduction of contaminants such as carbonaceous matter which then burned to leave soot deposits and the emission of gases during the firing process which created bubbles in the vitrified paste or the glaze. Serious blemishes resulted in

H. G. M. Edwards, *Welsh Armorial Porcelain*,
https://doi.org/10.1007/978-3-030-97439-8

porcelain items being unsaleable, whereas it was a common practice to mask minor blemishes, such as small spots of soot or little bubbles in an otherwise perfectly shaped piece, by the shrewd and strategic placement the in enamelling of small devices or motifs such as insects, butterflies or moths prior to a final glazing process taking place.

Body paste The wet mixture of raw material components which comprise the formulaic recipes for porcelain bodies. Texturally moist to the touch, the paste could be moulded or formed into shapes preparatory to the firing process taking place in the biscuit kiln. This term is also applied to the fired porcelain body prior to its chemical analysis, when of course it is then not a wet "paste" at all but rather a hard ceramic material.

Burnishing The gentle polishing of gilt decoration on a glazed ceramic surface to a highly polished reflective coating. Early gilding was accomplished using "honey gilding" whereby 24-carat gold leaf was applied to the surface with an adhesive medium of honey, or with the use of applied resins such as gum arabic, which were both replaced in the late eighteenth Century by mercury gilding using an amalgam of mercury which had been triturated with the gold leaf. During the final firing of the ceramic piece in the "glost" kiln at low temperatures the organic adhesive carrier component or the mercury in the amalgam was volatilised off into the atmosphere, leaving a dull golden finish which was then hand-polished by skilled gilders using a finely powdered jeweller's rouge, iron (III) oxide Fe_2O_3, to give a brilliant finish. The skill of the gilder was much admired in porcelain workshops as it offset and complemented the applied enamelling decoration but it was always an expensive part of the finishing process. Some service commissions received only a gilded decoration without accompanying enamelling decoration whereas others were not gilded at all. In some manufactories, such as Derby, the gilder was a recognised artistic master and was allowed to place an assigned numeral indicative of their work on the underside of the piece. In the third phase of the Nantgarw China Works, Young and Pardoe chose to eliminate the gilding on some of their work on remnant pieces from the Billingsley and Walker era to economise upon the decorating costs and the simple edge gilding usually employed on locally decorated pieces in the earlier era then comprised enamelled colours, especially a chocolate brown and deep green.

Chemical descriptors Care should be taken to differentiate between the descriptors of chemical compounds and their component elements. For example, silica is SiO_2 and contains silicon and oxygen: likewise, calcia is calcium oxide CaO (lime), alumina is aluminium oxide Al_2O_3, magnesia is magnesium oxide MgO, potash is potassium oxide K_2O and soda is sodium oxide Na_2O. Other chemicals are described precisely as they occur, such as phosphorus pentoxide P_2O_5 – also known as phosphorus (V) oxide or phosphorus pentoxide, relative to phosphorus (III) oxide P_2O_3—and sulfur dioxide SO_2. Iron oxide is troublesome as this can mean both ferrous oxide FeO and ferric oxide Fe_2O_3, where the iron is in the oxidation state (II) and (III), respectively, and the former is a green colour and the latter orange-red.

China A generic term for high-fired ceramics, which may be glazed or unglazed, and used for decorative or utilitarian purposes around the home. Believed to originate from descriptors of ceramics that were made in simulation of imported Chinese porcelain, this term covers a wide range of ceramics including porcelain, earthenwares, creamwares, stonewares, celadons, majolica and delftwares. The term "china" has been used freely to describe ceramic articles of a range of compositions since the importation of porcelain from China in the sixteenth Century and some of the artefacts do not contain china clay (kaolinite).

Chinese porcelain A generic term which historically describes the characteristic hard paste porcelain comprising *kaolinite* and *petuntse* as major raw materials fired at a high kiln temperature; termed "true porcelain" by some historians to differentiate it from "artificial porcelain" which is retained for the soft paste body variants which included synthetic components as raw materials, for example calcined bone ash, potash, lime and glass frit. During the early eighteenth Century, the Chinese perfected a soft paste body variant which has been recognised as a comparative rarity. Chinese porcelains have a range of compositions which varied both with their production sites geographically and also with the specified quality: the highest quality porcelains from the Imperial kilns at Jingdezhen were composed of an equal ratio of china clay (kaolin) and petuntse.

Cipher See Monogram

Coat-of-arms An officially registered heraldic device granted in Great Britain by Royal authority via the College of Arms in London and the Lord Lyon King of Arms in Scotland to members of a noble family or an arms-bearing organisation. Other European countries have their equivalent officers for the administering and granting of arms. This includes a shield (escutcheon) with dexter (right) and sinister (left) supporters, as viewed from the bearer of the shield – or by an observer from the front of the shield in a mirror image. The supporters are usually real or mythical animals or birds such as the lion, leopard, eagle, unicorn, dragon, wyvern and gryphon. Under the shield (base) a motto may be carried and also a badge of honour. On top of the shield (chief) can sit a coronet signifying the rank in the peerage of the bearer of the arms, a helm and visor accompanied by mantlings(lambrequins) and supporting a twisted silk torse upon which the crest sits. Atop the crest may appear a battle cry legend. This comprises the full coat-of-arms, which is very rarely encountered on the Welsh armorial porcelains as studied here but is encountered more frequently elsewhere.

Crested china A term reserved for a partial coat-of-arms depicted upon porcelain or earthenwares which may exhibit the crest solely, with or without its torse, and perhaps accompanied by a motto, usually placed in either the verge or reserve or centrally located on the artefact. It may be enamelled in polychrome pigments, or it is more usually found in gilt or occasionally and more rarely in a monochrome pigment.

Cullet Also known as glass frit, this is a glass additive in porcelain body recipes which is included to increase the transparency of the fired body. Generally purchased by porcelain manufactory proprietors from neighbouring glassworks (for example, as the "end of day glass" mixture) this was then added as a finely ground material to the second stage firing process of soft paste biscuit porcelain syntheses. Cullet is a rather indefinite term scientifically as it could contain either flint glass (a highly refractive glass containing up to 60% lead oxide, PbO) or crown glass (also known as soda glass, which contained either soda or potash, or both, as an alkaline flux, Na_2O and K_2O, but critically for analytical chemical interpretation of the elemental data, this soda glass is lead-free. Analytically the presence of a glass additive is important as in the absence of a lead – containing material from other sources in the recipes, the presence of a lead oxide signature in the analytical data derived from the body paste is a definitive indicator of the presence of a flint glass cullet additive. In several production notes from manufactories, proprietors did not refer to the addition of cullet in their raw materials composition, yet analytically this has been inferred from the presence of a lead oxide signature (from flint glass) or the presence of an abnormally high soda and potash content (arising potentially from a crown, or soda glass cullet additive). Also known as glass frit, a confusion occasionally occurs in terminology because of the use of the word "frit" which is used to also describe a ground calcined body paste after a first firing sequence, whether this contains glass cullet or not.

Electromagnetic spectrum This describes the range of wavelengths which define the characteristics of electromagnetic radiation from the high-energy, low wavelength, high frequency cosmic and X rays, through the ultraviolet (200–400 nm) and then the visible spectrum (400–700 nm) from the violet (through indigo, blue, green, yellow and orange) to the red into the infrared, and thence through to the low-energy, high wavelength, low frequency microwave radiation. Analytical techniques probe different ranges of the electromagnetic spectrum to interrogate the molecular and elemental compositions of ceramic specimens and derive quantitative and qualitative data which can be correlated with the materials used in the ceramic bodies, glazes and pigments. The unit of the wavelength used analytically to define the electromagnetic spectrum is the nanometre, 10^{-9} metre or nm. As the wavelength of the radiation increases, the associated energy becomes smaller, through from the Xray to the microwave region, according to the Planck-Einstein radiation law, $E = h\nu$, where E is the energy of the radiation, h is the Planck constant and ν is the frequency of the electromagnetic radiation, which is inversely proportional to the wavelength of the radiation, λ. Generally, the elemental composition is interrogated in the Xray region and the molecular composition is found from molecular energy transitions occurring in the visible and near infrared regions of the spectrum. Analysis of the elemental and molecular spectral data gives a complementary picture of the porcelain body, glaze and pigments used in the artefact.

Elemental oxides Usually these are determined from an SEM/EDAXS or XRF experiment which detects the key elements in fired porcelain such as silicon and aluminium and a range of others: a conversion factor hence needs to be applied to estimate the percentage of silica, alumina or other metal and nonmetal oxide present in the sample. From the ratio of the elemental relative atomic masses in the formula to the corresponding molecular weights of the oxides, the conversion factors can be listed as follows: Si to SiO_2 2.1, Al to Al_2O_3 1.9, Mg to MgO 1.7, Ca to CaO 1.4, Pb to PbO 1.1, Na to Na_2O 1.35, K to K_2O 1.2, Ti to TiO_2 1.7, Fe to Fe_2O_3 1.4 and P to P_2O_5 2.3. This means that an analytical determination of 10% elemental silicon in the experiment would correlate with a content of 21% silicon dioxide (silica) in the specimen. This does not necessarily imply that the porcelain in this case has a composition which required 21% of sand as the analytical result is a compound quantitation of all the components present which contained silica: these might include materials such as sand, flint, chert, smalt, feldspar, glass frit, china clay, secondary ball clays and steatite.

Frit The result of a fine grinding process which is applied to the components of a ceramic paste and especially used to describe the compositional mixtures of bone ash, china clay, flint glass (cullet) and soaprock. The necessity of very fine grinding to produce a homogeneous mixture of components which originally may have very tangible differences in hardness was appreciated by the earliest porcelain manufacturers and a fine frit usually then required just the addition of a single component or water, perhaps with some additional alkaline flux or powdered china clay, to effect a suitable and tractable paste mixture for moulding and firing in a biscuit kiln. The term frit is also applied to a first-stage preparatory firing process of a component mixture which has been fired at a high temperature then ground, mixed with new components and then re-fired to produce biscuit porcelain in the two-stage process that is typical of most soft paste porcelain syntheses. Care must be taken in the understanding of the meaning of frit in recipes as it was used to describe both the finely ground porcelain body after its initial firing and also the ground glass component used as an additive synthetic raw material prior to its making up as a body paste (see, Glass).

Gilding The application of a gilt decoration to fired porcelain involving 24-carat gold leaf in a carrier such as mercury, honey or an organic resin followed by burnishing, see under ***burnishing***. Gilding was a highly specialist task in decorator's workshops, the best gilders being permitted to sign their work with a numeral on the underside of the piece or being mentioned specifically in the service commission documentation and gilding was an expensive component of the decorative process.

Glass A supercooled vitreous material which is rigid but has not crystallised – glass is often referred to scientifically as a supercooled liquid material. Several raw material components are shared with porcelains, including sand and an alkaline flux comprising calcia, soda and potash. Glass may be crown glass, which is lead-free, or flint glass, which contains significant amounts of lead oxide, which usually has been added to the molten glass in the form of red lead, trilead tetroxide, Pb_3O_4. Pulverised glass residues in the form of cullet, obtained from neighbouring glassworks,who wished to dispose of surplus and residual "end-of-day glass", were frequently added to porcelains and the determination of an analytical signature for lead oxide in a fired porcelain body results in the conclusion that flint glass residues were a component in the body paste; a crown glass cullet additive is usually manifest in an increased soda, calcia and perhaps also a potash percentage. Glass firing was conducted over a similar temperature range as soft paste porcelains especially for their glazing procedure which was broadly accomplished at a temperature of between 950–1150 °C, and the aim was to achieve the cotectic region where the pressure, temperature and composition of the raw material components would result in a melting of the component raw materials without the separate crystallisation out of the individual species. Although normally one would classify a glass as transparent, certain types of glass, such as milk glass, had heavy metal oxides added as opacifiers which rendered them translucent, in keeping with porcelain.

Glaze The application of a coating to a fired porcelain biscuit body to seal the ceramic pores and to enhance the appearance with a glossy, smooth texture. Usually applied over the decorative pigments, it acted as a protectant for the decoration. Overglaze decoration, known technically as painting rather than enamelling in the ceramic workshops, was subject to paint loss and wear during usage. The glaze was usually applied as a slip, a thick suspension of clays, alkaline earths and other additives with lead oxide component (or a tin oxide additive, such as cassiterite, for a lead -free glaze). Fired at a lower temperature than the biscuit body, this final stage of preparation enhanced the beauty of the porcelain: it was essential that the glaze composition mechanically and chemically was compatible with the fired porcelain otherwise firing faults were created and an unsightly glaze "craquelure" could form, whereby the glaze physically separates from the porcelain substrate in contraction from the kiln upon cooling. Certain decorative pigments were favoured for their enhancement properties upon interaction with the glaze application, cobalt blue being one of these, and it was equally important that the pigment maintained its integrity and was able to resist "running" on application of the glaze.

Glost kiln Used in the final stage of porcelain preparation before and/or after applying the enamelled decoration and involving the application of an alkaline "slip" containing china clay, soda, potash and a lead oxide component usually in the form of a powdered flint glass, which at lower temperatures will form a hard, transparent glaze coating. Occasionally, a glost kiln was used for the calcination of raw material components at lower temperatures and for the drying of raw materials in which variable amounts of water were found to occur. Initially, glazes were lead-based but in the first decade of the nineteenth Century, when the toxicity of lead compounds was fully appreciated, reversion to a less toxic, tin-based glaze was effected (as patented by John Rose at the Coalport China Works in 1820, using cassiterite, SnO_2, instead of lead oxide, litharge or massicot, PbO). China sold from manufactories "in the white" could be unglazed or glazed for decoration in ateliers, with subsequent glazing being undertaken by the enamellers.

Goss china a type of armorial porcelain made *en masse* as souvenirs and mementoes for holidaymakers, tourists and visitors from the mid-nineteenth Century. The coats-of-arms of towns and places of historical interest featured on a variety of miniature bone china artefacts and numbering up to ten thousand different armorials on more than 2000 shapes.

Initials see Monogram.

Instrumental Chemical Analysis the use of instrumentation to detect qualitatively and to estimate quantitatively the presence of chemical elements and their compounds in ceramic bodies and glazes. The basis of instrumental analysis is the use of special signatures or properties of chemical species or chemical bonds such as silica, SiO_2, or the Si=O bonds in complex macromolecular structures, such as silicaceous ceramic bodies, from their behaviour towards the scattering or absorption of electromagnetic radiation in the electromagnetic spectrum from the low wavelength (high wavenumber) Xray region to the high wavelength (low wavenumber) microwave region. When coupled with a microscope, an analytical instrument is capable of interrogating very small specimens or domain regions in a larger specimen of the order of cubic microns or cubic nanometres. Instrumentation that has been commonly used in the assessment of the constitution of porcelain and ceramic bodies, glazes and pigments includes scanning electron microscopy/energy dispersive and X Ray spectrometry (SEM/EDAXS), X Ray fluorescence spectroscopy (XRF) for the elemental oxide composition and X Ray diffraction spectrometry (XRD), laser Raman Spectroscopy (RS) and Fourier-transform infrared spectroscopy (FTIR) for mineralogical and pigment identification.

Monogram Initials which are arranged in a design so that the removal of one initial compromises the design and renders one or more of the other initials indecipherable. A cipher is a design in which initials which are merely arranged together whereby the removal of one does not affect the others. Monograms, ciphers and initials all appear on applied ceramic decoration and when accompanied by an heraldic device such as a coronet or a crest then constitute an armorial representation. Otherwise, they are not strictly considered as armorial although historically they have often been included as such.

Moulding The shape and size of porcelain items can be a reliable stylistic indicator of a particular factory's output and a clue to the attribution of unmarked pieces. Mouldings were generally unique to each manufactory and comprised individual and idiosyncratic cross-sections of flatwares such as saucers and plates, the shapes of cups, the presence and pattern of footrims, the shape of spill vases (e.g. cylindrical, fluted, tulip-shape, trumpet-shaped, flared), the impressed moulded and embossed verges of plates and the number and type of indentations at the rim. Many factories characteristically used the application of moulded flowers which were integral to the item and also applied reliefs such as feet, the heads of animals, floral decoration and beads, as well as knops or handles on tureens and comports.

Named service A porcelain service which has been commissioned by a particular client or individual from the china factory or one of its agents, especially where factory records support the attribution and where occasionally a reason for its purchase may be stated, such as the *Duke of Cambridge* service in Nantgarw porcelain which was ordered from the Nantgarw agents, John Mortlock, Oxford Street, London, as a wedding present from George, the Prince Regent, for the marriage of his younger brother, Adolphus, Duke of Cambridge, to Princess Augusta, Landgravine of Hesse-Cassel, at Buckingham Palace in June, 1818.

The term also includes services which have been purchased from the later sale of effects or estates and have thereby acquired the name of their new owners by association. Examples of the former are the *Duke of Cambridge, Marquess of Anglesey* and *Baron Phipps of Normanby* services in Nantgarw porcelain, the *Lysaght* and *Gosforth Castle* services in Swansea porcelain, and of the latter type, the *Burdett-Coutts* and *Lady Seaton* services in Swansea porcelain and the *Mackintosh Edwards* and *Twyning* services in Nantgarw porcelain, respectively. Occasionally, the named services do not bear the name of the person who commissioned them but rather the place where they were to be used or perhaps referred to even the decoration adopted: an example of the former is the *Marino Ballroom* service in Swansea porcelain commissioned by John Vivian for his residence in Sketty Hall, Swansea, and of the latter, the *Garden Scenery* service of Swansea porcelain originally commissioned by Lewis Weston Dillwyn, proprietor of the Swansea China Works, which was later acquired and retained by the Dillwyn-Venables-Llewellyn family of Llysdinam, Powys.

Pigments In ceramics decoration, these are coloured minerals which are thermally stable at the operating temperatures adopted in the glost kiln after application of the glaze slip and firing up to 600 °C or slightly greater. Mineral pigments are usually metal oxides, sulfates and sulfides such as haematite, gypsum and orpiment. Care needs to be taken in the interpretation of old recipes for the decoration of ceramics as minerals were often confusingly assigned an incorrect nomenclature, such as *minium*, which has been used historically to describe both red lead, trilead tetroxide, Pb_3O_4, and cinnabar, mercury sulfide, HgS, and red oxide of lead can mean both red lead ($2PbO.PbO_2$) and litharge (PbO). Analytically, the interrogation of pigments on enamelled porcelains is a fertile area of study as manufacturers often changed their pigment sources and the composition of their pigments with time and the presence of trace quantities of unusual elements such as zinc, niobium, rubidium and copper can indicate valuable information about the timelines of production and sources of the pigment by the manufactory concerned. For example, the presence of a minor elemental component of bismuth in a cobalt blue pigment on an Iznik glazed tile from the sixteenth/seventeenth Century indicated that the source of the cobalt ore was European (from the Erzengebirge mine in Saxony) rather than the more usually encountered Asian mines. Arsenic traces in the cobalt blue pigment on Chinese eighteenth Century porcelains are indicative of the use of European cobalt ores in the preparation of the pigment rather than the Asian sources which are arsenic – free.

Porcelain types commencing with hard paste and soft paste porcelains historically (also termed rather confusingly as "true" and "artificial" porcelains), this inadequate description generated a further differentiation into the porcelain as hard paste (highly siliceous), phosphatic, bone china, magnesian, glassy and hybrid types. The categorisation of a porcelain type is dependent not only upon the body composition but also upon the process methodology and firing sequences adopted. However made, all types of porcelain must possess a degree of translucency to qualify for its appellation and inclusion in the category.

Raw Materials The raw materials are components comprising the body and glaze recipes used in porcelain syntheses. Mostly these were natural minerals and rocks which were sourced from precise mines and locations which gave rise to several alternative names. Natural raw materials include feldspars, china clays, ball clays and soapstone. Synthetic raw materials include bone ash, pearl ash, soda ash, smalt, lime and magnesia.

Service A porcelain service could be commissioned directly from the factory, from an appointed agent, bought at auction from "chinamen" or acquired through an import agency such as the Honourable East India Company who traded with Chinese porcelain agencies and their brokers or factors (Hongs) through the port of Canton (Guangzhou). A porcelain service was normally defined as a dinner, dessert or tea service, but combinations of dinner-dessert, tea-coffee and special breakfast services were quite commonly ordered. The usual dessert or dinner service comprised 42 pieces approximately, but special orders expanded this to over 100 pieces, with the inclusion of soup dishes, tureens, centre fruit comports and stands, ice pails, spill vases, muffin dishes, guglets (water bottles), vegetable dishes, egg-cups and carrier, sauce boats and trios (coffee can/cup, tea cup and saucer). There are instances of very large services being commissioned, such as those from Meissen in the eighteenth Century of over 1000 pieces with some 50 or 60 place settings. The service was usually decorated according to the specific requirements of the commission and the inclusion of a coat of arms or heraldic crest centrally or at the verge would confer upon it the description of an "armorial" service. An armorial service is especially valuable for provenancing and historical purposes as it facilitates the assignment of the commission to a particular person in an arms-bearing family and often enables one to attribute a definitive date for a commission, such as the granting of arms or perhaps a family succession to a title or elevation in the peerage which may have been celebrated accordingly with the ordering of a porcelain service.

Translucency This is perhaps the greatest achievable asset of porcelain manufacture to which every manufacturer subscribed and hoped to attain in emulation of the Chinese "eggshell" wares which were imported to Europe in the mid-eighteenth Century. It describes the transmission of visible radiation, or light, through a solid object: at one end of the spectral scale is glass, which is usually *transparent* (100 % transmission) and at the other is earthenware which is *opaque* (0 % transmission). However, see entry above for "Glass", where the occurrence of a milk glass is noted in which the transparency is less than 100% and the glass is then termed translucent. Porcelain is measured by its translucency, which is the clarity for transmission when viewed with background lighting. An intermediate descriptor is "semi-opaque" or even "semi-porcelain", which was applied to some china during the mid-nineteenth Century, which is rather indefinite and conveys little information about their true category. Other descriptors such as "vitreous china" and "ironstone china" likewise do little to assist in their classification. By definition, porcelain must be able to transmit light to some extent unlike stonewares, faience, terracotta and earthenwares.

References

References for Appendix I

A. Brogniart, *Traite des Artes Ceramiques, ou des Poteries, Considerees dans Leur Histoire, Leur Pratique et Leur Theorie*, vol 2 (Bechet Jeune, Libraire Editeur, Paris, 1854)

H.G.M. Edwards, *18th and 19th Century Porcelain Analysis: A Forensic Provenancing Assessment* (Springer-Nature, Dordrecht, 2020)

H.G.M. Edwards, *Porcelain Analysis and Its Role in the Forensic Attribution of Ceramic Specimens* (Springer-Nature, Dordrecht, 2021a)

J. Fairbairn, *Crests of the Families of Great Britain and Ireland and Their Kindred in Other Lands,* vols. I and II, ed. (T.C. and E.C. Jack, Edinburgh and London, 1905 and New Orchard Publishing, London, 1986a)

G.A. Godden, *An Illustrated Guide to Lowestoft Porcelain* (Barrie & Jenkins, London, 1969)

P. Pelliot, *Notes on Marco Polo*, vol 2 (Imprimerie Nationale & Libraire Adrian- Maisonneuve, Paris, 1959), pp. 810–811

N. Pine, *The Concise Encyclopaedia and Price Guide to Goss China*, 7th edn. (Milestone Publications, Portsmouth, 2000)

L. Pine, *Goss and Souvenir Heraldic China* (Shire Publications, Princes Risborough/Bloomsbury Books, London, 2005)

L. Pine, N. Pine, *William Henry Goss: The Story of the Staffordshire Family of Potters Who Invented Heraldic Porcelain* (Milestone Publications, Portsmouth, 1987)

References for Appendix II

W. Bemrose, *Bow, Chelsea and Derby Porcelain: Being Further Information Relating to the Factories Obtained from Original Documents Not Hitherto Published* (Bemrose & Sons, London, 1898)

P. Bradshaw, *Eighteenth Century English Porcelain Figures 1745–1795* (Antique Collectors Club, Woodbridge, 1981)

A. Dawson, *French Porcelain: A Catalogue of the British Museum Collection* (British Museum Publications, London, 1994)

H. G. M. Edwards, *Welsh Armorial Porcelain*,
https://doi.org/10.1007/978-3-030-97439-8

L.W. Dillwyn, *British Conferva or Coloured Figures and Description of British Plants Referred by Botanists to the Genus Conferva* (W. Phillips, George Yard, London, 1809)

L.W. Dillwyn, *Notes of Recipes for Porcelain Bodies and Glazes 1815–1817* (Library of the Victoria & Albert Museum, South Kensington, London, deposited by John Campbell Esq. in, 1920a)

H. Eccles, B. Rackham, *Analyses of English Porcelain in the Victoria and Albert Museum Collection* (South Kensington, Victoria & Albert Museum Publications, London, 1922a)

H.G.M. Edwards, *Swansea and Nantgarw Porcelains: A Scientific Reappraisal* (Springer – Nature, Dordrecht, 2017)

H.G.M. Edwards, *Nantgarw and Swansea Porcelains: An Analytical Perspective* (Springer – Nature, Dordrecht, 2018)

H.G.M. Edwards, *Porcelain to Silica Bricks: The Extreme Ceramics of William Weston Young, 1776–1847* (Springer-Nature, Dordrecht, 2019a)

H.G.M. Edwards, *Porcelain Analysis and Its Role in the Forensic Attribution of Ceramic Specimens* (Springer-Nature, Dordrecht, 2021b)

F. Gambon, *Porslen Abertawe a Nantgarw (Swansea and Nantgarw Porcelain)* (Oriel Plas Glyn-y-Weddw, Llanbedrog/Abersoch/Gwynedd, 2016a)

Gartre'n Ol, Coming Home – An Exhibition of the Finest Nantgarw Porcelain, Nantgarw China Works Museum, (Tyla Gwyn, Nantgarw, 2019a)

J. Gray, *The Cambrian Company, Swansea Pottery in London, 1806–1808*, 2012.

J. Haslem, *The Old Derby China Factory: The Workmen and their Productions. Containing Details of their Chief Artist Workmen, the Various Marks Used, Fac-Similes Copied from the Old Derby Pattern Books, the Original Price List of More than 400 Figures and Groups, Etc., Etc.*; George Bell & Sons Covent Garden, London, 1876: reprinted by (E.P. Publishing, Wakefield, 1973)

W. Elis Jenkins, *Swansea Porcelain* (D. Brown & Sons Ltd., Cowbridge, 1970)

L. Jewitt, *The Ceramic Art of Great Britain from the Prehistoric Times Down to the Present Day*, vol 1 and 2 (Virtue & Co. Ltd., Paternoster Row, London, 1878)

W.D. John, *William Billingsley* (Ceramic Book Company, Newport, 1968a)

A.E. Jones, S.L. Joseph, *Swansea Porcelain: Shapes and Decoration* (David Brown Publishers, Cowbridge, 1988a)

R. Massey, An Unrecorded Derby Figure. *Transactions of the English Ceramic Circle* **20**(1), 91–94 (2008)

E. Morton Nance, *The Pottery and Porcelain of Swansea and Nantgarw* (Batsford, London, 1942a)

H. Owen, *Two Centuries of Ceramic Art in Bristol* (Bell & Daldy, London, 1873)

R. de Plinval de Guillebon, *Paris Porcelain, 1770–1850,* translated from the French by R. Charleston, (Barrie & Jenkins, London, 1972)

W.J. Pountney, *Old Bristol Potteries* (J.W. Arrowsmith, Bristol, 1920), pp. 115–116

W. Turner, *The Ceramics of Swansea and Nantgarw: A History of the Factories* (Bemrose & Sons Ltd., London, 1897a)

J. Twitchett, *Derby Porcelain: An Illustrated Guide, 1748–1848* (The Antique Collectors Club, Woodbridge, 2002)

I.J. Williams, *The Nantgarw Pottery and its Products: An Examination of the Site* (The National Museum of Wales and the Press Board of the University of Wales, Cardiff, 1932)

W.W. Young, *The Diaries of William Weston Young (1776–1847), 1802–1843,* 30 volumes (West Glamorgan Archives Service, Swansea, SA13SN). https://archiveshub.jisc.sc.uk/data/gb216-d/dxch/ddxch/i/hub

References for Appendix III

L.W. Dillwyn, *Notes of Recipes for Porcelain Bodies and Glazes 1815–1817* (Library of the Victoria & Albert Museum, South Kensington, London, deposited by John Campbell Esq. in, 1920b)

H. Eccles, B. Rackham, *Analyses of English Porcelain in the Victoria and Albert Museum Collection* (South Kensington, Victoria & Albert Museum Publications, London, 1922b)

R. Edmondson, "Billingsley, Randall and Rose", in *Welsh Ceramics in Context*, vol. 1, ed. J. Gray (The Royal Institution of South Wales, Swansea, 2003), p. 203

G. Edwards, *A Natural History of Uncommon Birds: And of Some Other Rare and Undescribed Animals, Quadrupeds, Fishes, Reptiles, Insects, &c., Exhibited in Two Hundred and Ten Copper-Plates, from Designs Copied Immediately from Nature and Curiously Coloured After Life*, vol. 4, (The Royal College of Physicians, Warwick Lane, London 1743–1751): and

G. Edwards, *Gleanings of Natural History, Exhibiting Figures of Quadrupeds, Birds, Insects, Plants &c.*, vols. 1 and 2, (The Royal College of Physicians, London, 1758–1760)

H.G.M. Edwards, *Porcelain to Silica Bricks: The Extreme Ceramics of William Weston Young, 1776–1847* (Springer-Nature, Dordrecht, 2019b)

H.G.M. Edwards, *Porcelain Analysis and Its Role in the Forensic Attribution of Ceramic Specimens* (Springer-Nature, Dordrecht, 2021c)

O. Fairclough, "The London China Trade, 1800–1830", *Transactions of the English Ceramics Circle*, **16/2**, pp. 209 and ff. (1997)

F. Gambon, *Porslen Abertawe a Nantgarw (Swansea and Nantgarw Porcelain)* (Oriel Plas Glyn-y-Weddw, Gwynedd, 2016b)

Gartre'n Ol*: Coming Home – An Exhibition of the Finest Nantgarw Porcelain, Nantgarw China Works Museum*, (Tyla Gwyn, Nantgarw, 2019b)

W.D. John, *Nantgarw Porcelain* (Ceramic Book Co., Newport, Gwent, 1948a)

W.D. John, K. Coombes, G.J. Coombes, *Nantgarw Porcelain Album* (Ceramic Book Company, Newport, Gwent, 1975a)

A.E. Jones, Sir Leslie Joseph, *Swansea Porcelain: Shapes and Decoration*, David Brown Publishers, Cowbridge, 1988.

E. Morton Nance, *The Pottery and Porcelain of Swansea and Nantgarw* (Batsford, London, 1942b)

W. Turner, *The Ceramics of Swansea and Nantgarw: A History of the Factories* (Bemrose & Sons Ltd., London, 1897b)

W.W. Young, *The Diaries of William Weston Young (1776–1847), 1802–1843*, 30 volumes (West Glamorgan Archives Service, Swansea, SA13SN). https://archiveshub.jisc.sc.uk/data/gb216-d/dxch/ddxch/i/hub.

References for Appendix IV

Sir A.H. Church, *English Porcelain: A Handbook to the China Made in England During the 18th Century as Illustrated by Specimens Chiefly in the National Collection*, (A South Kensington Museum Handbook, Chapman & Hall Ltd., London, 1885 and 1894)

P. Colomban, H.G.M. Edwards, C. Fountain, Raman spectroscopic and SEM/EDAXS analyses of highly transparent Nantgarw soft-paste porcelain. J. Eur. Ceram. Soc. **40**, 4664–4675 (2020)

H. Eccles, B. Rackham, *Analysed Specimens of English Porcelain in the Victoria & Albert Museum Collection* (Victoria and Albert Museum Publications, London, 1922c)

H.G.M. Edwards, *Porcelain to Silica Bricks: The Extreme Ceramics of William Weston Young, 1776–1847* (Springer Publishing, Dordrecht, 2019c)

C. Jacob-Hanson, "Charlotte Schreiber – The Unforgotten Grand Dame of the "*Chasse*"", Northern Ceramics Newsletter, No. 166, June 2012, pp 50–61, (2012)

E. Morton Nance, *The Pottery and Porcelain of Swansea and Nantgarw* (London, 1942c)

V. Owens, *Lady Charlotte Guest: The Exceptional Life of a Female Industrialist* (Pen & Sword Books, Barnsley, South Yorkshire, 2020)

B. Rackham, *The Catalogue of The Schreiber Collection of English Porcelain, Pottery, Glass and Enamels collected by Charles Schreiber MP and the Lady Charlotte Elizabeth Schreiber and Presented to The South Kensington Museum in 1884, Volume I: Porcelain, Volume II: Earthenwares, Volume III: Enamels and Glass* (Victoria & Albert Museum Publications, London, 1915)

Lady C. Schreiber, *The Lady Charlotte Schreiber Journals: Confidences of a Collector of Ceramics and Antiques Through Britain, France, Holland, Belgium, Spain, Portugal, Turkey, Austria and Germany from the Year 1869–1885*, ed. by M. John Guest (1839–1909), (John Lane, London/New York, 1911)

W. W. Young, *The Diaries of William Weston Young, 1776–1847(1802–1843)*, 30 volumes (West Glamorgan Archive Service, Swansea, SA1 3SN). https://arcgiveshub.jisc.sc.uk/data/gb216-d/dxch/ddxch/i/hub

References for Appendix V

H.G.M. Edwards, *Porcelain to Silica Bricks: The Extreme Ceramics of William Weston Young, 1776–1847* (Springer-Nature, Dordrecht, 2019d)

H.G.M. Edwards, *Porcelain Analysis and Its Role in the Forensic Attribution of Ceramic Specimens* (Springer-Nature, Dordrecht, 2021d)

J. Fairbairn, *Crests of the Families of Great Britain and Ireland and Their Kindred in Other Lands*, vols. I and II, (T.C. and E.C. Jack, Edinburgh and London, 1905 and New Orchard Publishing, London, 1986b)

Gentlemen's Magazine and Historical Chronicle, "Obituary with Anecdotes of Remarkable Persons", vol. **LXXIV**, 7/ii, (Nichols Son & Bentley, Cicero's Head, Fleet Street, London, 1814), pp 604–605

W. Turner, *The Ceramics of Swansea and Nantgarw. A History of The Factories with Biographical Notices of the Artists and Others, Notes on the Merits of the Porcelain, the Marks Thereon Etc.* (Bemrose & Sons Ltd, London, 1897c)

W. W. Young, *The Diaries of William Weston Young, 1776–1847(1802–1843)*, 30 volumes (West Glamorgan Archive Service, Swansea, SA1 3SN). https://arcgiveshub.jisc.sc.uk/data/gb216-d/dxch/ddxch/i/hub.

References for Appendix VI

R.J.H. Griffiths, *A Mortlock Encyclopedia*, (Havant, 2000–2019)

Selected Bibliography for Welsh Armorial Porcelains

Sir J. Bernard Burke, *Burke's General Armory—The General Armory of England, Scotland, Ireland and Wales: Comprising a Registry of Armorial Bearings from the Earliest to the Present Time* (Harrison & Sons, London, 1884)

H.G.M. Edwards, *Porcelain to Silica Bricks: The Extreme Ceramics of William Weston Young, 1776–1847* (Springer -Nature, Dordrecht, 2019e)

H.G.M. Edwards, *Porcelain Analysis and Its Role in the Forensic Attribution of Ceramic Specimens* (Springer -Nature, Dordrecht, 2021e)

C.N. Elvin, *A Handbook of Mottoes Borne by the Nobility, Gentry, Cities, Public Companies Etc.*, Bell & Daldy, Fleet Street, London, 1860. Reprinted with Supplement and an Index by R. Pinches, (Heraldry Today, Beauchamp Place, London and Genealogical Publishers, Baltimore, 1971)

J. Fairbairn, *Crests of the Families of Great Britain and Ireland and Their Kindred in Other Lands*, vols. I and II, (T.C. and E.C. Jack, Edinburgh and London, 1905 and New Orchard Publishing, London, 1986c)

O. Fairclough, The use of set pattern decoration at the Swansea China Works, in *Welsh Ceramics in Context*, ed. by J. Gray, vol. II, (The Royal Institution of South Wales, Swansea, 2005), pp. 192–208

A.C. Fox-Davies, *Armorial Families: A Complete Peerage, Baronetage and Knightage, and a Directory of Some Gentlemen Coat -Armour and being the First Attempt to Show Which Arms are in Use at the Moment are Borne by Some Legal Authority*, vol. 1 and 2, T.C., E.C. Jack, (Grange Publishing Works, Edinburgh, 1895)

A.C. Fox-Davies, *A Complete Guide to Heraldry: with Illustrations by G. Johnston, Herald Painter to the Lyon Court*, T.C., E.C. Jack, (Edinburgh and London, 1909)

F. Gambon, *Porslen Abertawe a Nantgarw Swansea and Nantgarw Porcelain*, (Oriel Plas Glyn-y-Weddw, Llanbedrog, Abersoch, Gwynedd, 2016c)

Gartre'n Ol *Coming Home Exhibition Catalogue – An Exhibition of the Finest Nantgarw Porcelain, (Nantgarw China Works Museum, Tyla Gwyn, Nantgarw,* 2019c)

W.D. John, *Nantgarw Porcelain* (Ceramic Book Company, Newport, 1948b)

W.D. John, *Swansea Porcelain* (Ceramic Book Company, Newport, 1958)

W.D. John, *William Billingsley* (Ceramic Book Company, Newport, 1968b)

W.D. John, K. Coombes, G.J. Coombes, *Nantgarw Porcelain Album* (Ceramic Book Company, Newport, 1975b)

A.E. Jones, S.L. Joseph, *Swansea Porcelain: Shapes and Decoration* (D. Brown, Cowbridge, 1988b)

K.S. Meager, *Swansea and Nantgarw Potteries: Catalogue of the Collection of Welsh Pottery and Porcelain on Exhibition at the Glynn Vivian Art Gallery, Swansea* (Swansea Corporation, Swansea, 1949)

E. Morton Nance, *The Pottery and Porcelain of Swansea and Nantgarw* (Batsford, London, 1942d)

D. Phillips, Heraldic Swansea Emblems. *Nantgarw & Swansea Newsletter* **18** (2006)

C. Scott-Giles, *Boutell's Heraldry* (Frederick Warne & Co. Ltd, London, 1954)

W. Turner, *The Ceramics of Swansea and Nantgarw. A History of The Factories with Biographical Notices of the Artists and Others, Notes on the Merits of the Porcelain, the Marks Thereon Etc.* (Bemrose & Sons Ltd, London, 1897d)

Index

H. G. M. Edwards, *Welsh Armorial Porcelain*,
https://doi.org/10.1007/978-3-030-97439-8